Lecture Notes in Physics

Edited by H. Araki, Kyoto, J. Ehlers, München, K. Hepp, Zürich
R. L. Jaffe, Cambridge, MA, R. Kippenhahn, Göttingen, D. Ruelle, Bures-sur-Yvette
H. A. Weidenmüller, Heidelberg, J. Wess, Karlsruhe and J. Zittartz, Köln

Managing Editor: W. Beiglböck

384

Mitchell D. Smooke (Ed.)

Reduced Kinetic Mechanisms and Asymptotic Approximations for Methane-Air Flames

A Topical Volume

With Contributions by R. W. Bilger H. K. Chelliah J.-Y. Chen
R. W. Dibble M. B. Esler V. Giovangigli J. Göttgens
D. A. Goussis S. H. Lam N. Peters B. Rogg K. Seshadri
M. D. Smooke S. H. Starner C. Treviño F. A. Williams

Springer-Verlag
Berlin Heidelberg GmbH

Editor

Mitchell D. Smooke
Department of Mechanical Engineering
Yale University, New Haven, CT 06520, USA

ISBN 978-3-662-13854-0   ISBN 978-3-540-47496-8 (eBook)
DOI 10.1007/978-3-540-47496-8

© Springer-Verlag Berlin Heidelberg 1991
Originally published by Springer-Verlag Berlin Heidelberg New York in 1991
Softcover reprint of the hardcover 1st edition 1991

2153/3140-543210 – Printed on acid-free paper

Preface

Combustion models that are used in the simulation of pollutant formation and ignition phenomena and in the study of chemically controlled extinction limits often combine detailed chemical kinetics with complicated transport phenomena. As the number of chemical species and the geometric complexity of the computational domain increases, the modeling of such systems becomes computationally prohibitive on even the largest supercomputer. The difficulty centers primarily on the number of chemical species and on the size of the different length scales in the problem. If one can reduce the number of species in a reaction network while still retaining the predictive capabilities of the mechanism, potentially larger problems could be solved with existing technology. In addition, current large scale problems could be moved to smaller workstation computers.

Although the concept of reducing a chemical reaction mechanism (both in the number of species and reactions) has been known by chemists for some time, it is only within the last few years that it has been applied with considerable success to combustion systems. As Peters discusses in Chap. 3 of this volume, the first applications of these ideas were surprisingly, not to hydrogen-air systems, but rather to hydrocarbon flames. Since the mid-1980s research in this area has increased dramatically. Scientists have continued to investigate both the generation and the subsequent application of reduced chemical kinetic mechanisms to a variety of combustion systems. Some of the work has been primarily numerical in nature while other studies have been more analytical.

In spite of the large amount of work performed in this area, there are still a number of questions that remain unanswered. First, how do the reduced chemistry numerical and asymptotic solutions compare? For example, are the computed flame speeds, reaction rates, and extinction conditions in good agreement? Also, how do the results compare with experiments? Some of these questions have been answered already on a mechanism by mechanism basis but nowhere to date has a systematic comparison been made on a set of well-defined premixed and nonpremixed test problems.

These questions formed the basis of a number of discussions at the XIth ICDERS meeting in Warsaw, Poland, during the summer of 1987. At that meeting several of the eventual contributors to this book decided that a systematic study of these issues for premixed and nonpremixed methane-air combustion was needed. In particular, it was decided that numerical solutions of a set of premixed and nonpremixed test problems (Chaps. 1 and 2) would be generated for standardized chemistry and transport approximations. These solutions and the data bases needed in generating them would then be made available to anyone interested in studying reduced chemistry methane-air combustion. This approach would allow a more systematic comparison between reduced chemistry numerical and analytical work.

Formulation of the test problem model together with the transport and chemistry approximations required several intermediate workshops—Sydney, Australia (January 1988) and New Haven, Connecticut (March 1988). Once a consensus was reached regarding these issues, it was decided that a meeting would be held the following year at UCSD in La Jolla, California, in March 1989. The chapters contained in this volume are based upon work first presented at that meeting and then subsequently refined in the following year.

The book is essentially divided into two parts. The first four chapters form a small tutorial on the procedures used in formulating the test problems and in reducing reaction mechanisms by applying steady-state and partial equilibrium approximations. The final six chapters discuss various aspects of the reduced chemistry problem for premixed and nonpremixed combustion. Chapter 5 compares premixed and nonpremixed four-step solutions with the skeletal chemistry solutions. Chapters 6 and 7 examine asymptotically the structure of premixed and nonpremixed flames, respectively, using a four-step mechanism. Chapter 8 applies a first-order sensitivity analysis to laminar methane-air flames using both a full and a reduced reaction mechanism. The last two chapters take a somewhat different approach. Chapter 9 examines the application of reduced chemistry to turbulent nonpremixed combustion and in Chap. 10 an innovative procedure for mechanism reduction is described using numerical singular perturbation analysis. It is our hope that this text will be the first of a number of similar contributions devoted to this important subject.

Finally, this volume would not have been possible without the help and co-operation of all the authors and the people at Springer-Verlag.

Mitchell D. Smooke
New Haven, Connecticut
November 1990

Contents

CHAPTER 1

FORMULATION OF THE PREMIXED AND NONPREMIXED TEST PROBLEMS

Mitchell D. Smooke
Department of Mechanical Engineering
Yale University
New Haven, Connecticut

and

Vincent Giovangigli
Ecole Polytechnique and CNRS
Centre de Mathématiques Appliquées
91128, Palaiseau cedex, France

1. Introduction

Combustion models that are used in the simulation of pollutant formation, ignition phenomena and in the study of chemically controlled extinction limits often combine detailed chemical kinetics with complicated transport phenomena. As the number of chemical species and the geometric complexity of the computational domain increases, the modeling of such systems becomes computationally prohibitive on even the largest supercomputer. While parallel architectures and algorithmic improvements have the potential of enhancing the level of problems one can solve, the modeling of three-dimensional time-dependent systems with detailed transport and finite rate chemistry will remain beyond the reach of combustion researchers for several years to come. The situation is even less promising if one wants to consider direct numerical simulation of turbulence with finite rate chemistry. While some applications can be studied effectively by lowering the dimensionality of the computational domain, there are many systems in which this is neither feasible nor scientifically sound.

The difficulty centers primarily on the number of chemical species and on the size of the different length scales in the problem. While local mesh refinement will ultimately solve the length scale problem, the size of the chemical system depends in large part upon the fuel one considers. For matrix based solution methods such as Newton's method, the cost of a flame computation scales with the square of the number of species. If one doubles the number of species, then the cost of the calculation quadruples. For field by field solution methods the scaling is linear in the number of species though the number of iterations required for convergence is usually large enough to produce a less efficient algorithm than the matrix based method. More importantly, however, since post-processing via first-order sensitivity analysis is essential in any flame study, a Jacobian matrix will have to be formed ultimately even if field by field solution methods are employed.

It is clear from these arguments that, if one can reduce the number of species in a reaction network while still retaining the predictive capabilities of the mechanism, potentially larger problems could be solved with existing technology. In addition, current large scale problems could be moved to smaller workstation computers. Reaction mechanism reduction (both in the number of elementary species and in the number of reactions) have been studied, for example, by Peters [1], Peters and Kee [2], Peters and Williams [3] and Bilger and Kee [4].

In spite of the large amount of work performed in this area, there are still a number of questions that remain unanswered. First, how do the reduced chemistry numerical and asymptotic solutions compare? For example, are the computed flame speeds, reaction rates, and extinction conditions in good agreement? Also, how do the results compare with experiments? Some of these questions have been answered already on a mechanism by mechanism basis but nowhere to date has a systematic comparison been made on a set of well defined premixed and nonpremixed test problems. In this volume we will attempt to answer these questions by using a carefully formulated set of flames with simplified transport and skeletal chemistry for methane-air combustion. (The term skeletal chemistry will refer to a mechanism that is larger than the reduced mechanisms reported in [1-4] but substantially smaller than the full mechanisms used in detailed chemistry studies, e.g., [7], [13], [16], [20-21]). In particular, for various equivalence ratios, pressures and strain rates we will generate a reference data base which can then be used in assessing the results of the mechanism reductions.

The two models we will consider are the laminar premixed and the laminar counterflow diffusion flame. In both cases the governing equations can be reduced to a set of coupled nonlinear two-point boundary value problems with separated boundary conditions. Numerical solution of these problems have been studied by a number of researchers [5-31].

In the laboratory nonadiabatic laminar premixed flames are obtained by flowing a premixed fuel and oxidizer through a cooled porous plug burner. As the gases emerge from the burner they are ignited and a steady flame sits above the burner surface (see Figure 1). In general, there is a positive temperature gradient at the burner surface. In the adiabatic (freely propagating) case, the flame is contained in an infinite domain with zero gradients of the dependent variables at either end. The premixed flame problem is appealing due to its simple flow geometry and it has been used by kineticists in understanding elementary reaction mechanisms in the oxidation of fuels [7], [13-16], [20-21].

Counterflow diffusion flames have played an important role in recent models of turbulent nonpremixed combustion. The reacting surface in these models can be viewed as being composed of a number of thin, laminar, diffusion flamelets (see, e.g., [32-35]) Using the flamelet concept, researchers have been able to include the effects of complex chemistry and detailed transport in the modeling of turbulent reacting flows [36]. In addition to their use in flamelet models, counterflow diffusion flames are recognized as an important tool with which to study the complex transport and chemical kinetic interactions that occur in nonpremixed combustion [37-40].

Several counterflow configurations have been studied experimentally. In one case two cylindrical axisymmetric, coaxial, jets are employed [41]. Fuel is introduced from one jet and air from the other. A flat counterflow diffusion flame sits between the two jets with a stagnation plane located near the point of the peak temperature. In another configuration fuel is blown radially outward from a porous cylinder into an oncoming stream of air (see Figure 2). A free stagnation line parallel to the cylinder axis forms in front of the cylinder's porous surface. Combustion occurs within a thin flame zone near the stagnation line where the fuel and oxidizer are in stoichiometric proportion [37], [40]. This "Tsuji" configuration will be employed in all of the counterflow test studies contained in this volume.

In this chapter we will first formulate the premixed and diffusion flame problems. We will then consider several modifications of the governing equations that will simplify the

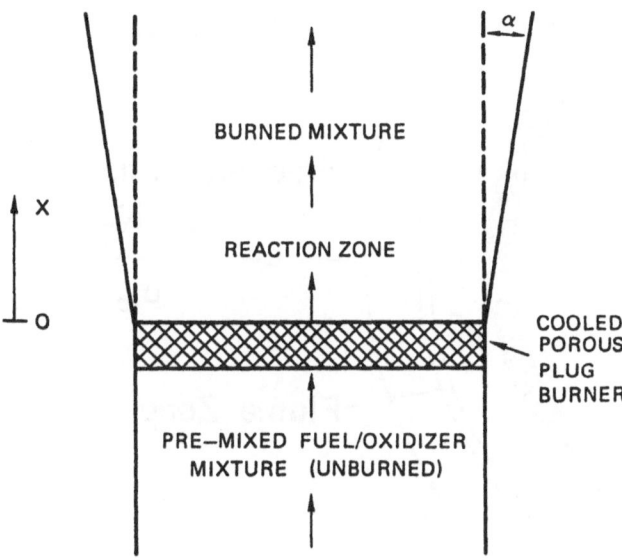

Figure 1. Schematic of a burner stabilized premixed laminar flame.

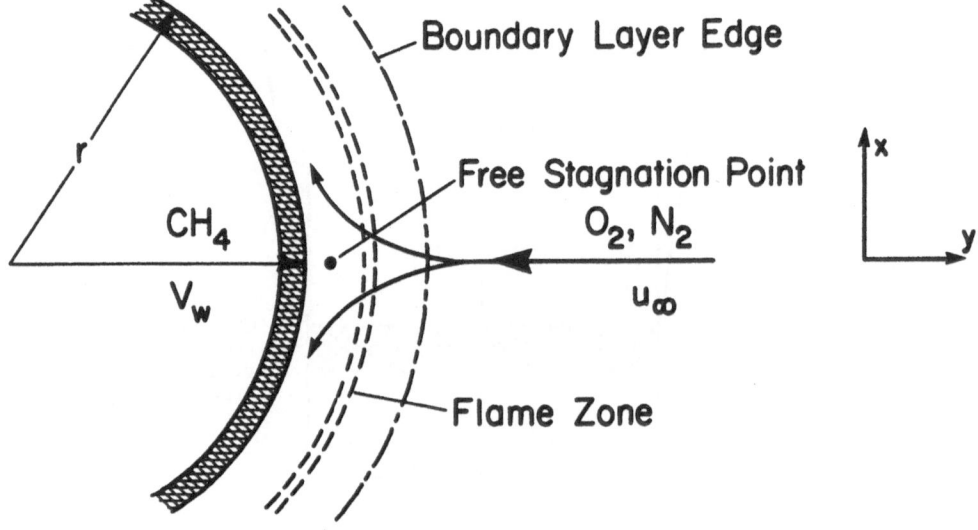

Figure 2. Schematic of a "Tsuji" counterflow diffusion flame.

numerical and asymptotic study of these systems without compromising the predictive capabilities of the models. In this way we will be able to facilitate a more accurate comparison between the numerical and analytic solutions of the test problems.

2. Problem Formulations

Premixed Flame Problem

The formulation of the premixed flame problem we consider closely follows the one originally proposed by Hirschfelder and Curtiss [42]. Our goal is to predict theoretically the mass flow rate (adiabatic flame speed), the mass fractions of the species and the temperature as functions of the independent coordinate x. Upon neglecting viscous effects, body forces, radiative heat transfer and the diffusion of heat due to concentration gradients, the equations governing the structure of a steady, one-dimensional, isobaric flame (with expansion angle $\alpha = 0$) are

$$\dot{M} = \rho u = \text{constant}, \tag{2.1}$$

$$\dot{M}\frac{dY_k}{dx} + \frac{d}{dx}(\rho Y_k V_k) - \dot{w}_k W_k = 0, \qquad k = 1, 2, \ldots, K, \tag{2.2}$$

$$c_p \dot{M}\frac{dT}{dx} - \frac{d}{dx}\left(\lambda \frac{dT}{dx}\right) + \sum_{k=1}^{K} \rho Y_k V_k c_{p_k}\frac{dT}{dx} + \sum_{k=1}^{K} \dot{w}_k h_k W_k = 0, \tag{2.3}$$

$$\rho = \frac{p\overline{W}}{RT}. \tag{2.4}$$

In these equations x denotes the independent spatial coordinate; $\dot{M}$, the mass flow rate; T, the temperature; Y_k, the mass fraction of the kth species; p, the pressure; u, the velocity of the fluid mixture; ρ, the mass density; W_k, the molecular weight of the kth species; $\overline{W}$, the mean molecular weight of the mixture; R, the universal gas constant; λ, the thermal conductivity of the mixture; c_p, the constant pressure heat capacity of the mixture; c_{p_k}, the constant pressure heat capacity of the kth species; $\dot{w}_k$, the molar rate of production of the kth species per unit volume; h_k, the specific enthalpy of the kth species; and V_k, the diffusion velocity of the kth species. The form of the diffusion velocities, the transport coefficients and the chemical production rates is described in detail in [43-44].

The boundary conditions for the problem in the reactant stream are

$$T(-\infty) = T_u, \tag{2.5}$$

$$Y_k(-\infty) = \epsilon_k(\phi), \qquad k = 1, \ldots, K, \tag{2.6}$$

where T_u is the cold reactant stream temperature and $\epsilon_k(\phi)$ is the known incoming mass flux fraction of the kth species which depends on the reactant stream equivalence ratio ϕ. In the hot stream, we have

$$\frac{dT}{dx}(+\infty) = 0, \tag{2.7}$$

$$\frac{dY_k}{dx}(+\infty) = 0, \qquad k = 1, 2, \ldots, K. \tag{2.8}$$

In a freely propagating flame, the mass flow rate M is not known; it is an eigenvalue to be determined. Calculation of the flow rate proceeds by introducing the trivial differential equation

$$\frac{d\dot{M}}{dx} = 0, \tag{2.9}$$

and an additional boundary condition to the system in (2.1-2.8). The particular choice of the extra boundary condition is somewhat arbitrary. It must be chosen, however, to insure that the spatial gradients of both the temperature and the mass fractions are vanishingly small as x approaches the two boundaries. Typically, we impose the temperature at one additional point, e.g., at the origin $x = 0$

$$T(0) = T_i, \tag{2.10}$$

which also removes the translational invariance of the problem.

From a computational viewpoint, the infinite spatial domain is not convenient to implement numerically. A preferable set of boundary conditions on a finite domain are obtained by first integrating the species conservation equations between $-\infty$ and 0, and then assuming that T_i is sufficiently low so that the production terms $\dot{\omega}_k$ are negligible for $T \leq T_i$. This produces the familiar mixed species boundary conditions at $x = 0$

$$\dot{M}Y_k(0) + \rho(0)Y_k(0)V_k(0) = \dot{M}\epsilon_k(\phi), \quad k = 1, 2, \ldots, K. \tag{2.11}$$

If the enthalpy conservation equation is then integrated between $-\infty$ and 0 and we use (2.10) and (2.11), we can write [45]

$$\lambda(0)\frac{dT}{dx}(0) = \sum_{k=1}^{K} \epsilon_k(\phi)[h_k(T(0)) - h_k(T_u)]. \tag{2.12}$$

The computational domain is then further reduced to $[0, L]$ by replacing the boundary conditions at $+\infty$ by the relations

$$\frac{dT}{dx}(L) = 0, \tag{2.13}$$

$$\frac{dY_k}{dx}(L) = 0, \qquad k = 1, \ldots, K, \tag{2.14}$$

where the choice of L must be large enough to ensure vanishingly small gradients at $x = L$. The premixed flame model then consists of equations (2.1-2.4) together with (2.10-2.14).

Counterflow Diffusion Flame Problem

We consider counterflow diffusion flames in a Tsuji configuration. Our model for these flames assumes a laminar, stagnation point flow. The one-dimensional governing equations can be derived by considering a boundary layer model in which the equations for mass, momentum, chemical species and energy are written in the form

$$\frac{\partial(\rho u)}{\partial x} + \frac{\partial(\rho v)}{\partial y} = 0, \tag{2.15}$$

$$\rho u \frac{\partial u}{\partial x} + \rho v \frac{\partial u}{\partial y} + \frac{\partial p}{\partial x} - \frac{\partial}{\partial y}\left(\mu \frac{\partial u}{\partial y}\right) = 0, \tag{2.16}$$

$$\rho u \frac{\partial Y_k}{\partial x} + \rho v \frac{\partial Y_k}{\partial y} + \frac{\partial}{\partial y}\left(\rho Y_k V_{ky}\right) - \dot{w}_k W_k = 0, \quad k = 1, 2, \ldots, K, \tag{2.17}$$

$$\rho u c_p \frac{\partial T}{\partial x} + \rho v c_p \frac{\partial T}{\partial y} - \frac{\partial}{\partial y}\left(\lambda \frac{\partial T}{\partial y}\right) + \sum_{k=1}^{K} \rho Y_k V_{ky} c_{pk} \frac{\partial T}{\partial y} + \sum_{k=1}^{K} \dot{w}_k W_k h_k = 0, \tag{2.18}$$

The system is closed with the ideal gas law. In addition to the quantities already defined, x and y denote independent spatial coordinates in the tangential and transverse directions, respectively; u and v the tangential and the transverse components of the velocity, respectively; μ the viscosity of the mixture; and V_{ky} is the diffusion velocity of the kth species in the y direction.

The free stream (tangential) velocity at the edge of the boundary layer is given by $u_\infty = ax$ where a is the strain rate. We introduce the notation

$$U = \frac{u}{u_\infty}, \tag{2.19}$$

$$V = \rho v, \tag{2.20}$$

where U is related to the derivative of a modified stream function (see e.g., [30]). Using these expressions, the boundary layer equations can be transformed into a system of ordinary differential equations valid along the stagnation-point streamline $x = 0$. For a system in rectangular coordinates we have

$$\frac{dV}{dy} + a\rho U = 0, \tag{2.21}$$

$$V \frac{dU}{dy} - \frac{d}{dy}\left(\mu \frac{dU}{dy}\right) - a(\rho_\infty - \rho U^2) = 0, \tag{2.22}$$

$$V \frac{dY_k}{dy} + \frac{d}{dy}\left(\rho Y_k V_k\right) - \dot{w}_k W_k = 0, \quad k = 1, 2, \ldots, K, \tag{2.23}$$

$$c_p V \frac{dT}{dy} - \frac{d}{dy}\left(\lambda \frac{dT}{dy}\right) + \sum_{k=1}^{K} \rho Y_k V_k c_{pk} \frac{dT}{dy} + \sum_{k=1}^{K} \dot{w}_k W_k h_k = 0. \tag{2.24}$$

The boundary conditions for the Tsuji configuration are given by

$$V(0) = V_w, \tag{2.25}$$

$$U(0) = 0, \tag{2.26}$$

$$V_w Y_k(0) + \rho Y_k(0) V_k = V_w \epsilon_k, \quad k = 1, 2, \ldots, K, \tag{2.27}$$

$$T(0) = T_w, \tag{2.28}$$

at the cylinder wall ($y = 0$) and

$$U = 1, \tag{2.29}$$

$$Y_k = Y_{k\infty}, \quad k = 1, 2, \ldots, K, \tag{2.30}$$

$$T = T_\infty, \tag{2.31}$$

as $y \to \infty$ The mass flux, temperature and the incoming mass flux fractions (V_w, T_w and ϵ_k) at the wall are specified, as are the mass fractions of the species and the temperature (Y_{k_∞} and T_∞) at the edge of the boundary layer.

3. Model Simplifications

One of the goals of this volume is to be able to compare asymptotic solutions of premixed and nonpremixed flames with corresponding numerical computations employing reduced chemistry. In many of the detailed transport/finite rate chemistry studies of flames, the transport coefficients are formed using kinetic theory expressions with tabulated values of the appropriate collision integrals (see, e.g., [20] and [44]). This type of formulation poses no difficulty for computational models since the solution is ultimately represented as a discrete set of numbers. It is not, however, the most convenient representation from which to develop a closed form asymptotic solution. As a result, we will introduce a number of simplifying assumptions that can aid the comparison process between the numerical and the analytical solutions. Specifically, we will focus on modifications to the energy equation and on the simplification of the transport model.

Enthalpy Flux Terms

Both the premixed and nonpremixed energy equations contain a term of the form

$$H = \sum_{k=1}^{K} c_{p_k} \rho Y_k V_k \frac{dT}{dx}. \tag{3.1}$$

We point out that if all the species heat capacities are equal, i.e., $c_{p_k} = c =$ constant, then

$$H = c\rho \frac{dT}{dx} \sum_{k=1}^{K} Y_k V_k, \tag{3.2}$$

which is identically equal to zero since

$$\sum_{k=1}^{K} Y_k V_k = 0. \tag{3.3}$$

In practice, we do not expect all of the heat capacities to be equal but we anticipate that their variation will be small so that $|H|$ will be small compared to the other three terms in the energy equation. To investigate these ideas, we have plotted the convective term, the conduction term, the chemistry term and the enthalpy flux term in the energy equation as a function of the independent spatial coordinate for three premixed flames and one counterflow flame. The computations were performed with the model formulations described in the previous section. The transport coefficients were evaluated as in [20], [44] and the reaction mechanism was the skeletal mechanism listed in Table II. In Figure 3 we illustrate the four terms for an atmospheric pressure, premixed, methane-air flame with an equivalence ratio $\phi = 1.0$. We note that the enthalpy flux term is negligible in all regions of the flame. Similar results hold for a one atmosphere premixed flame with $\phi = 0.6$ (Figure 4) and for a stoichiometric flame at a pressure of 30.0 atmospheres (Figure 5). The situation is almost identical for a counterflow flame at a strain rate of 100 sec^{-1} (Figure 6). Based upon these results, we will eliminate the enthalpy flux term from our premixed and nonpremixed models.

Thermal Diffusivity Approximation

In detailed flame models the heat capacity of the mixture is often approximated by a polynomial fit in the temperature to the JANAF data [43]. The thermal conductivity is

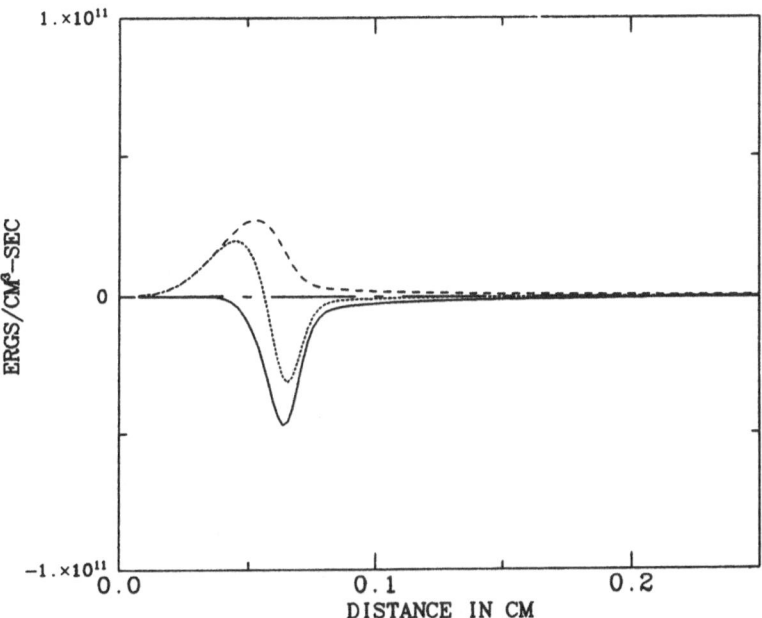

Figure 3. Comparison of the size of the diffusive (dot), convective (dash), enthalpy flux (solid-dash) and chemistry terms (solid) in the energy equation for an atmospheric pressure, stoichiometric, premixed methane-air flame with detailed transport and skeletal chemistry.

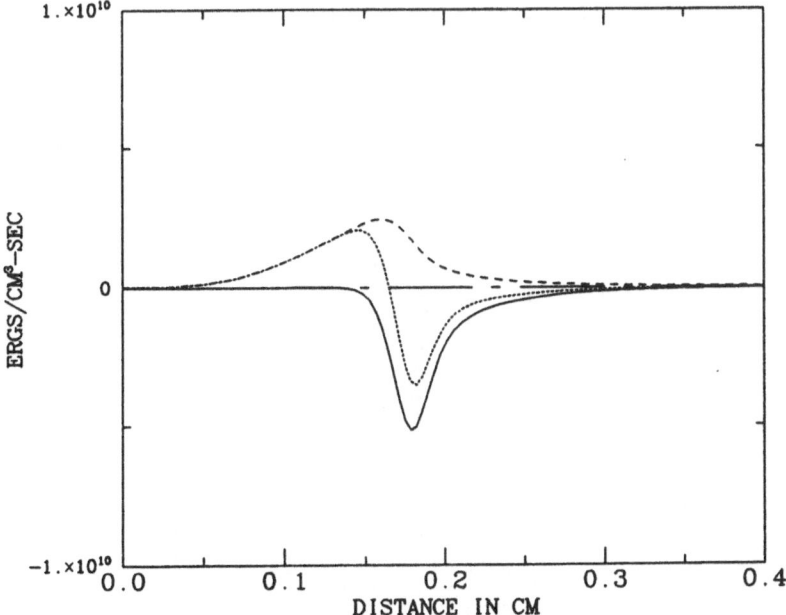

Figure 4. Comparison of the size of the diffusive (dot), convective (dash), enthalpy flux (solid-dash) and chemistry terms (solid) in the energy equation for a $\phi = 0.6$, atmospheric pressure, premixed methane-air flame with detailed transport and skeletal chemistry.

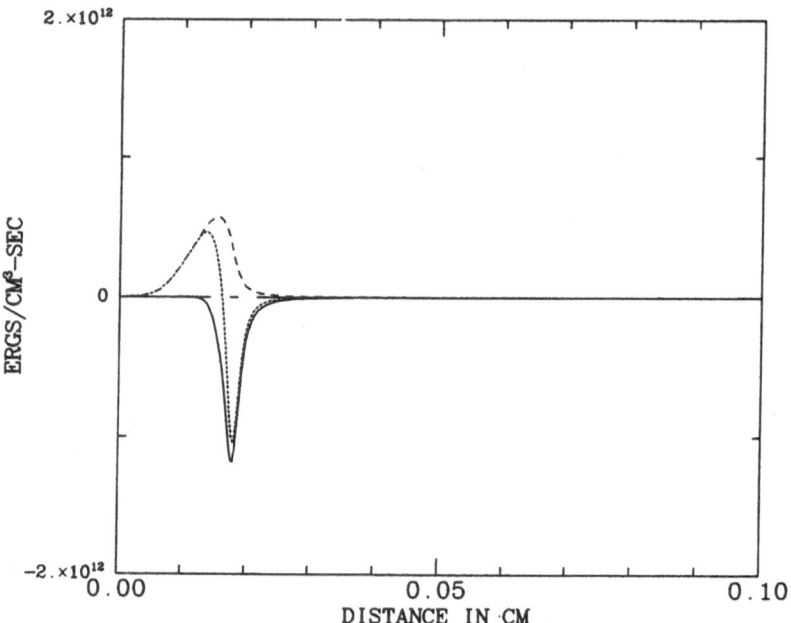

Figure 5. Comparison of the size of the diffusive (dot), convective (dash), enthalpy flux (solid-dash) and chemistry terms (solid) in the energy equation for a 30 atmosphere, stoichiometric, premixed methane-air flame with detailed transport and skeletal chemistry.

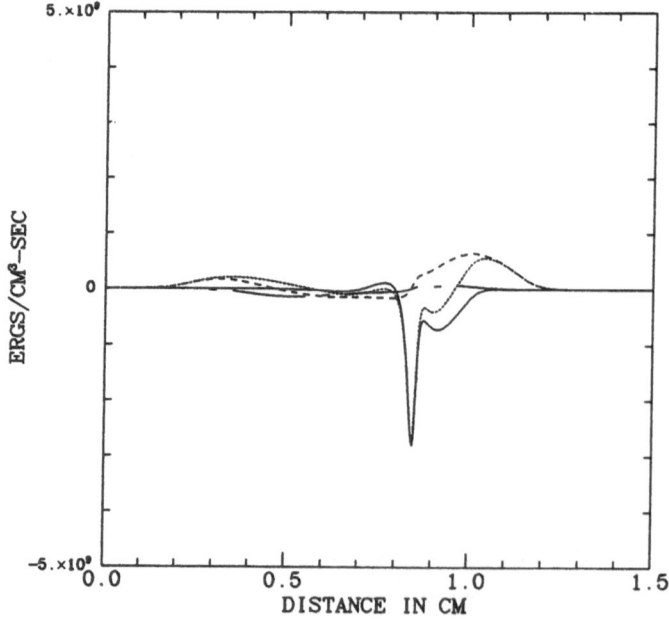

Figure 6. Comparison of the size of the diffusive (dot), convective (dash), enthalpy flux (solid-dash) and chemistry terms (solid) in the energy equation for a methane-air, counterflow diffsuion flame ($a = 100 \text{ sec}^{-1}$) with detailed transport and skeletal chemistry.

approximated by a complex relation involving translational, vibrational and rotational factors [44]. While each approximation by itself can be fairly complicated, the ratio of the two quantities can often be approximated by a simple expression. To investigate this possibility, we postulate the relation

$$\frac{\lambda}{c_p} = A \left(\frac{T}{T_o}\right)^r,$$ (3.4)

for a constant A, a reference temperature $T_o = 298\text{K}$ and an exponent r. Using the detailed transport-finite rate skeletal chemistry model employed in the enthalpy flux evaluation, we generated vaues of λ/c_p as a function of the independent spatial coordinate for a stoichiometric, one atmosphere, premixed flame. We then determined the parameters A and r with a nonlinear least squares fitting procedure. Specifically, we found

$$A = 2.58 \times 10^{-4} \text{ g/cm-sec} \text{ and } r = 0.7.$$ (3.5)

In Figure 7 we compare the detailed transport-finite rate skeletal chemistry values of λ/c_p with the simplified relation in (3.4-3.5) as a function of temperature for a one atmosphere, stoichiometric, premixed flame. Similar comparisons for a one atmosphere $\phi = 0.6$ flame and a stoichiometric 30 atmosphere flame are illustrated in Figures 8-9. In all cases we find excellent agreement over the entire temperature range. Results for the counterflow diffusion flame with $a = 100 \text{ sec}^{-1}$ are contained in Figure 10. We note, however, that the simple power law is only able to approximate the detailed transport form of λ/c_p on the fuel lean side. A two zone model could be developed in an analagous fashion but, for simplicity and consistency, we will only use the relations in (3.4) and (3.5) in the test problems.

Heat Conduction Term

The premixed and nonpremixed energy equation contain terms of the form

$$C = \frac{1}{c_p}\frac{d}{dx}\left(\lambda\frac{dT}{dx}\right).$$ (3.6)

Since we have combined the thermal conductivity and the heat capacity of the mixture into a single ratio, it would be reasonable to rewrite the heat conduction term using this expression, i.e., we would like to write

$$C = \frac{d}{dx}\left(\frac{\lambda}{c_p}\frac{dT}{dx}\right).$$ (3.7)

However, we note that

$$\frac{1}{c_p}\frac{d}{dx}\left(\lambda\frac{dT}{dx}\right) = \frac{d}{dx}\left(\frac{\lambda}{c_p}\frac{dT}{dx}\right) + \frac{\lambda}{c_p^2}\frac{dc_p}{dx}\frac{dT}{dx}.$$ (3.8)

We observe that an additional term in the energy equation is generated by this reformulation. To investigate whether this term can be neglected, we compare in Figures 11-14 the size of the two terms on the right-hand side of (3.8) as a function of distance for three premixed flames and a counterflow diffusion flame with $a = 100 \text{ sec}^{-1}$. While the term containing the derivative of the heat capacity of the mixture is not dominant, it can account for as much as 20% of the contribution to (3.6). If the term containing the derivatives of the heat capacity is neglected, the burning velocity for the one atmosphere, stoichiometic flame becomes 27.07 cm/sec and the counterflow diffusion flame extinction strain rate becomes $a = 212 \text{ sec}^{-1}$. These numbers differ significantly from

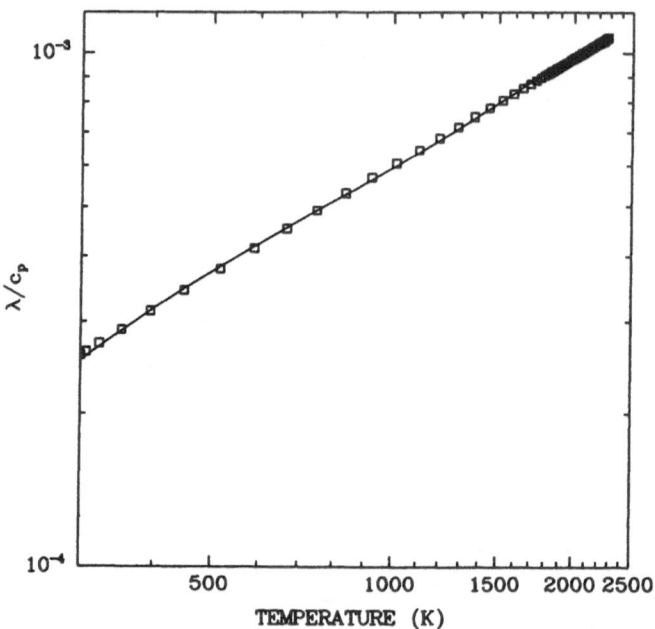

Figure 7. A comparison between the detailed transport-skeletal chemistry value of λ/c_p ($\square$) and the least squares fit (solid) as a function of the temperature for an atmospheric pressure, stoichiometric, premixed methane-air flame.

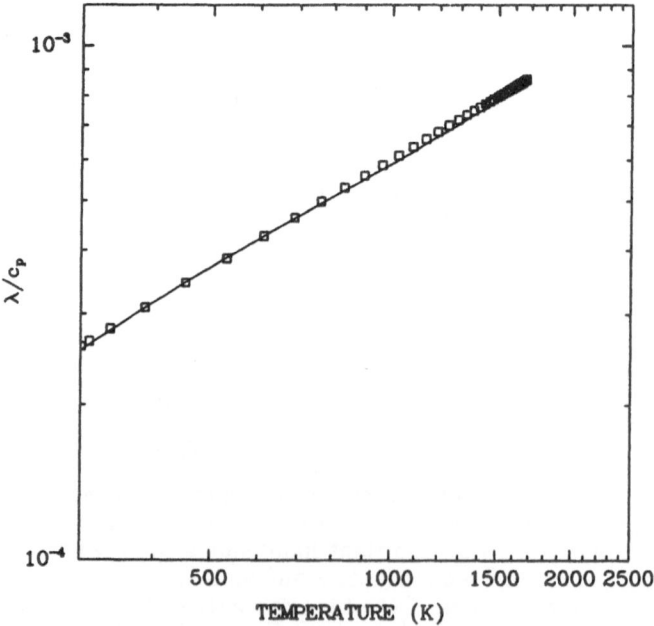

Figure 8. A comparison between the detailed transport-skeletal chemistry value of λ/c_p ($\square$) and the least squares fit (solid) as a function of the temperature for a $\phi = 0.6$, atmospheric pressure, premixed methane-air flame.

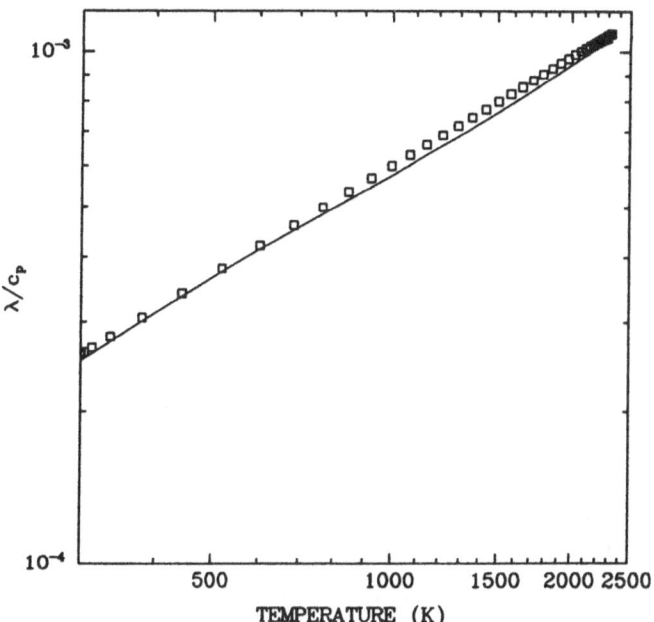

Figure 9. A comparison between the detailed transport-skeletal chemistry value of λ/c_p ($\square$) and the least squares fit (solid) as a function of the temperature for a 30 atmosphere, stoichiometric, premixed methane-air flame.

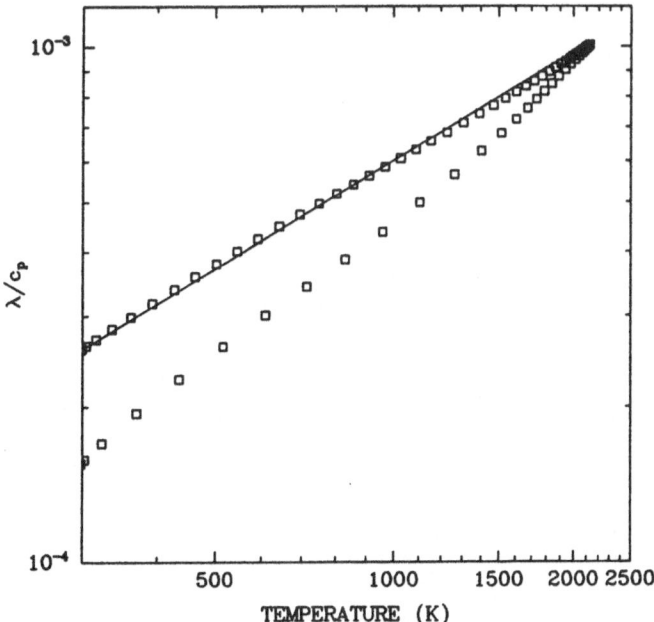

Figure 10. A comparison between the detailed transport-skeletal chemistry value of λ/c_p ($\square$) and the least squares fit (solid) as a function of the temperature for a methane-air, counterflow diffusion flame with $a = 100 \ \text{sec}^{-1}$.

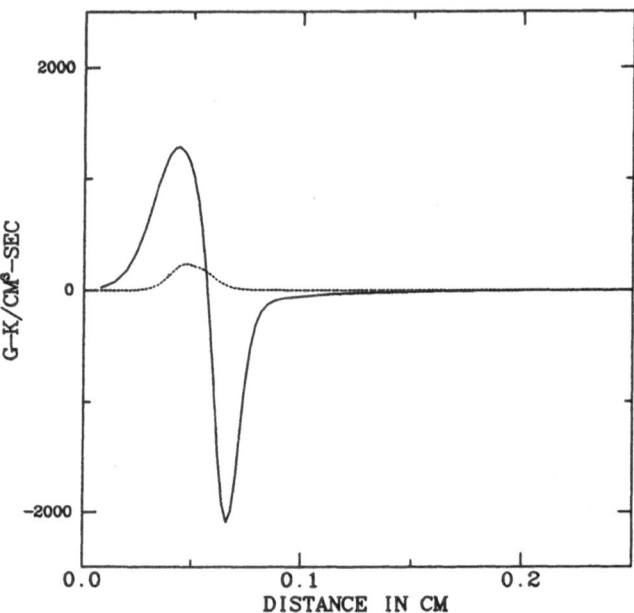

Figure 11. A comparison between the size of the heat conduction term in Eqn. (3.7) (solid) and the second term on the right-hand side of Eqn. (3.8) (dot) for an atmospheric pressure, stoichiometric, premixed methane-air flame.

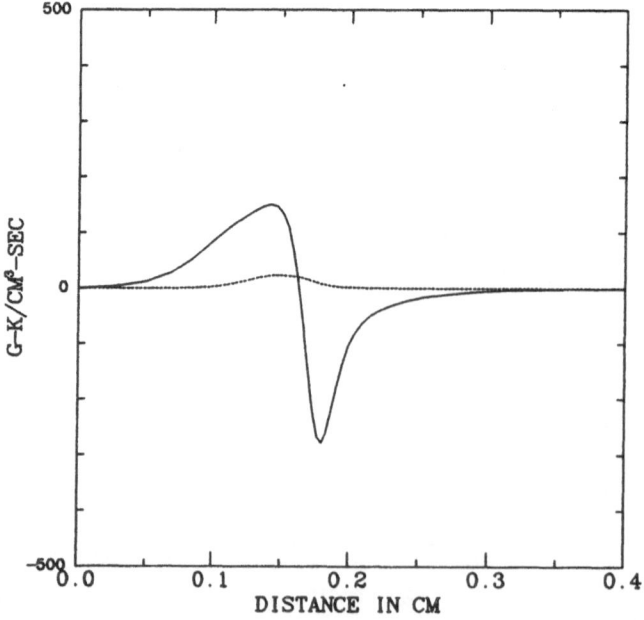

Figure 12. A comparison between the size of the heat conduction term in Eqn. (3.7) (solid) and the second term on the right-hand side of Eqn. (3.8) (dot) for a $\phi = 0.6$, atmospheric pressure, premixed methane-air flame.

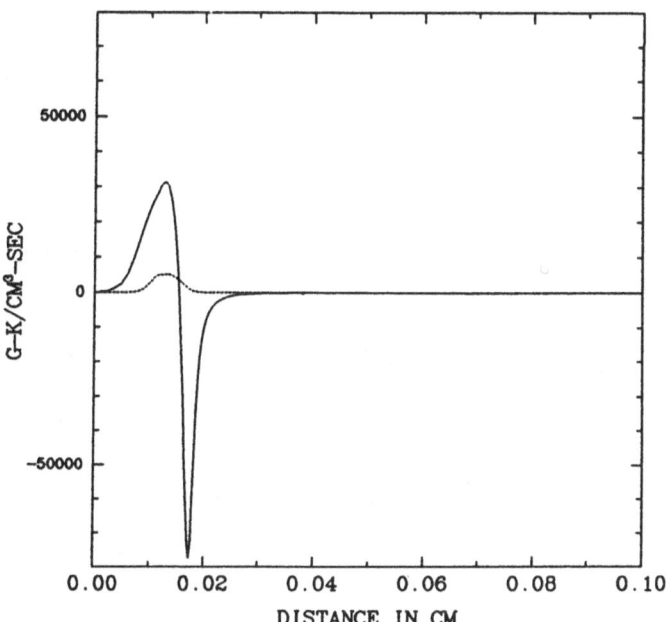

Figure 13. A comparison between the size of the heat conduction term in Eqn. (3.7) (solid) and the second term on the right-hand side of Eqn. (3.8) (dot) for a 30 atmosphere, stoichiometric, premixed methane-air flame.

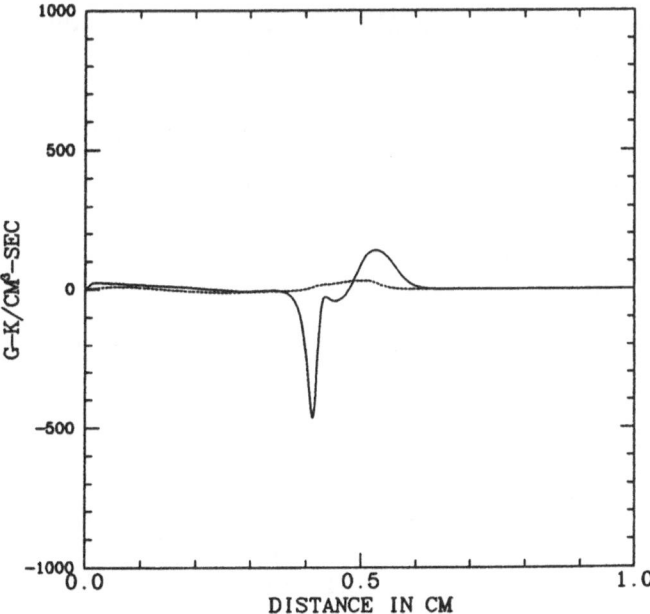

Figure 14. A comparison between the size of the heat conduction term in Eqn. (3.7) (solid) and the second term on the right-hand side of Eqn. (3.8) (dot) for a methane-air, counterflow diffusion flame with $a = 100$ sec^{-1}. par

the experimentally measured flame speed of 37-40 cm/sec and extinction strain rate of $a = 350$ sec^{-1}. As we will see in the next chapter, by keeping the heat capacity derivative term the computed flame speed becomes 37.67 cm/sec and the extinction strain rate becomes $a = 353$ sec^{-1}. Based upon these results, the heat capacity derivative term in the energy equation will be retained.

Lewis Number Approximations

We define the Lewis number as the ratio of the thermal diffusivity to the mass diffusivity, i.e.,

$$\text{Le}_k = \frac{\lambda}{\rho D_k c_p}. \tag{3.9}$$

Our transport model can be simplified dramatically if the Lewis numbers of the various species in our kinetic model can be taken to be approximately constant. Using the detailed transport-finite rate skeletal chemistry model for a one atmosphere stoichiometric premixed flame and a counterflow diffusion flame with $a = 100$ sec^{-1}, we generated the spatially dependent Lewis numbers for each of the 16 chemical species. This data was then fit to a constant. The results for the two flames are contained in Figures 15-22. We note that, with the exception of the low temperature regions, the fits are remarkably good. Table I lists the Lewis numbers we computed for each of the 16 species.

Diffusion Velocity Approximations

The species balance equations contain terms of the form

$$D = \frac{d}{dx}\left(\rho Y_k V_k\right), \tag{3.10}$$

where V_k is the diffusion velocity of the kth species. To relate the diffusion velocity to the mass diffsion coefficients, we employ the approximation

$$Y_k V_k \approx -D_k \frac{dY_k}{dx}, \tag{3.11}$$

where Y_k is the mass fraction of the kth species. Hence, we can write

$$\frac{d}{dx}\left(\rho Y_k V_k\right) \approx -\frac{d}{dx}\left(\rho D_k \frac{dY_k}{dx}\right). \tag{3.12}$$

In some detailed transport combustion models a conservation equation is solved for each elementary species [22]. If the relation

$$\sum_{k=1}^{K} Y_k V_k = 0, \tag{3.13}$$

is not satisfied from the diffusion velocity model, then a correction velocity v_c is often introduced so that

$$\sum_{k=1}^{K}(V_k + v_c)Y_k = 0. \tag{3.14}$$

In our simplified model we will not solve for all of the species. Instead, we will solve for only $K - 1$ chemical species with the mass fraction of N_2 determined from mass

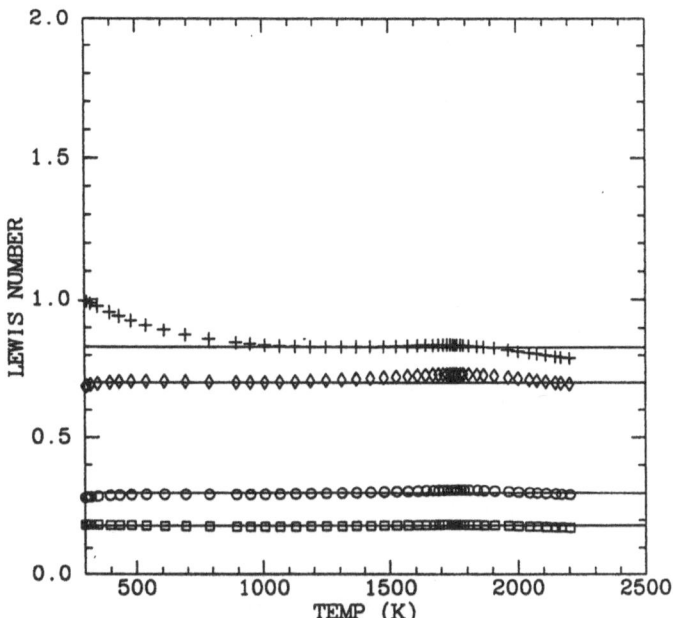

Figure 15. A comparison between the detailed transport-skeletal chemistry Lewis numbers and the least squares fits (solid) as a function of the temperature for $H_2O(+)$, $O(\diamond)$, $H_2(\circ)$ and $H(\square)$ for an atmospheric pressure, stoichiometric, premixed methane-air flame.

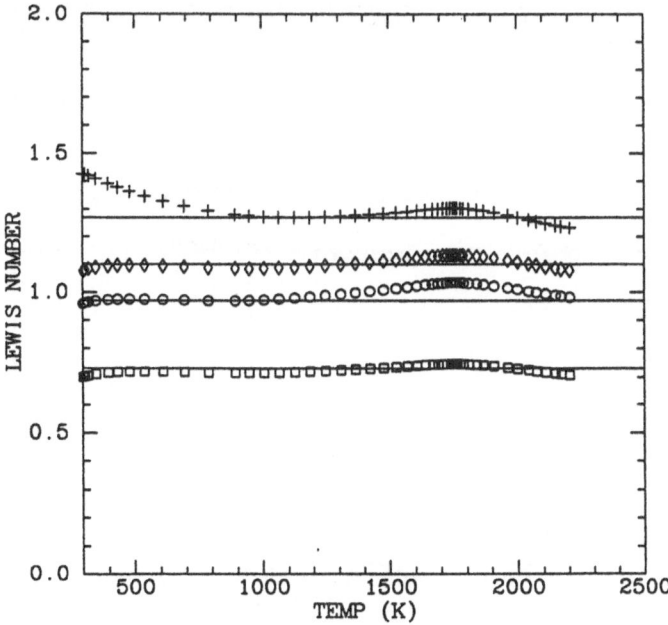

Figure 16. A comparison between the detailed transport-skeletal chemistry Lewis numbers and the least squares fits (solid) as a function of the temperature for $CH_2O(+)$, $HO_2(\diamond)$, $CH_4(\circ)$ and $OH(\square)$ for an atmospheric pressure, stoichiometric, premixed methane-air flame.

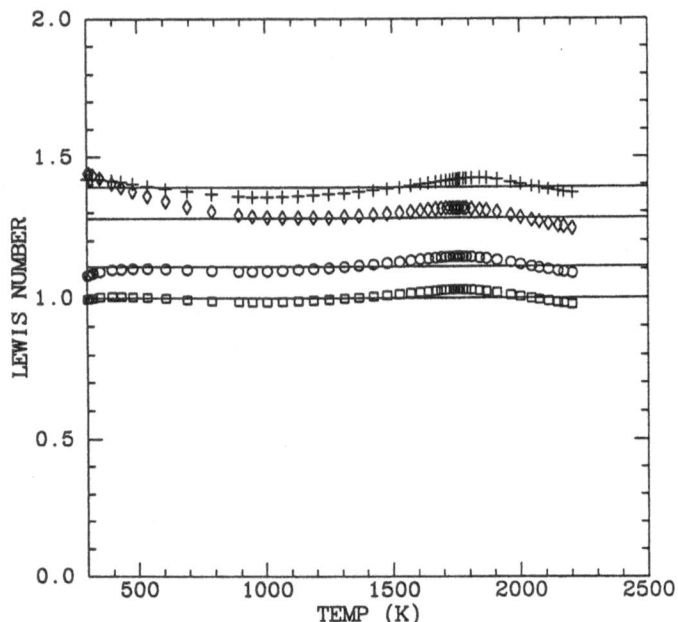

Figure 17. A comparison between the detailed transport-skeletal chemistry Lewis numbers and the least squares fits (solid) as a function of the temperature for $CO_2(+)$, $HCO(\diamond)$, $CO(\circ)$ and $CH_3(\square)$ for an atmospheric pressure, stoichiometric, premixed methane-air flame.

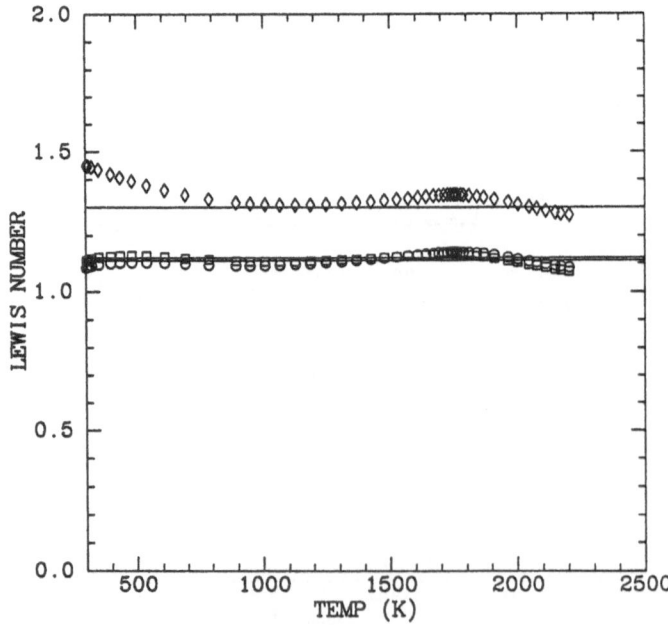

Figure 18. A comparison between the detailed transport-skeletal chemistry Lewis numbers and the least squares fits (solid) as a function of the temperature for $CH_3O(\diamond)$, $O_2(\circ)$ and $H_2O_2(\square)$ for an atmospheric pressure, stoichiometric, premixed methane-air flame.

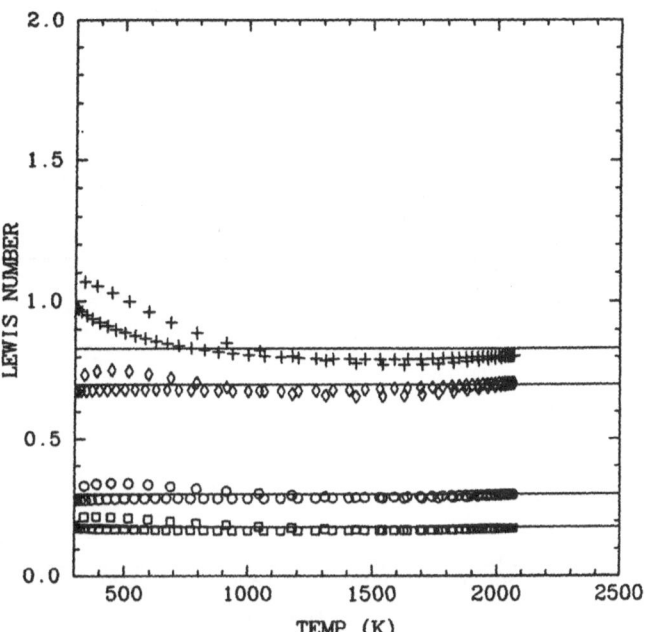

Figure 19. A comparison between the detailed transport-skeletal chemistry Lewis numbers and the least squares fits (solid) as a function of the temperature for $H_2O(+)$, $O(\diamond)$, $H_2(\circ)$ and $H(\square)$ for a methane-air, counterflow diffusion flame with $a = 100$ sec^{-1}.

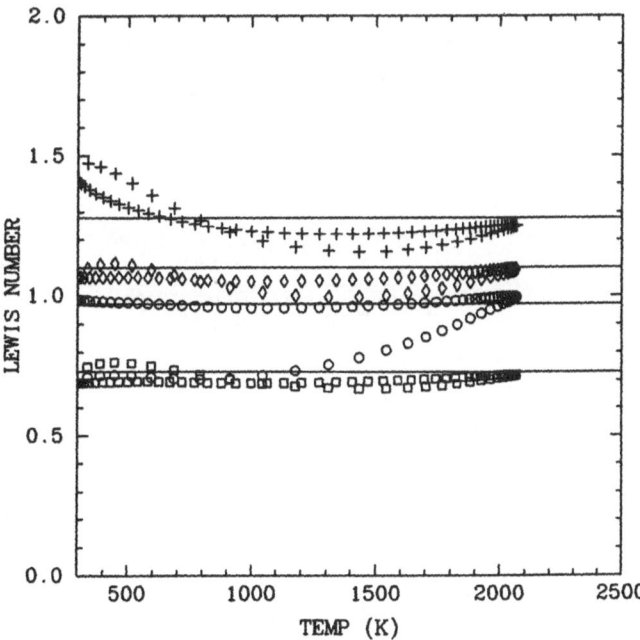

Figure 20. A comparison between the detailed transport-skeletal chemistry Lewis numbers and the least squares fits (solid) as a function of the temperature for $CH_2O(+)$, $HO_2(\diamond)$, $CH_4(\circ)$ and $OH(\square)$ for a methane-air, counterflow diffusion flame with $a = 100$ sec^{-1}

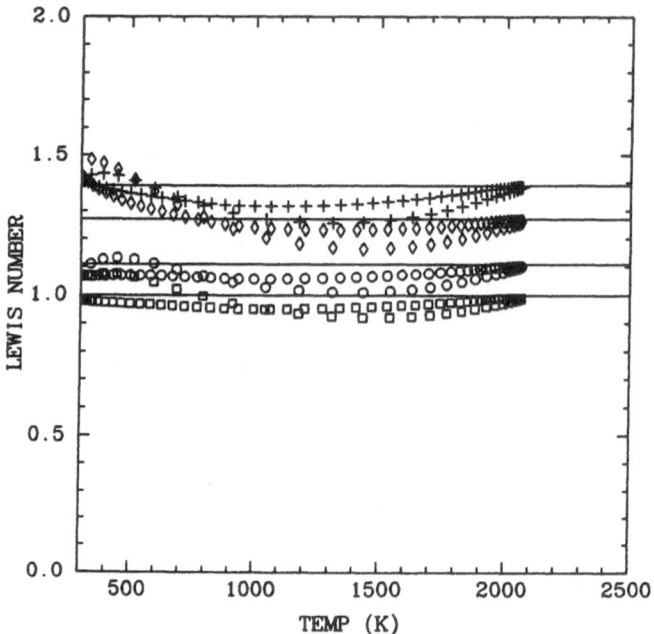

Figure 21. A comparison between the detailed transport-skeletal chemistry Lewis numbers and the least squares fits (solid) as a function of the temperature for $CO_2(+)$, $HCO(\diamond)$, $CO(\circ)$ and $CH_3(\square)$ for a methane-air, counterflow diffusion flame with $a = 100 \ \text{sec}^{-1}$.

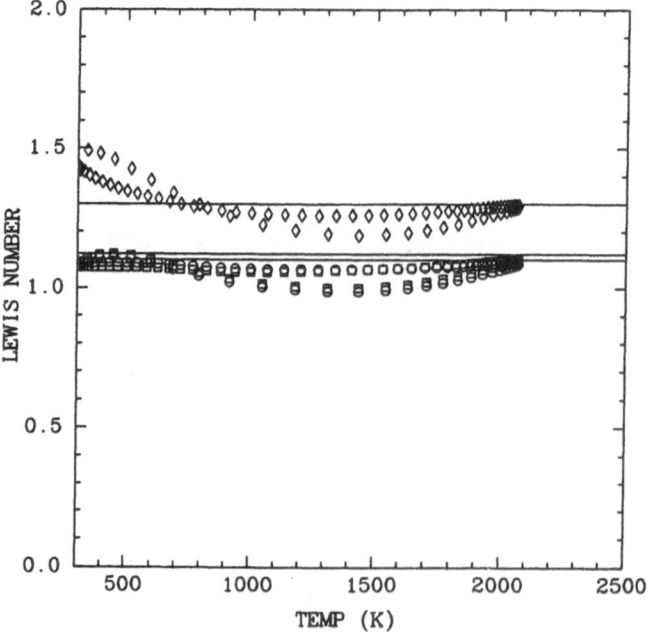

Figure 22. A comparison between the detailed transport-skeletal chemistry Lewis numbers and the least squares fits (solid) as a function of the temperature for $CH_3O(\diamond)$, $O_2(\circ)$ and $H_2O_2(\square)$ for a methane-air, counterflow diffusion flame with $a = 100 \ \text{sec}^{-1}$.

conservation, i.e.,

$$Y_{N_2} = 1 - \sum_{k=1}^{K-1} Y_k. \tag{3.15}$$

TABLE I

Simplified Transport Model Lewis Numbers

Species	Value
CH_4	0.97
O_2	1.11
H_2O	0.83
CO_2	1.39
H	0.18
O	0.70
OH	0.73
HO_2	1.10
H_2	0.30
CO	1.10
H_2O_2	1.12
HCO	1.27
CH_2O	1.28
CH_3	1.00
CH_3O	1.30
N_2	1.00

If needed, the diffusion velocity of N_2 can be back calculated from (3.13). With this formulation, the mass diffusivity can be related to the transport model in (3.4-3.5) via the Lewis numbers, i.e.,

$$\rho D_k = \frac{1}{Le_k} \left(\frac{\lambda}{c_p} \right) = \left(\frac{2.58 \times 10^{-4}}{Le_k} \right) \left(\frac{T}{298} \right)^{0.7}. \tag{3.16}$$

Viscosity Model

The momentum equation in the counterflow model contains a term involving the viscosity of the mixture. From the definition of the Prandtl number we have

$$Pr = \frac{\mu c_p}{\lambda} \tag{3.17}$$

By selecting a constant Prandtl number equal to 0.75 we can write

$$\frac{d}{dy}\left(\mu\frac{dU}{dy}\right) = \Pr\frac{d}{dy}\left(\frac{\lambda}{c_p}\frac{dU}{dy}\right). \tag{3.18}$$

4. Chemistry Model

The skeletal mechanism from which the reduction process will begin is listed in Table II. It contains 10 reversible and 15 irreversible reactions. Of particular importance is the reaction

$$CH_4 + (M) \rightleftharpoons CH_3 + H + (M) \tag{4.1}$$

This reaction is known to be pressure dependent. To account for the functional dependence on the pressure we have employed a Lindemann approximation for the forward and reverse rate constants. Specficially, the rates are given by the relation

$$k = \frac{k_\infty}{1 + k_{fall}/[M]}, \tag{4.2}$$

where k_∞ is the appropriate high pressure rate constant, k_{fall} is the fall-off rate and $[M]$ is the third body concentration equal to p/RT.

In the course of performing the test problem calculations with the skeletal mechanism and the model simplifications discussed above, it was found that flame speeds at one atmosphere were 42.5 cm/sec and the extinction limit for the counterflow flame was $a = 391$ sec^{-1}. This is higher than the experimental flame speed measurements which fall in the 37-40 cm/sec neighborhood and it is higher than the measured extinction strain rate of 350 sec^{-1}. Warnatz [46] has employed a Kassel sum formulation to approximate the pressure dependence of (4.1). He has tabulated values for this rate at several pressures. If the Warnatz rate for the reverse reaction of (4.1) at one atmosphere is compared with the fall-off rate at one atmosphere as a function of temperature (Figure 23), we see that the Warnatz rate is higher than the fall-off rate in the region between 500-2500 K. It turns out that these differences are critical in the calculation of both premixed flame speeds and in the calculation of the extinction limit for the counterflow flame. To incorporate the preferred Warnatz rate within the Lindemann form, we include a correction factor α such that at one atmosphere (p_{atm}) the Warnatz rate (k_W) is equal to the value defined by (4.2) as a function of temperature. Specifically, we have

$$k_W = \frac{\alpha k_\infty}{1 + k_{fall}/[M]}. \tag{4.3}$$

Solving for α we find

$$\alpha = (1 + k_{fall}\frac{RT}{p_{atm}})\left(\frac{k_W}{k_\infty}\right). \tag{4.4}$$

The inclusion of this factor is critical in obtaining excellent agreement with experiments for both premixed and nonpremixed flames. For the forward direction of (4.1) we employ

$$k_{W_f} = 2.3 \times 10^{38}(T^{-7}) \exp\left(-114360/RT\right), \tag{4.5}$$

and for the reverse direction

$$k_{W_b} = 1.9 \times 10^{36}(T^{-7}) \exp\left(-9050/RT\right). \tag{4.6}$$

TABLE II

Skeletal Methane-Air Reaction Mechanism
Rate Coefficients in the Form $k_f = AT^\beta \exp(-E_0/RT)$.
Units are moles, cubic centimeters, seconds, Kelvins and calories/mole.

	REACTION	A	β	E
1f.	$H + O_2 \rightarrow OH + O$	2.000E+14	0.000	16800.
1b.	$OH + O \rightarrow H + O_2$	1.575E+13	0.000	690.
2f.	$O + H_2 \rightarrow OH + H$	1.800E+10	1.000	8826.
2b.	$OH + H \rightarrow O + H_2$	8.000E+09	1.000	6760.
3f.	$H_2 + OH \rightarrow H_2O + H$	1.170E+09	1.300	3626.
3b.	$H_2O + H \rightarrow H_2 + OH$	5.090E+09	1.300	18588.
4f.	$OH + OH \rightarrow O + H_2O$	6.000E+08	1.300	0.
4b.	$O + H_2O \rightarrow OH + OH$	5.900E+09	1.300	17029.
5.	$H + O_2 + M \rightarrow HO_2 + M^a$	2.300E+18	-0.800	0.
6.	$H + HO_2 \rightarrow OH + OH$	1.500E+14	0.000	1004.
7.	$H + HO_2 \rightarrow H_2 + O_2$	2.500E+13	0.000	700.
8.	$OH + HO_2 \rightarrow H_2O + O_2$	2.000E+13	0.000	1000.
9f.	$CO + OH \rightarrow CO_2 + H$	1.510E+07	1.300	-758.
9b.	$CO_2 + H \rightarrow CO + OH$	1.570E+09	1.300	22337.
10f.	$CH_4 + (M) \rightarrow CH_3 + H + (M)^b$	6.300E+14	0.000	104000.
10b.	$CH_3 + H + (M) \rightarrow CH_4 + (M)^b$	5.200E+12	0.000	-1310.
11f.	$CH_4 + H \rightarrow CH_3 + H_2$	2.200E+04	3.000	8750.
11b.	$CH_3 + H_2 \rightarrow CH_4 + H$	9.570E+02	3.000	8750.
12f.	$CH_4 + OH \rightarrow CH_3 + H_2O$	1.600E+06	2.100	2460.
12b.	$CH_3 + H_2O \rightarrow CH_4 + OH$	3.020E+05	2.100	17422.
13.	$CH_3 + O \rightarrow CH_2O + H$	6.800E+13	0.000	0.
14.	$CH_2O + H \rightarrow HCO + H_2$	2.500E+13	0.000	3991.
15.	$CH_2O + OH \rightarrow HCO + H_2O$	3.000E+13	0.000	1195.
16.	$HCO + H \rightarrow CO + H_2$	4.000E+13	0.000	0.
17.	$HCO + M \rightarrow CO + H + M$	1.600E+14	0.000	14700.
18.	$CH_3 + O_2 \rightarrow CH_3O + O$	7.000E+12	0.000	25652.
19.	$CH_3O + H \rightarrow CH_2O + H_2$	2.000E+13	0.000	0.
20.	$CH_3O + M \rightarrow CH_2O + H + M$	2.400E+13	0.000	28812.
21.	$HO_2 + HO_2 \rightarrow H_2O_2 + O_2$	2.000E+12	0.000	0.
22f.	$H_2O_2 + M \rightarrow OH + OH + M$	1.300E+17	0.000	45500.
22b.	$OH + OH + M \rightarrow H_2O_2 + M$	9.860E+14	0.000	-5070.
23f.	$H_2O_2 + OH \rightarrow H_2O + HO_2$	1.000E+13	0.000	1800.
23b.	$H_2O + HO_2 \rightarrow H_2O_2 + OH$	2.860E+13	0.000	32790.
24.	$OH + H + M \rightarrow H_2O + M^a$	2.200E+22	-2.000	0.
25.	$H + H + M \rightarrow H_2 + M^a$	1.800E+18	-1.000	0.

[a] Third body efficiencies: $CH_4 = 6.5, H_2O = 6.5, CO_2 = 1.5, H_2 = 1.0, CO = 0.75, O_2 = 0.4, N_2 = 0.4$ All other species $=1.0$

[b] Lindemann form, $k = k_\infty/(1 + k_{fall}/[M])$ where $k_{fall} = .0063 \exp(-18000/RT)$.

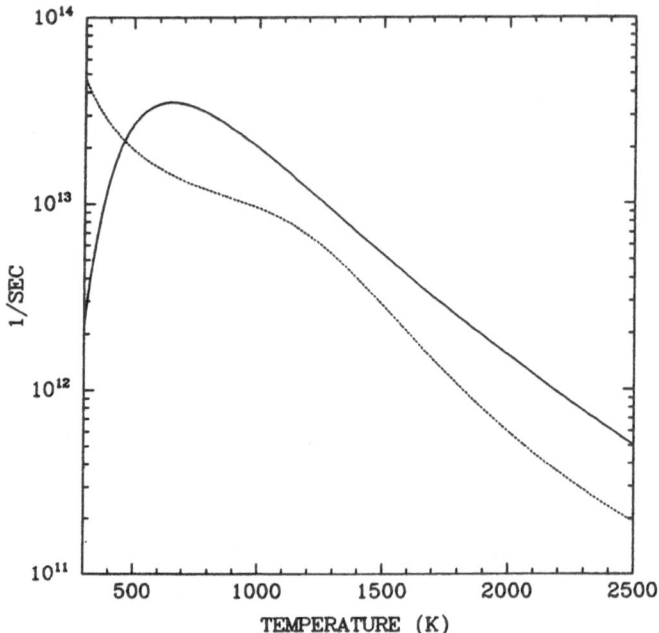

Figure 23. A comparison between the rate in Eqn. (4.6) (solid) and the corresponding fall-off rate (dot) for reaction 10b (Table II) as a function of the temperature.

5. Final Form of the Governing Systems

With the approximations and simplifications made in Sections 3 and 4, the final form of the governing equations for the premixed system becomes

$$\frac{d\dot{M}}{dx} = 0, \tag{5.1}$$

$$\dot{M}\frac{dY_k}{dx} - \frac{1}{\mathrm{Le}_k}\frac{d}{dx}\left(\frac{\lambda}{c_p}\frac{dY_k}{dx}\right) - \dot{w}_k W_k = 0, \qquad k = 1, 2, \ldots, K-1, \tag{5.2}$$

$$\dot{M}\frac{dT}{dx} - \frac{d}{dx}\left(\frac{\lambda}{c_p}\frac{dT}{dx}\right) - \frac{\lambda}{c_p^2}\frac{dc_p}{dx}\frac{dT}{dx} + \frac{1}{c_p}\sum_{k=1}^{K}\dot{w}_k h_k W_k = 0, \tag{5.3}$$

$$Y_{N_2} = 1 - \sum_{k=1}^{K-1} Y_k, \tag{5.4}$$

$$\rho = \frac{p\overline{W}}{RT}. \tag{5.5}$$

with the boundary conditions at $x = 0$ given by

$$T(0) = T_i, \tag{5.6}$$

$$\dot{M}Y_k(0) + \rho(0)Y_k(0)V_k(0) = \dot{M}\epsilon_k(\phi), \quad k = 1, 2, \ldots, K, \tag{5.7}$$

$$\lambda(0)\frac{dT}{dx}(0) = \sum_{k=1}^{K}\epsilon_k(\phi)\Big(h_k(T(0)) - h_k(T_u)\Big), \tag{5.8}$$

and at $x = L$ by

$$\frac{dT}{dx}(L) = 0,$$ (5.9)

$$\frac{dY_k}{dx}(L) = 0, \qquad k = 1,\ldots,K.$$ (5.10)

The corresponding form of the counterflow diffusion flame equations is given by

$$\frac{dV}{dy} + a\rho U = 0,$$ (5.11)

$$V\frac{dU}{dy} - \Pr\frac{d}{dy}\left(\frac{\lambda}{c_p}\frac{dU}{dy}\right) - a(\rho_\infty - \rho U^2) = 0,$$ (5.12)

$$V\frac{dY_k}{dy} - \frac{1}{\mathrm{Le}_k}\frac{d}{dy}\left(\frac{\lambda}{c_p}\frac{dY_k}{dy}\right) - \dot{w}_k W_k = 0, \qquad k = 1,2,\ldots,K-1,$$ (5.13)

$$V\frac{dT}{dy} - \frac{d}{dy}\left(\frac{\lambda}{c_p}\frac{dT}{dy}\right) - \frac{\lambda}{c_p^2}\frac{dc_p}{dy}\frac{dT}{dy} + \frac{1}{c_p}\sum_{k=1}^{K}\dot{w}_k h_k W_k = 0,$$ (5.14)

$$Y_{N_2} = 1 - \sum_{k=1}^{K-1} Y_k,$$ (5.15)

$$\rho = \frac{p\overline{W}}{RT},$$ (5.16)

with the boundary conditions at $y = 0$ given by

$$V(0) = V_w,$$ (5.17)
$$U(0) = 0,$$ (5.18)
$$V_w Y_k(0) + \rho Y_k(0) V_k = V_w \epsilon_k, \quad k = 1,2,\ldots,K,$$ (5.19)
$$T(0) = T_w,$$ (5.20)

and as $y \to L$ by

$$U = 1,$$ (5.21)
$$Y_k = Y_{k_\infty}, \quad k = 1,2,\ldots,K,$$ (5.22)
$$T = T_\infty.$$ (5.23)

References

[1] Peters, N., "Numerical and Asymptotic Analysis of Systematically Reduced Reaction Schemes for Hydrocarbon Flames," in Numerical Simulation of Combustion Phenomena, R. Glowinski et al., Eds., Lecture Notes in Physics, Springer-Verlag, (1985), p. 90.

[2] Peters, N. and Kee, R. J., "The Computation of Stretched Laminar Methane-Air Diffusion Flames Using a Reduced Four-Step Mechanism," *Comb. and Flame*, **68**, (1987), p. 17.

[3] Peters, N. and Williams, F. A., "The Asymptotic Structure of Stoichiometric Methane Air Flames," *Comb. and Flame*, **68**, (1987), p. 185.

[4] Bilger, R. W., Starner, S. H. and Kee, R. J., "On Reduced Mechanisms for Methane-Air Combustion in Non-Premixed Flames," to appear in *Comb. and Flame*, (1990).

[5] Spalding, D. B., "The Theory of Flame Phenomena with a Chain Reaction," *Phil. Trans. Roy. Soc. London*, **249A** (1956), p. 1.

[6] Adams, G. K. and Cook, G. B., "Mechanism and Speed of the Hydrazine Decomposition Flame," *Comb. and Flame*, **4** (1960), p. 9.

[7] Dixon-Lewis, G., "Flame Structure and Flame Reaction Kinetics I. Solution of Conservation Equations and Application to Rich Hydrogen-Oxygen Flames," *Proc. Roy. Soc. London*, **298A**, (1967), p. 495.

[8] Dixon-Lewis, G., "Kinetic Mechanism, Structure, and Properties of Premixed Flames in Hydrogen-Oxygen-Nitrogen Mixtures," *Phil. Trans. of the Royal Soc. London*, **292**, (1979), p. 45.

[9] Spalding, D. B., Stephenson D. L. and Taylor, R. G., "A Calculation Procedure for the Prediction of Laminar Flame Speeds," *Comb. and Flame.*, **17**, (1971), p. 55.

[10] Wilde, K. A., "Boundary-Value Solutions of the One-Dimensional Laminar Flame Propagation Equations," *Comb. and Flame*, **18**, (1972), p. 43.

[11] Bledjian, L., "Computation of Time-Dependent Laminar Flame Structure," *Comb. and Flame*, **20**, (1973) p. 5.

[12] Margolis, S. B., "Time-Dependent Solution of a Premixed Laminar Flame," *J. Comp. Phys.*, **27**, (1978), p. 410.

[13] Warnatz, J., "Calculation of the Structure of Laminar Flat Flames I; Flame Velocity of Freely Propagating Ozone Decomposition Flames," *Ber. Bunsenges. Phys. Chem.*, **82**, (1978), p. 193.

[14] Warnatz, J., "Calculation of the Structure of Laminar Flat Flames II; Flame Velocity and Structure of Freely Propagating Hydrogen-Oxygen and Hydrogen-Air-Flames," *Ber. Bunsenges. Phys. Chem.*, **82**, (1978), p. 643.

[15] Warnatz, J., "Calculation of the Structure of Laminar Flat Flames III; Structure of Burner-Stabilized Hydrogen-Oxygen and Hydrogen-Fluorine Flames," *Ber. Bunsenges. Phys. Chem.*, **82**, (1978), p. 834.

[16] Westbrook C. K. and Dryer, F. L., "A Comprehensive Mechanism for Methanol Oxidation," *Comb. Sci. and Tech.*, **20**, (1979), p. 125.

[17] Westbrook, C. K. and Dryer, F. L., "Prediction of Laminar Flame Properties of Methanol Air Mixtures," *Comb. and Flame*, **37**, (1980), p. 171.

[18] Coffee T. P. and Heimerl, J. M., "The Detailed Modeling of Premixed, Laminar Steady-State-Flames. I. Ozone," *Comb. and Flame*, **39**, (1980) p. 301.

[19] Coffee, T. P. and Heimerl, J. M., "Transport Algorithms for Premixed, Laminar, Steady-State Flames," *Comb. and Flame*, **43**, (1981), p. 273.

[20] Miller, J. A., Mitchell, R. E., Smooke, M. D., and Kee, R. J., "Toward a Comprehensive Chemical Kinetic Mechanism for the Oxidation of Acetylene: Comparison of Model Predictions with Results from Flame and Shock Tube Experiments," Nineteenth Symposium (International) on Combustion, Reinhold, New York, (1982), p. 181.

[21] Miller, J. A., Smooke, M. D., Green, R. M. and Kee, R. J., "Kinetic Modeling of the Oxidation of Ammonia in Flames," *Comb. Sci. and Tech.*, **34**, (1983), p. 149.

[22] Miller, J. A., Kee, R. J., Smooke, M. D. and Grcar, J. F., "The Computation of the Structure and Extinction Limit of a Methane-Air Stagnation Point Diffusion Flame," Paper # WSS/CI 84-10 presented at the 1984 Spring Meeting of the Western States Section of the Combustion Institute, University of Colorado, Boulder, CO, April 2-3, 1984.

[23] Smooke, M. D., "Solution of Burner Stabilized Premixed Laminar Flames by Boundary Value Methods," *J. Comp. Phys.*, **48**, (1982), p. 72.

[24] Smooke, M. D., Miller, J. A. and Kee, R. J., "On the Use of Adaptive Grids in Numerically Calculating Adiabatic Flame Speeds," Numerical Methods in Laminar Flame Propagation, N. Peters and J. Warnatz (Eds.). Friedr. Vieweg and Sohn, Wiesbaden, (1982).

[25] Smooke, M. D., Miller, J. A. and Kee, R. J., "Determination of Adiabatic Flame Speeds by Boundary Value Methods," *Comb. Sci. and Tech.*, **34**, (1983), p. 79.

[26] Smooke, M. D., Miller, J. A. and Kee, R. J., "Solution of Premixed and Counter-flow Diffusion Flame Problems by Adaptive Boundary Value Methods," Numerical Boundary Value ODEs, U. M. Ascher and R. D. Russell, eds., Birkhäuser, Boston, (1985), p. 303.

[27] Smooke, M. D., "On the Use of Adaptive Grids in Premixed Combustion," *AIChE J.*, **32**, (1986), p. 1233.

[28] Hahn, W. A. and Wendt, J. O. L., "NO_x Formation in Flat, Laminar, Opposed Jet Methane Diffusion Flames," Eighteenth Symposium (International) on Combustion, Reinhold, New York, (1981), p. 121.

[29] Sato, J. and Tsuji, H., "Extinction of Premixed Flames in a Stagnation Flow Considering General Lewis Number," *Comb. Sci. and Tech.*, **33**, (1983), p. 193.

[30] Dixon-Lewis, G., David, T., Haskell, P. H., Fukutani, S., Jinno, H., Miller, J. A., Kee, R. J., Smooke, M. D., Peters, N., Effelsberg, E., Warnatz, J. and Behrendt, F., "Calculation of the Structure and Extinction Limit of a Methane-Air Counterflow Diffusion Flame in the Forward Stagnation Region of a Porous Cylinder," Twentieth Symposium (International) on Combustion, Reinhold, New York, (1985), p. 1893.

[31] Giovangigli, V. and Smooke, M. D., "Extinction of Strained Premixed Laminar Flames with Complex Chemistry," *Comb. Sci. and Tech.*, **53**, (1987), p. 23.

[32] Williams, F. A., "Recent Advances in Theoretical Descriptions of Turbulent Diffusion Flames," in Turbulent Mixing in Non-Reactive and Reactive Flows, S. N. B. Murthy, ed., Plenum Press, New York, (1975), p. 189.

[33] Peters, N., "Laminar Diffusion Flamelet Models in Nonpremixed Turbulent Combustion," *Prog. Energy Comb. Sci.* **10**, (1984), p. 319.

[34] Liew, S. K., Bray, K. N. C. and Moss, J. B., "A Flamelet Model of Turbulent Non-Premixed Combustion," *Comb. Sci. and Tech.*, **27**, (1981), p. 69.

[35] Liew, S. K., Bray, K. N. C. and Moss, J. B., "A Stretched Laminar Flamelet Model of Turbulent Nonpremixed Combustion," *Comb. and Flame*, **56**, (1984), p. 199.

[36] Rogg, B., Behrendt, F. and Warnatz, J., "Turbulent Non-Premixed Combustion in Partially Premixed Diffusion Flamelets with Detailed Chemistry," Twenty-First Symposium (International) on Combustion, Reinhold, New York, (1986), p. 1533.

[37] Tsuji, H. and Yamaoka, I., "The Counterflow Diffusion Flame in the Forward Stagnation Region of a Porous Cylinder," Eleventh Symposium (International) on Combustion, Reinhold, New York, (1967), p. 979.

[38] Tsuji, H. and Yamaoka, I., "The Structure of Counterflow Diffusion Flames in the Forward Stagnation Region of a Porous Cylinder," Twelfth Symposium (International) on Combustion, Reinhold, New York, (1969), p. 997.

[39] Tsuji, H. and Yamaoka, I., "Structure Analysis of Counterflow Diffusion Flames in the Forward Stagnation Region of a Porous Cylinder," Thirteenth Symposium (International) on Combustion, Reinhold, New York, (1971), p. 723.

[40] Tsuji, H., "Counterflow Diffusion Flames," *Progress in Energy and Comb.*, **8**, (1982), p. 93.

[41] Smooke, M. D., Puri, I. K. and Seshadri, K., "A Comparison Between Numerical Calculations and Experimental Measurements of the Structure of a Counterflow Diffusion Flame Burning Diluted Methane in Diluted Air," Twenty-First Symposium (International), (1986), p. 1783.

[42] Curtiss, C. F. and Hirschfelder, J. O., "Transport Properties of Multicomponent Gas Mixtures," *J. Chem. Phys.*, **17**, (1949), p. 550.

[43] Kee, R. J., Miller, J. A. and Jefferson, T. H., "CHEMKIN: A General-Purpose, Transportable, Fortran Chemical Kinetics Code Package," Sandia National Laboratories Report, SAND80-8003, (1980).

[44] Kee, R. J., Warnatz, J., and Miller, J. A., "A Fortran Computer Code Package for the Evaluation of Gas-Phase Viscosities, Conductivities, and Diffusion Coefficients," Sandia National Laboratories Report, SAND83-8209, (1983).

[45] Giovangigli, V. and Smooke, M. D., "Application of Continuation Methods to Premixed Laminar Flames," submitted to *Comb. Sci. and Tech.*, (1990).

[46] Warnatz, J., "The Mechanism of High Temperature Combustion of Propane and Butane," *Comb. Sci. and Tech.* **34**, (1983), p. 177.

CHAPTER 2

PREMIXED AND NONPREMIXED
TEST PROBLEM RESULTS

Mitchell D. Smooke
Department of Mechanical Engineering
Yale University
New Haven, Connecticut

and

Vincent Giovangigli
Ecole Polytechnique and CNRS
Centre de Mathématiques Appliquées
91128, Palaiseau cedex, France

1. Introduction

In this chapter we present the results of applying the model simplifications discussed in the previous chapter with skeletal methane-air chemistry to a sequence of premixed and nonpremixed flames. We focus our results on adiabatic flames speeds, extinction strain rates, temperature and species profiles along with reaction rate data. Corresponding results for full transport calculations are often provided as a means of verifying the simplified transport model. Results are reported in the independent spatial coordinate, a normalized spatial coordinate and the mixture fraction.

2. Solution Method

In both the premixed and nonpremixed computations we are interested in following the solution as a system parameter is varied. For example, in the premixed problems we are interested in allowing the equivalence ratio to change as the pressure is held fixed. Similarly, we are also interested in varying the pressure for a fixed equivalence ratio. For the diffusion flames, the strain rate is the parameter of interest. While we could compute a single flame with specified values of these parameters and then use this computed solution as a starting estimate for a new problem with different parameter values, this is extremely inefficient. Instead, we apply an arclength continuation method such that the grid and the solution smoothly change as the parameter is varied.

Specifically, the solution algorithm we implement proceeds with a phase-space, pseudo-arclength continuation method with Newton-like iterations and global adaptive gridding [1-2]. After we replace the continuous spatial derivatives by finite difference expressions, the premixed model in Eqns. (5.1-5.10) of Chapter 1 and the diffusion flame model in Eqns. (5.11-5.23) of Chapter 1 reduce to a system of the form

$$\mathcal{F}(\mathcal{X},\gamma) = 0, \qquad (2.1)$$

where $\mathcal{X}$ is the solution vector and γ is a system parameter (such as the equivalence ratio, the pressure or the strain rate). The solutions $(\mathcal{X},\gamma)$ in (2.1) form a one-dimensional manifold which, as a result of the presence of turning points, cannot be parameterized in the form $(X(\gamma), \gamma)$. The upper part of the manifold denotes the stable solutions and the lower part the unstable ones assuming there are no Hopf bifurcations.

To generate this solution set, (X, γ) is reparameterized into $(X(s), \gamma(s))$ where s is a new independent parameter and γ becomes an eigenvalue. The system in (2.1) can now be written

$$\mathcal{F}(X(s), \gamma(s)) = 0, \tag{2.2}$$

and the dependence of s on the augmented solution vector (X, γ) is specified by an extra scalar equation

$$\mathcal{N}(X(s), \gamma(s), s) = 0, \tag{2.3}$$

which is chosen such that s approximates the arclength of the solution branch in a given phase space (see [1-2]). Rather than solving the coupled system in (2.2) and (2.3), we replace γ by a function Γ of x and we let the unknown $Z = (X^T, \Gamma)^T$ be the solution of a three-point limit value problem

$$\mathcal{H}(Z, s) = \begin{bmatrix} \mathcal{F}(X, \Gamma) \\ \dfrac{d\Gamma}{dx} \\ \mathcal{N}(X(\hat{x}), \Gamma(\hat{x}), s) \end{bmatrix} = 0, \tag{2.4}$$

where $\hat{x}$ is a given point in $[0, L]$. The main advantage of considering (2.4) is that, if (2.2) can be discretized in a block-tridiagonal form, then (2.4) has the same property. The numerical Jacobian matrix obtained from (2.4) has a block-tridiagonal structure and one can prove easily that it is nonsingular at simple turning points. Specially developed block tridiagonal linear equation solvers can then be used to invert the corresponding linear systems.

The system in (2.4) is solved by combining a first-order Euler predictor and a corrector step involving Newton-like iterations and adaptive gridding. To resolve the high activity regions of the dependent solution components, the mesh must be refined in the continuation calculations. Specifically, after solving the governing equations on a previously determined mesh, we determine a new equidistributed mesh

$$Z^* = \{0 = x_1^* < x_2^* < \ldots < x_{n^*}^* = L\}, \tag{2.5}$$

by imposing the condition

$$\int_{x_i^*}^{x_{i+1}^*} w \; dx = 1, \quad 1 \le i \le n^* - 1, \tag{2.6}$$

where the weight function w depends upon the gradient and curvature of the dependent solution components and on mesh regularity properties [2]. The governing equations are solved on this new mesh and then another continuation step is taken. It optimizes the number of points and determines the new grid in only one pass. In addition, the method also takes into account every component of the solution in forming a new mesh.

Once a solution to the flame equations has been obtained, it is often useful to investigate those parameters to which the solution is most "sensitive". Sensitivity analysis is an important tool for the physical investigation and validation of mathematical models (see, e. g., [4]). In the past, however, the major obstacle in obtaining sensitivity information systematically was the additional amount of computation required for solving the sensitivity equations. An advantage of solving the premixed and counterflow flames by a Newton based method is that first-order sensitivity information can be obtained at a fraction of the cost of the total calculation [5].

Sensitivity analysis enables the investigator to help predict the variation of a parameter vector β on the dependent variables in the problem. The vector β may contain rate constants, mass flux fractions, etc.. In particular, the quantities of interest are the first order sensitivity coefficients $\partial X / \partial \beta_k$, $k = 1, 2, \ldots, N$. The appropriate equations for these quantities can be derived by differentiating (2.1) with respect to β_k. We have

$$\frac{d}{d\beta_k}\left(\mathcal{F}(X;\beta)\right) = \frac{\partial \mathcal{F}}{\partial X}\frac{\partial X}{\partial \beta_k} + \frac{\partial \mathcal{F}}{\partial \beta_k} = 0, \quad k = 1, 2, \ldots, N. \tag{2.7}$$

Upon recalling that the Jacobian matrix $J = \partial \mathcal{F} / \partial X$, we can re-arrange (2.7) in the form

$$J\frac{\partial X}{\partial \beta_k} = -\frac{\partial \mathcal{F}}{\partial \beta_k}, \quad k = 1, 2, \ldots, N. \tag{2.8}$$

We remark that although (2.8) can be solved at any step of the Newton iteration and at any level of grid refinement, we ordinarily solve the sensitivity equations on the finest grid with the last Jacobian formed. In practice, it is only at this stage of the calculation that the numerical approximation to J can represent adequately the analytic Jacobian. In addition, since the Jacobian matrix on the final grid is already formed and factored, each sensitivity coefficient can be calculated by performing relatively inexpensive back substitutions.

3. Premixed Flames

The first set of calculations we consider focuses on the variation of flame structure as a function of the equivalence ratio ϕ. In Chapter 1 we indicated that we are interested in using the skeletal mechanism reported in Table II to compute flame structure for stoichiometric to lean methane-air flames at one atmosphere. Specifically, flames with equivalence ratios of $\phi = 1.0, 0.9, 0.8, 0.7$, and 0.6 are to be computed. By reparameterizing the premixed flame problem so that the equivalence ratio is the free parameter, we can apply the arclength continuation method discussed above to investigate premixed flame structure as the equivalence ratio changes. In this way we can accurately follow the movement of the flame while adaptively refining the flame front. This is in distinction to simply computing a flame for a given value of ϕ and then using this solution as the starting estimate for another computation with a different ϕ. Ordinarily, this approach will result in a mesh which could include the union of the two grids for the two flames. If this process is carried out over the entire equivalence ratio range, a very inefficient computation could result.

Three sets of equivalence ratio calculations were performed. One with the simple transport and the skeletal chemistry (denoted hereafter as "simple transport") described in the previous chapter, another with kinetic theory transport and skeletal chemistry (denoted hereafter as "full transport") and another with simple transport and the reduced four-step chemistry (Case C) discussed in the next chapter. In Figures 1 and 2 we illustrate the peak temperature and flame speed variation as ϕ is changed. The solid line corresponds to the simple transport calculations, the dashed line to the full transport calculations and the dotted line to the reduced chemistry computations. We note exceptional agreement over the entire range of ϕ considered. Peak temperatures are almost indistinguishable among the three cases and the flame speeds rarely differ by more than a couple of cm/sec. In Table I we report specific flame speeds for the three models for the five flames under study.

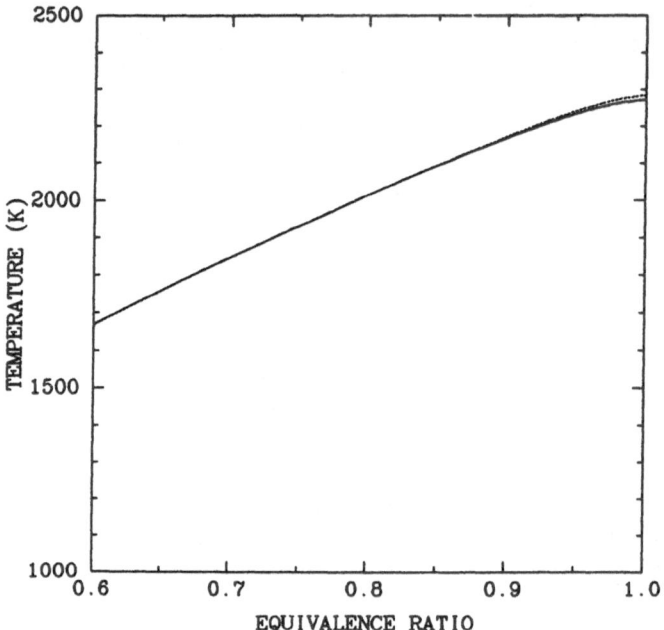

Figure 1. Variation of the peak temperature as a function of the equivalence ratio for simple transport-skeletal chemistry (solid), full transport-skeletal chemistry (dash) and simple transport-reduced chemistry (dot) models of premixed methane-air flames.

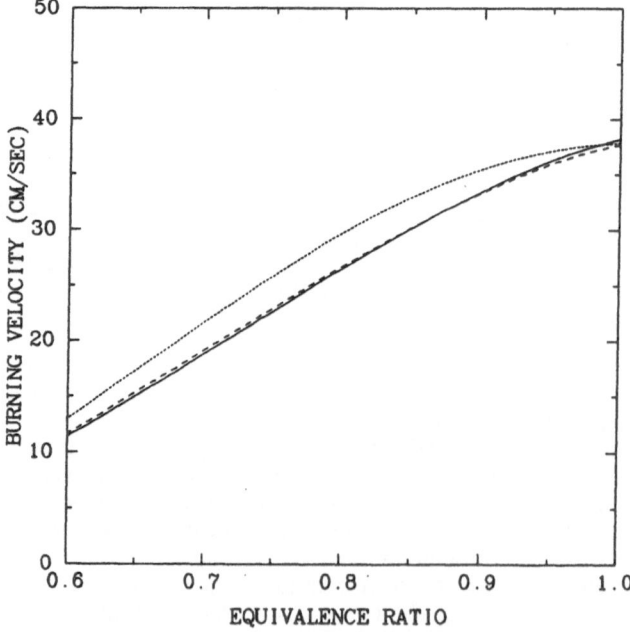

Figure 2. Variation of the burning velocity as a function of the equivalence ratio for simple transport-skeletal chemistry (solid), full transport-skeletal chemistry (dash) and simple transport-reduced chemistry (dot) models of premixed methane-air flames.

TABLE I

Premixed Flame Speeds (cm/sec)
Variation with Equivalence Ratio
p = 1 atm

Equivalence Ratio	Simple Transport	Full Transport	Reduced Chemistry
$\phi = 1.0$	37.67	38.48	37.90
$\phi = 0.9$	33.59	33.54	35.60
$\phi = 0.8$	26.50	26.62	29.76
$\phi = 0.7$	19.17	18.74	21.60
$\phi = 0.6$	11.59	11.40	13.02

The arclength continuation procedure outlined above can be modified to allow the pressure to become the free parameter in question. In this way we can investigate flame structure for variable pressure. This is particularly useful for flame studies in which the thermodynamic pressure varies by several atmospheres. The results in Figures 3 and 4 illustrate the variation of the peak temperature and the adiabatic flame speed as a function of the pressure. Computations are reported for the 1-30 atmosphere regime for the skeletal chemistry cases employing both simple (solid) and full transport (dash) approximations. No steady-state reduced chemistry solutions (dot) were obtainable above 10.5 atmospheres. As was the case for the equivalence ratio computations, the flame speed results are in excellent agreement across the full range of comparable pressures. Peak temperatures show some deviation (about 50K in the worst situation). Specific flame speeds are reported in Table II.

TABLE II

Premixed Flame Speeds (cm/sec)
Variation with Pressure
$\phi = 1.0$

Pressure in Atms	Simple Transport	Full Transport	Reduced Chemistry
p = 1.0	37.67	38.48	37.90
p = 5.0	18.75	18.20	15.39
p = 10.0	11.29	10.74	9.75
p = 20.0	6.81	6.43	-
p = 30.0	5.40	5.10	-

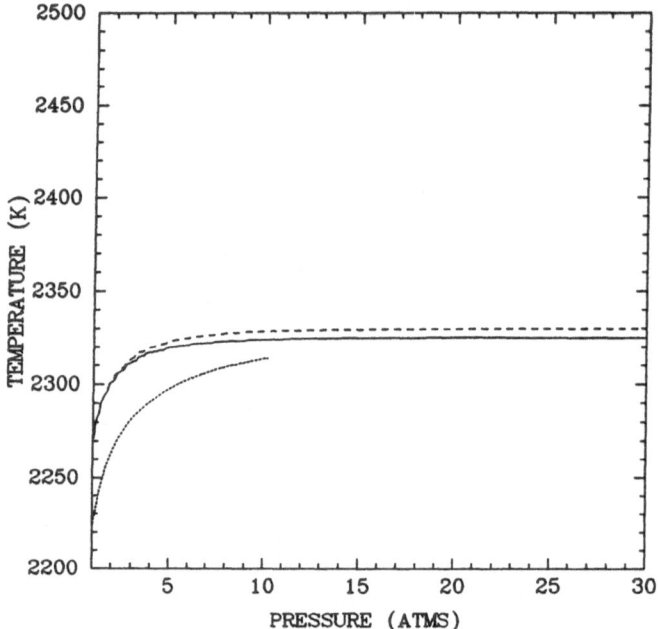

Figure 3. Variation of the peak temperature as a function of the pressure for simple transport-skeletal chemistry (solid), full transport-skeletal chemistry (dash) and simple transport-reduced chemistry (dot) models of premixed methane-air flames. No steady-state reduced chemistry solutions were obtained above 10.5 atmospheres.

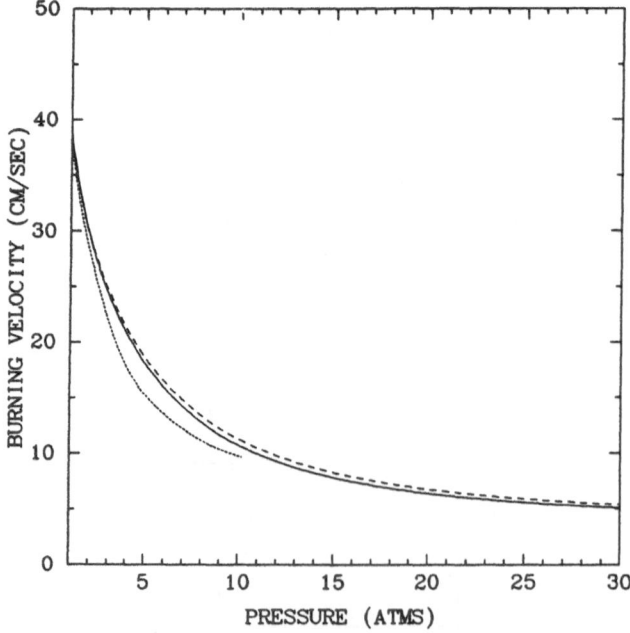

Figure 4. Variation of the burning velocity as a function of the pressure for simple transport-skeletal chemistry (solid), full transport-skeletal chemistry (dash) and simple transport-reduced chemistry (dot) models of premixed methane-air flames. No steady-state reduced chemistry solutions were obtained above 10.5 atmospheres.

Premixed flame structure for an atmospheric pressure, stoichiometric flame with simple transport and both skeletal and reduced chemistry are illustrated in Figures 5-7. The profiles are reported in terms of a normalized distance $\hat{x}$ through the flame. We define

$$\hat{x} = \dot{M}(x - x^o)c_p/\lambda, \tag{3.1}$$

where $\dot{M}$ is the mass flux through the flame, x^o is the location of the maximum fuel consumption and λ/c_p is evaluated from Eqns. (3.4-3.5) of Chapter 1. We observe excellent agreement in terms of peak heights and the general shape of the profiles as a function of the normalized distance coordinate. Specific tabulations of premixed flame structure for three of the flames with simple transport and both skeletal and reduced chemistry are given in Tables III and IV, respectively. T^o represents the temperature at the point of maximum fuel consumption.

TABLE III

Numerical Solutions for Premixed Flames
Simple Transport-Skeletal Chemistry
Pressure in Atmospheres

Flame Parameters	$\phi = 1.0, p = 1.0$	$\phi = 0.6, p = 1.0$	$\phi = 1.0, p = 10.0$
Flame Speed (cm/sec)	37.67	11.59	11.29
Peak H (mole)	7.38×10^{-3}	7.14×10^{-4}	1.04×10^{-3}
T^o (K)	1621	1360	1921
T_{max}(K)	2272	1668	2329

TABLE IV

Numerical Solutions for Premixed Flames
Simple Transport-Reduced Chemistry
Pressure in Atmospheres

Flame Parameters	$\phi = 1.0, p = 1.0$	$\phi = 0.6, p = 1.0$	$\phi = 1.0, p = 10.0$
Flame Speed (cm/sec)	37.90	13.02	9.75
Peak H (mole)	7.39×10^{-3}	8.81×10^{-4}	9.68×10^{-4}
T^o (K)	1577	1195	2037
T_{max}(K)	2285	1668	2314

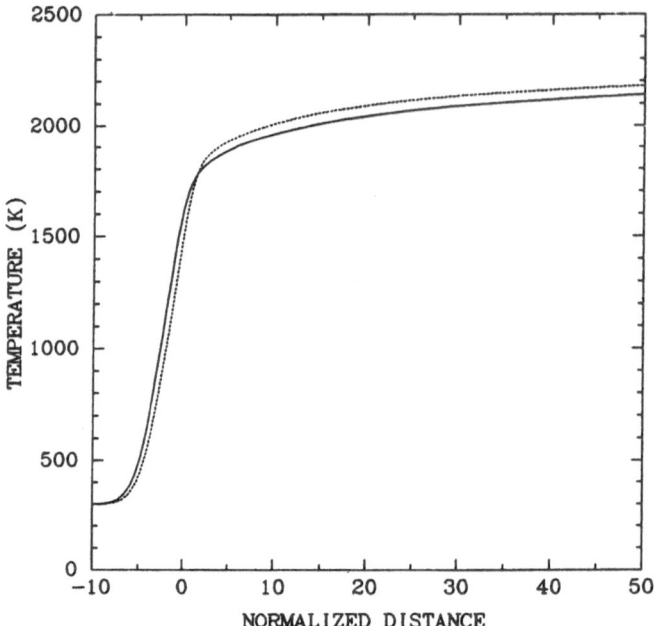

Figure 5. Computed temperature profiles as a function of the normalized distance for an atmospheric pressure, stoichiometric, methane-air flame employing a simple transport-skeletal chemistry (solid) and a simple transport-reduced chemistry (dot) model.

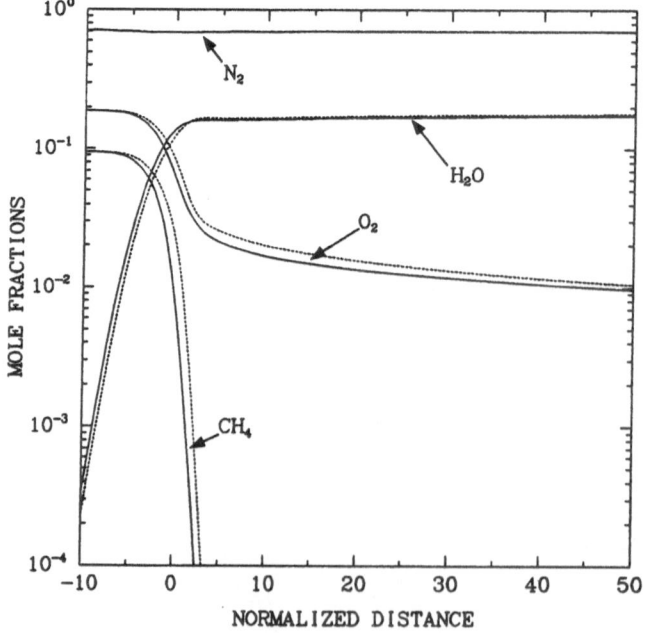

Figure 6. Computed major species profiles as a function of the normalized distance for an atmospheric pressure, stoichiometric, methane-air flame employing a simple transport-skeletal chemistry (solid) and a simple transport-reduced chemistry (dot) model.

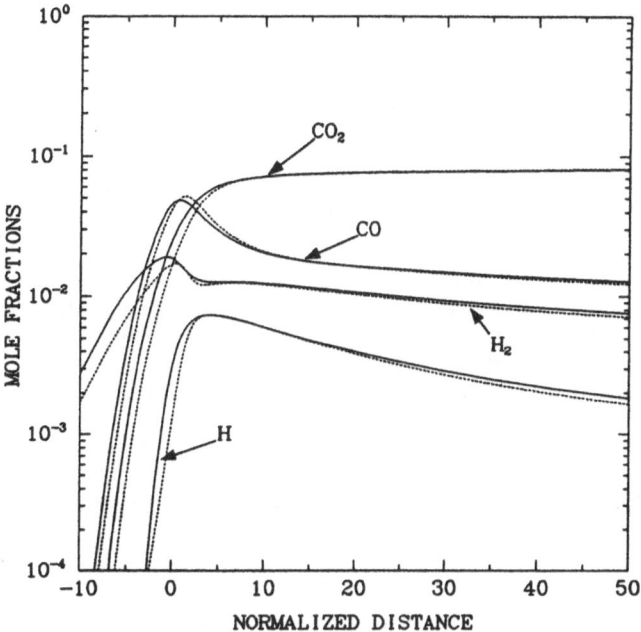

Figure 7. Computed major species profiles as a function of the normalized distance for an atmospheric pressure, stoichiometric, methane-air flame employing a simple transport-skeletal chemistry (solid) and a simple transport-reduced chemistry (dot) model.

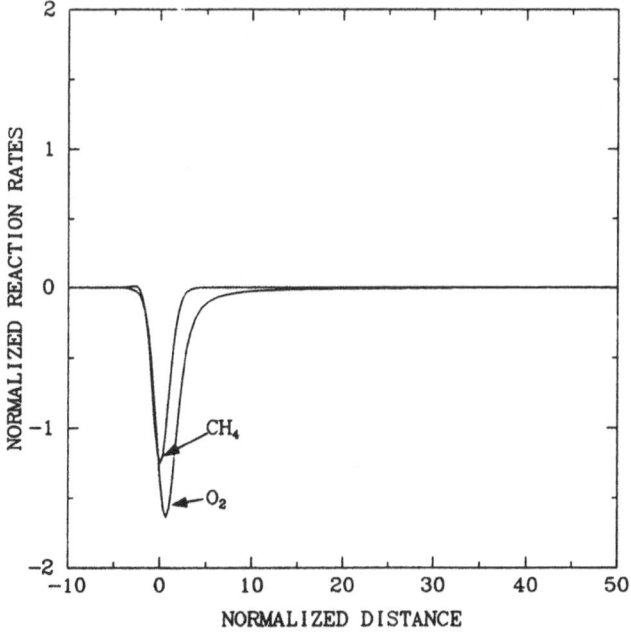

Figure 8. Computed normalized molar production rates as a function of the normalized distance for the major species in an atmospheric pressure, stoichiometric, methane-air flame employing a simple transport-skeletal chemistry model.

Molar production rates for an atmospheric pressure stoichiometric flame with simple transport and skeletal chemistry are illustrated in Figures 8-10. We nondimensionalize the production rate by dividing by the factor

$$\rho \dot{M} S_u \left(\frac{Y_{O_2}}{W_{O_2}} \right) \left(\frac{c_p}{\lambda} \right), \tag{3.2}$$

where ρ is the local mass density, S_u is the adiabatic flame speed and the other quantities are all evaluated at the unburnt conditions.

The ten largest log-normalized [5] flame speed sensitivity coefficients are reported in Table V (see also Chapter 8). The sensitivity implementation is such that each reaction is allowed to be reversible, i.e., each pair of forward and reverse reactions are combined into a single reversible reaction and each irreversible reaction is made into a reversible reaction. We then perturb the forward rate while keeping the equilibrium constant fixed. As a result, the reverse rate is perturbed also. If only the forward rate were perturbed, the sensitivity analysis would be performed with a different equilibrium constant which could affect the results. A positive sensitivity coefficient indicates that the flame speed will increase if the corresponding forward rate is increased. The opposite holds for a negative sensitivity coefficient.

TABLE V

Premixed Flame Speed Sensitivity Coefficients
$\phi = 1.0, p = 1.0$ atm

Reaction Number	Sensitivity Coefficient
1	1.110
9	0.329
5	-0.290
10	-0.200
11	-0.160
13	0.160
16	-0.140
17	0.140
3	0.089
15	0.030

4. Counterflow Diffusion Flames

Counterflow flames in the Tsuji configuration were studied from low strain rates until extinction using arclength continuation methods. As in the premixed case, computations were carried out for simple and full transport approximations with skeletal chemistry

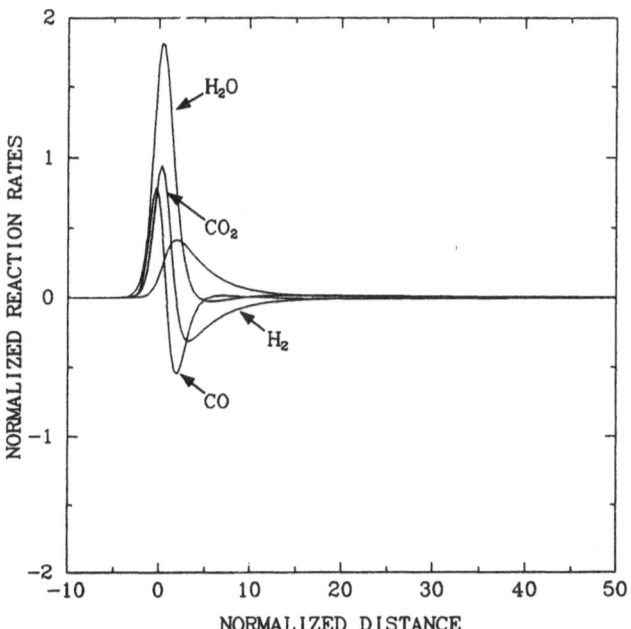

Figure 9. Computed normalized molar production rates as a function of the normalized distance for the major species in an atmospheric pressure, stoichiometric, methane-air flame employing a simple transport-skeletal chemistry model.

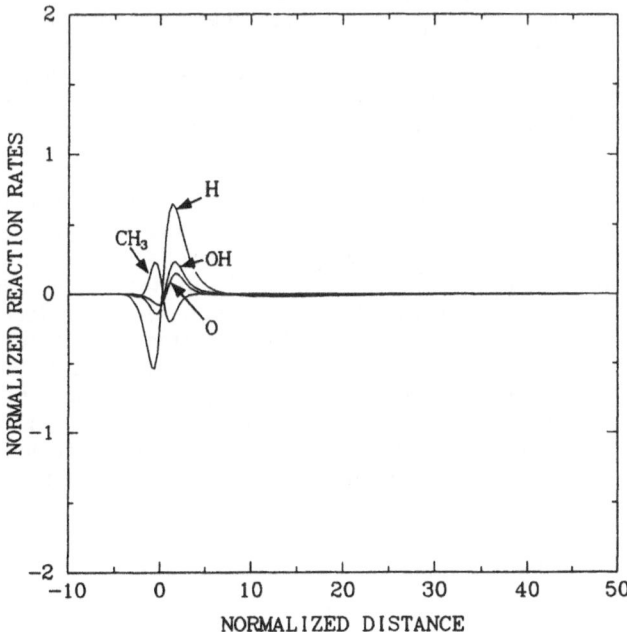

Figure 10. Computed normalized molar production rates as a function of the normalized distance for the major species in an atmospheric pressure, stoichiometric, methane-air flame employing a simple transport-skeletal chemistry model.

and for a simple transport-reduced chemistry model (Chapter 3, Case C). In Figure 11 we illustrate the "C-shaped" extinction curves for the three flame models. The solid line is for simple transport and skeletal chemistry, the dashed line is for full transport and skeletal chemistry and the dotted line is for simple transport and reduced chemistry. The two skeletal chemistry models produced almost identical results except for the region near extinction. Extinction for the simple transport flame occurred at $a = 353$ sec^{-1} and at $a = 361$ sec^{-1} for the full transport model. The upper portion of the reduced chemistry curve compares quite favorably with the results of the other two models. In particular, extinction occurs at $a = 323$ sec^{-1}. Significant variations exist in the lower unphysical branch compared with the skeletal chemistry solutions.

Temperature and major species profiles for a counterflow flame ($a = 100$ sec^{-1}) with simple transport and both skeletal and reduced chemistry are reported in Figures 12-14 in terms of the mixture fraction ξ where

$$\xi = \frac{2Z_C/W_C + .5Z_H/W_H + (Z_{O,O} - Z_O)/W_O}{2Z_{C,F}/W_F + .5Z_{H,F}/W_H + Z_{O,O}/W_O}, \tag{4.1}$$

which is defined in terms of the corresponding element mass fractions. Here $Z_{j,F}$ is the mass fraction at any location of element j contained in the fuel stream and $Z_{j,O}$ is the mass fraction at any location of element j contained in the oxidizer stream. The subscripts C H and O denote, respectively, the elements carbon, hydrogen and oxygen. We can write

$$Z_j = \sum_{i=1}^{K}(a_{ij}W_jY_i)/W_i, \tag{4.2}$$

where a_{ij} is a stoichiometric coefficient denoting the number of atoms of element j in molecule i. We notice exceptional agreement between the two profiles of the temperature, $CH_4, O_2, N_2, H_2O, CO_2$ and H. The major differences between the two chemistry models appear in the peak heights of CO and H_2. The reduced chemistry model predicts somewhat higher values of these species compared to the skeletal chemistry model. Reaction rate data for the simple transport-skeletal chemistry flame ($a = 100$ sec^{-1}) is contained in Figures 15-17. Here the molar production rate is divided by the local mass density.

A plot of the difference between the mixture fraction ξ_{st} at stoichiometric conditions and the mixture fraction ξ^o at the point of maximum fuel consumption is plotted as a function of the strain rate in Figure 18 for the simple transport-skeletal chemistry flame. The lower branch corresponds to physical solutions. For laboratory flames the variation in $\Delta\xi = \xi_{st} - \xi^o$ is almost constant. The variation in the value of the molecular oxygen, molecular hydrogen and carbon monoxide values at ξ^o for this model are reported in Figure 19 as a function of the strain rate (the upper branch corresponds to unphysical solutions). In Figure 20 we illustrate the oxygen leakage at $\xi = 0.1$ (the upper branch corresponds to unphysical solutions) as a function of the strain rate. Finally, in Tables VI and VII we summarize some of the results of the counterflow computations for simple transport-skeletal chemistry (Table VI) and simple transport-reduced chemistry (Table VII) flames.

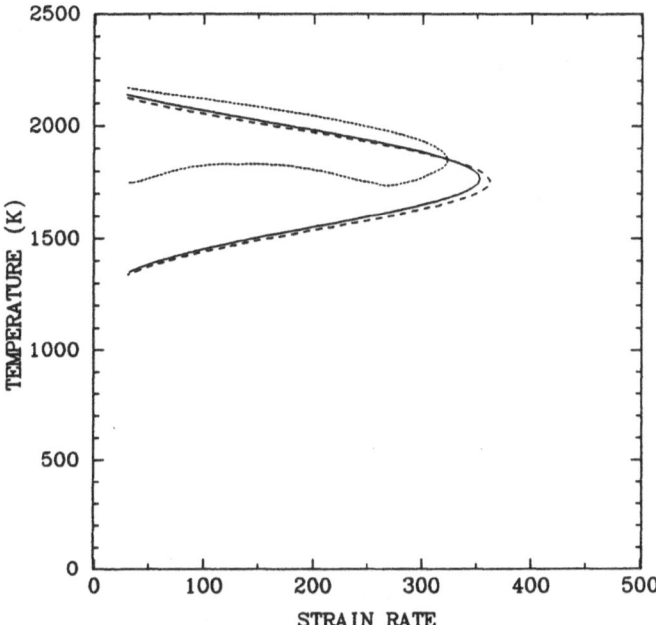

Figure 11. C-shaped extinction curves for the simple transport-skeletal chemistry (solid), the full transport-skeletal chemistry (dash) and the simple transport-reduced chemistry (dot) model of a counterflow, methane-air, diffusion flame. The upper branch corresponds to physical solutions and the lower branch to unphysical solutions.

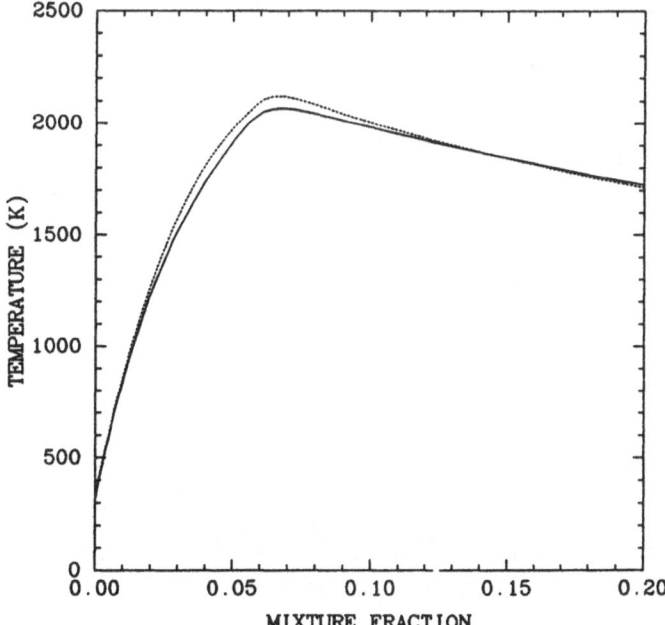

Figure 12. Computed temperature profiles as a function of the mixture fraction for a counterflow, methane-air, diffusion flame ($a = 100$ sec^{-1}) employing a simple transport-skeletal chemistry (solid) and a simple transport-reduced chemistry model (dot).

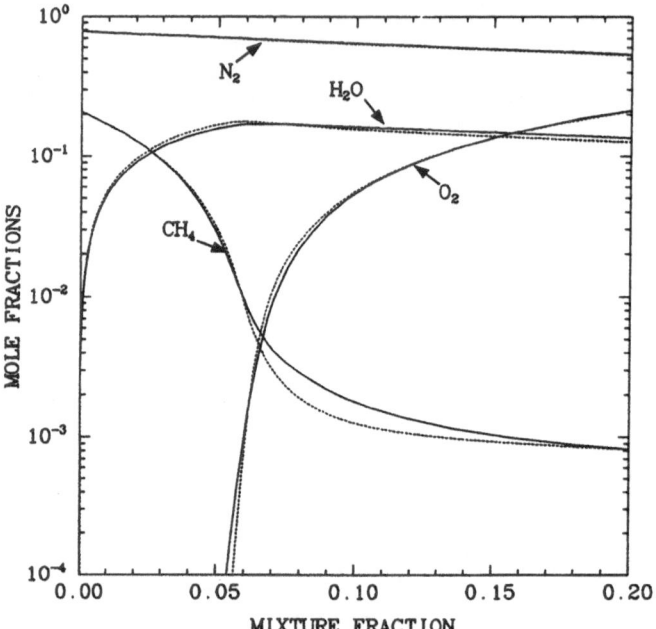

Figure 13. Computed major species profiles as a function of the mixture fraction for a counterflow, methane-air, diffusion flame ($a = 100$ sec^{-1}) employing a simple transport-skeletal chemistry (solid) and a simple transport-reduced chemistry model (dot).

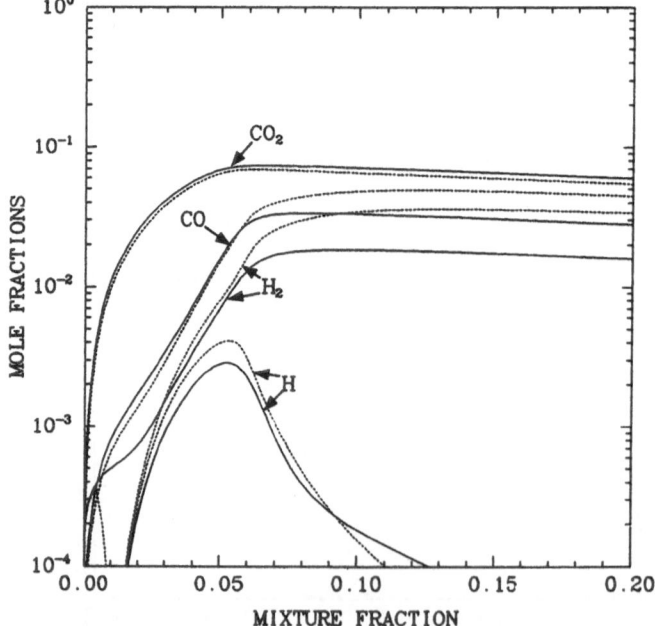

Figure 14. Computed major species profiles as a function of the mixture fraction for a counterflow, methane-air, diffusion flame ($a = 100$ sec^{-1}) employing a simple transport-skeletal chemistry (solid) and a simple transport-reduced chemistry model (dot).

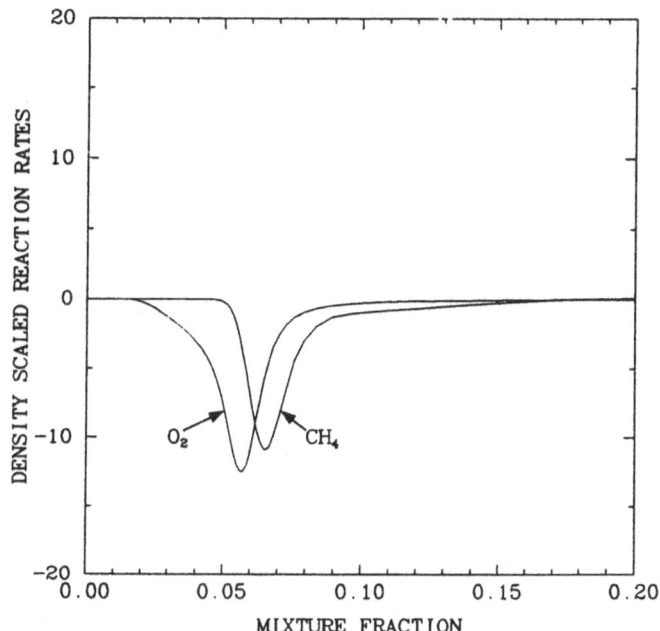

Figure 15. Density weighted molar production rates as a function of the mixture fraction for the major species in a counterflow, methane-air, diffusion flame ($a = 100$ sec^{-1}) employing a simple transport-skeletal chemistry model.

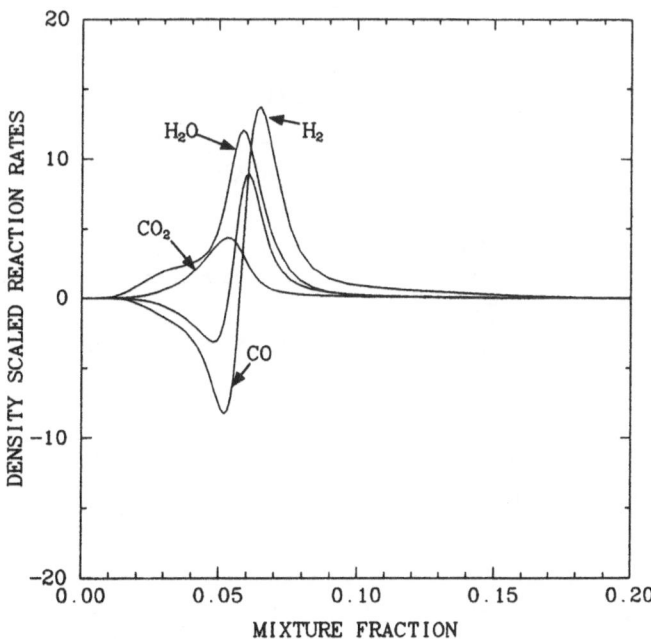

Figure 16. Density weighted molar production rates as a function of the mixture fraction for the major species in a counterflow, methane-air, diffusion flame ($a = 100$ sec^{-1}) employing a simple transport-skeletal chemistry model.

TABLE VI

Numerical Solutions for Counterflow Flames
Simple Transport-Skeletal Chemistry

Flame Parameters	$a = 30$	$a = 100$	$a = 300$	$a_{ext} = 353$
Peak T (K)	2137	2067	1887	1766
Peak H (mole)	2.50×10^{-3}	2.87×10^{-3}	2.70×10^{-3}	7.88×10^{-5}
Peak H_2 (mole)	.02124	.0186	.0157	1.04×10^{-3}
Peak CO (mole)	.0325	.0334	.0414	.0470
Peak CO_2 (mole)	.0779	.0734	.0581	.0785
Peak H_2O (mole)	.1736	.1711	.1636	.1040
$\xi_{st} - \xi^o$	-.0088	-.0100	-.0088	-.0050
T^o(K)	2137	2065	1883	1760
O^o_2 (mole)	2.17×10^{-3}	5.48×10^{-3}	.0204	.0354
H^o_2 (mole)	.0169	.0158	.0136	.0115
CO^o (mole)	.0293	.0316	.0395	.0424
O_2 Leakage	1.53×10^{-4}	1.89×10^{-3}	.0141	.0280

TABLE VII

Numerical Solutions for Counterflow Flames
Simple Transport-Reduced Chemistry

Flame Parameters	$a = 30$	$a = 100$	$a = 300$	$a_{ext} = 323$
Peak T (K)	2171	2120	1936	1847
Peak H (mole)	4.07×10^{-3}	4.127×10^{-3}	2.78×10^{-3}	2.22×10^{-3}
Peak H_2 (mole)	.0544	.0358	.0138	.0102
Peak CO (mole)	.0639	.0491	.0425	.0432
Peak CO_2 (mole)	.0675	.0689	.0579	.0510
Peak H_2O (mole)	.1878	.1781	.1631	.1560
$\xi_{st} - \xi^o$	-.0086	-.0067	.0001	.0022
T^o(K)	2171	2108	1886	1799
O^o_2 (mole)	1.81×10^{-3}	6.74×10^{-3}	.0303	.0409
H^o_2 (mole)	.0253	.0192	.0096	7.72×10^{-3}
CO^o (mole)	.0669	.0331	.0355	.0375
O_2 Leakage	9.30×10^{-5}	1.25×10^{-3}	.0141	.0234

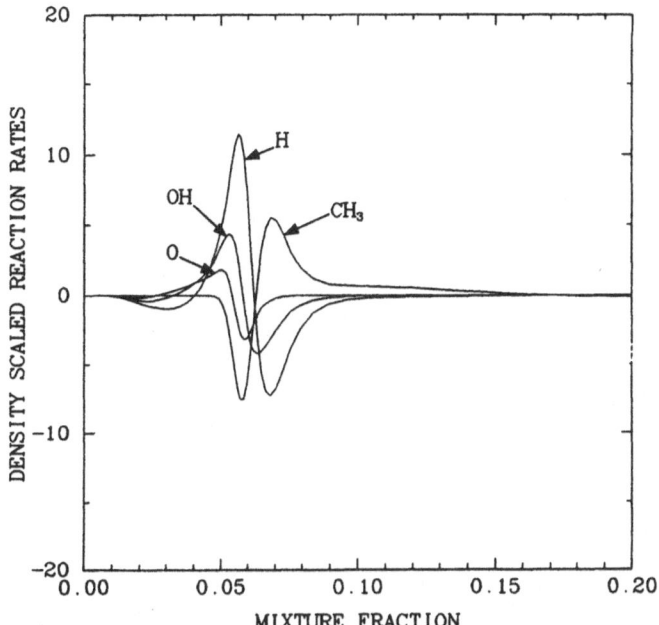

Figure 17. Density weighted molar production rates as a function of the mixture fraction for the major species in a counterflow, methane-air, diffusion flame ($a = 100$ sec^{-1}) employing a simple transport-skeletal chemistry model.

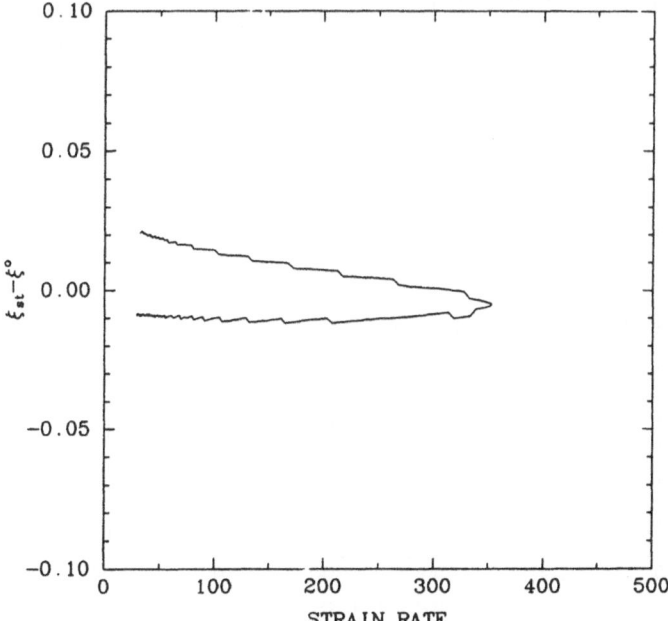

Figure 18. A plot of $\xi_{st} - \xi^o$ as a function of the strain rate for counterflow, methane-air, diffusion flames employing a simple transport-skeletal chemistry model.

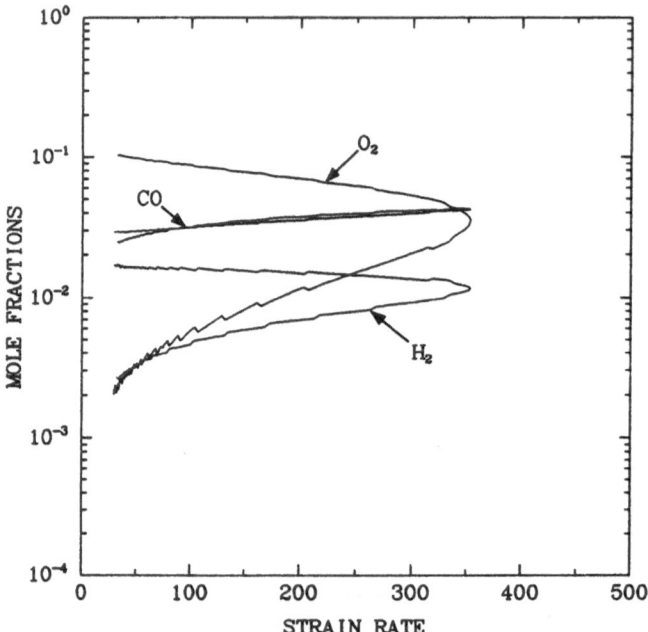

Figure 19. The variation in the values of O_2^0, CO^0 and H_2^0 as a function of the strain rate for counterflow, methane-air, diffusion flames employing a simple transport-skeletal chemistry model.

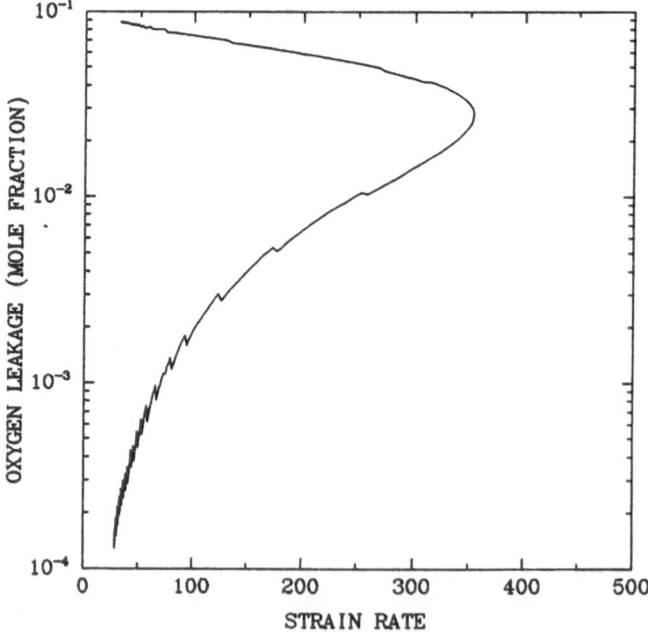

Figure 20. The variation in the oxygen leakage at $\xi = 0.1$ as a function of the strain rate for counteflow, methane-air, diffusion flames employing a simple transport-skeletal chemistry model.

References

[1] Giovangigli, V. and Smooke, M. D., "Extinction of Strained Premixed Laminar Flames with Complex Chemistry," *Comb. Sci. and Tech.*, **53**, (1987), p. 23.

[2] Giovangigli, V. and Smooke, M. D., "Adaptive Continuation Algorithms with Application to Combustion Problems,", *App. Num. Math.*, **5**, (1989), p. 305.

[3] Keller, H. B., "Numerical Solution of Bifurcation and Nonlinear Eigenvalue Problems," in Applications of Bifurcation Theory, P. Rabinowitz, Ed., Academic Press, New York, (1977), p. 359.

[4] Frank, P. M., Introduction to System Sensitivity Theory, Academic Press, New York, (1978).

[5] Smooke, M. D., Rabitz, H., Reuven, Y. and Dryer, F. L., "Application of Sensitivity Analysis to Premixed Hydrogen-Air Flames," *Comb. Sci. and Tech.*, **56**, (1988).

CHAPTER 3

REDUCING MECHANISMS

Norbert Peters
Institut für Technische Mechanik
RWTH Aachen, D-5100 Aachen
Germany

1. Introduction

While simplifications of the transport mechanism and numerical solutions were described in the preceeding two chapters, this chapter will treat the basic ideas and procedures involved in a systematic reduction of kinetic mechanisms. Reduced mechanisms for hydrocarbon flames are useful for at least two applications:

1. they reduce the computational effort in numerical calculations of flames by replacing a number of differential equations for intermediate species—those that are assumed as being in steady state—by algebraic relations;

2. they allow to study the flame structure by asymptotic methods and by that help to identify the relatively few kinetic parameters that mainly influence global properties such as the burning velocity or extinction strain rates.

While the general idea of reducing complex kinetic schemes by the introduction of steady state assumptions has been known to chemists for a long time [1], it has become fruitful for combustion applications only very recently. Interestingly enough, it was the application to hydrocarbon flames [2]–[4] rather than to the much simpler hydrogen flames that first showed the full potential of the methodology. Previous attempts for hydrogen flames had been discouraging mainly for two reasons: first the level of radicals as candidates for the steady state approximation in these flames is too high to justify such an assumption. Secondly, since only a small portion of the total number of species in the hydrogen-oxygen system could be assumed in steady state, the gain in reducing the numerical effort outrules the algebraic and numerical complications involved.

The key to the success in hydrocarbon flames lies in the fact that the chemistry for most of the hydrocarbon species proceeds in reaction chains, where each intermediate species is produced and consumed by only a few major reactions. This allows to derive explicit algebraic expressions for most of these species from their steady state relations. If these expressions involve other steady state species, a non-linear system of algebraic equations results. Consequently, its solution is not unique and among all the possible roots of this system, the right one must be singled out. Fortunately, this coupling occurs only rarely between intermediate hydrocarbon species. But it does occur between species like atomic oxygen O or the hydroxyl radical OH (which can often be assumed in steady state) and some steady state hydrocarbon species, if the reaction between these species is a dominant one. An example is the very fast reaction $CH_3 + O \rightarrow CH_2O + H$ in methane flames.

A way to overcome the difficulties introduced by the non-uniqueness of the system of algebraic equations is the truncation of some steady state relations. The consequences of a truncation will be addressed below. But before entering these details, some general aspects of the reduction strategy shall be presented. These concern the basis of the steady state assumption in terms of the magnitude of the formation and consumption rates, the choices that lead to a particular global reaction mechanism, and the consequences for subsequent asymptotic analyses.

2. Steady state approximations as an asymptotic limit

Steady state approximations for intermediate species can be justified in many different ways. They first were derived for zero dimensional homogeneous systems that depend only on time, and the term "steady state" was introduced because the time derivative of these species is set to zero

$$\frac{d[X_i]}{dt} = 0 = \sum_{k=1}^{r} \nu_{ik} w_k \,. \tag{2.1}$$

Here, $[X_i]$ denotes the molar density of species i, called the concentration, t the time, ν_{ik} the stoichiometric coefficient of species i in reaction k and w_k the reaction rate. The justification for this approximation is generally provided in physical terms by stating that the rate at which species i is consumed is much faster than the rate by which it is produced. Therefore its concentration always stays much smaller than those of the initial reactants and the final products. Since the concentration always stays small, its time derivative also stays small compared to the time derivatives of the other species, as eq. (2.1) implies.

As an example, one may look at the well-known Zeldovich mechanism for thermal production of NO

A $O + N_2 \rightarrow N + NO$
B $N + O_2 \rightarrow O + NO \,.$

Here, we assume that the level of atomic oxygen O is given as a result of the oxidation reactions in a combustion system. Now we assume that atomic nitrogen N is in steady state because reaction B is faster than reaction A. One then can add both reactions, and cancel N to obtain the global reaction

(I) $N_2 + O_2 = 2NO \,.$

In this case the O also cancels, but this is fortuitous. The rate of the overall reaction is that of the first reaction that is slow and therefore rate-determining. Since two moles of NO are formed according to reaction I, the time change of NO is

$$\frac{d[X_{NO}]}{dt} = 2k_A(T)[X_O][X_{N_2}] \,. \tag{2.2}$$

This shall now be derived in a more systematic way. The balance equations for NO and N are

$$\frac{d[X_{NO}]}{dt} = k_A[X_O][X_{N_2}] + k_B[X_N][X_{O_2}]$$
$$\frac{d[X_N]}{dt} = k_A[X_O][X_{N_2}] - k_B[X_N][X_{O_2}] \,. \tag{2.3}$$

These equations will be non-dimensionalized by introducing reference values for all concentrations and the temperature. We define

$$c_{NO} = [X_{NO}]/[X_{NO}]_{ref}, \quad c_N = [X_N]/[X_N]_{ref} \tag{2.4}$$

and a non-dimensional time as

$$\tau = t k_A(T_{ref})[X_O]_{ref}[X_{N_2}]_{ref}/[X_{NO}]_{ref} \,. \tag{2.5}$$

For simplicity, we assume the temperature and the concentrations of O_2, O and N_2 to be constant equal to their reference value. Then the reference value for N must be chosen as

$$[X_N]_{ref} = \frac{k_B(T_{ref})[X_O]_{ref}[X_{N_2}]_{ref}}{k_A(T_{ref})[X_{O_2}]_{ref}} \qquad (2.6)$$

in order to obtain the non-dimensional equations

$$\frac{dc_{NO}}{d\tau} = 1 + c_N$$
$$\varepsilon\frac{dc_N}{d\tau} = 1 - c_N. \qquad (2.7)$$

Here, ε denotes a small parameter defined by

$$\varepsilon = \frac{[X_N]_{ref}}{[X_{NO}]_{ref}} = \frac{k_A(T_{ref})[X_O]_{ref}[X_{N_2}]_{ref}}{k_B(T_{ref})[X_{O_2}]_{ref}[X_{NO}]_{ref}}. \qquad (2.8)$$

The two parts of this equation suggest that ε may be assumed small based on two different kinds of reasoning:

1. the concentration of the intermediate species in eq. (2.7) is small compared to the typical concentration of the product, which is NO in this case,

or

2. the rate constant k_A by which the intermediate N is formed is much smaller than the rate k_B at which it is consumed. This argument assumes that the ratio of the reference concentrations is of order unity.

The solution of the system (2.7) is readily obtained as

$$c_N = 1 - \exp(-\tau/\varepsilon)$$
$$c_{NO} = 2\tau + \varepsilon(\exp(-\tau/\varepsilon) - 1) \qquad (2.9)$$

showing that there are two time scales in this problem, namely τ and τ/ε. In the limit $\varepsilon \to 0$ the solution simplifies to

$$c_N = 1, \quad c_{NO} = 2\tau. \qquad (2.10)$$

This is equivalent to setting $[X_N]$ equal to the reference solution, eq. (2.6). Then $[X_{NO}]$ is in dimensional terms

$$[X_{NO}] = 2tk_A(T_{ref})[X_O]_{ref}[X_{N_2}]_{ref}, \qquad (2.11)$$

which is equivalent to the solution, which is obtained by integrating eq. (2.2).

It should be noted that eq. (2.10) does not satisfy the initial conditions $c_N = 0$ at $\tau = 0$. Therefore, the steady state solution breaks down in an initial layer of thickness ε, where the short time scale τ/ε is of order unity.

For the case $\varepsilon = 0.2$ the solution eq. (2.9) has been plotted in Fig. 1. It is seen that the concentration of c_N grows in the initial layer to the steady state value $c_N = 1$, which then is valid for large times. The concentration of c_{NO} grows initially slower than the linear time dependence of the steady state solution. There also is an order $O(\varepsilon)$ difference that remains in the solution for long times due to neglecting of the second

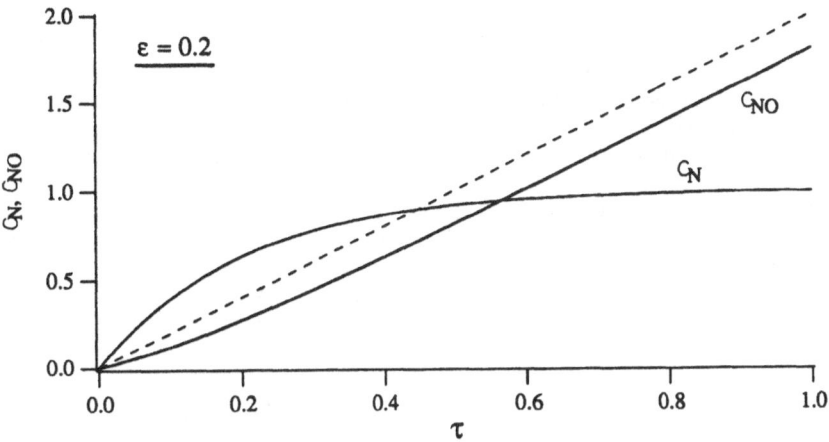

Fig. 1: Solution of eq. (2.9) for the non-dimensional concentrations of N and NO as a function of the non-dimensional time. The steady state solution for c_{NO}, eq. (2.10), is also shown (dashed line).

term in the solution for c_{NO}. This is again due to the initial boundary condition for c_N, which is not satisfied by the steady state solution.

This example illustrates that steady state assumption may be analysed rigorously by asymptotic methods and how the error that they introduce may be estimated. In a more complicated chemical system, where many steady state assumptions apply, this methodology can become cumbersome, since many small parameters that relate rate constants to each other, will appear. These parameters then must be ordered in a specific way to obtain a reasonable and self-consistent result. Very often it is easier to analyse numerical results from a complete solution and compare the magnitude of the concentration of the intermediates to the concentrations of the initial reactants or the final products. This corresponds to the reasoning associated with the first of equations (2.8). The error introduced by each steady state assumptions is then typically of the order of this concentration ratio. For many engineering purposes it will be acceptable to assume those intermediate species in steady state, whose concentration is significantly less than 10% of the initial fuel concentration.

Additional arguments for choosing the steady state species in flames, where diffusion plays a significant role, will be given below.

3. Global reaction mechanisms

The steady state assumption for a species i leads to an algebraic equation between reaction rates. Therefore each of these equations can be used to eliminate rates in the remaining balance equations for the non-steady state species. The stoichiometry of the resulting balance equations defines the global mechanism between the non-steady state species. Therefore the global mechanism depends on the choice of the reaction rates that were eliminated. The rule is that one should choose for each species the fastest rate by which it is consumed. Although this choice may be arbitrary sometimes, it has no consequence as far as the balance equations for the non-steady state species are concerned. We will illustrate this for the case of a hydrogen-oxygen mechanism involving only the first 8 reactions in Table II (conf. Chapter 1). The balance equations are

$$
\begin{aligned}
L([H]) &= -w_1 + w_2 + w_3 - w_5 - w_6 - w_7 \\
0 = L([OH]) &= w_1 + w_2 - w_3 - 2w_4 + 2w_6 - w_8 \\
0 = L([O]) &= w_1 - w_2 + w_4 \\
L([H_2]) &= -w_2 - w_3 + w_7 \\
L([O_2]) &= -w_1 - w_5 + w_7 + w_8 \\
L([H_2O]) &= w_3 + w_4 + w_8 \\
0 = L([HO_2]) &= w_5 - w_6 - w_7 - w_8 .
\end{aligned}
\tag{3.1}
$$

Here, $L([X_i])$ denotes a linear differential operator which may contain not only the time derivative as the one on the l.h.s. of eq. (2.3) but also, for a non-homogeneous system, convective and diffusive terms. The specific form for flames will be introduced below. The species OH, O and HO_2 were assumed in steady state in eq. (3.1) and the corresponding L-operators were set equal to zero that leads to three algebraic equations between the reaction rates w_k. After choosing to eliminate rate w_2 for O, w_3 for OH and w_7 for HO_2 as their respective fastest consumption rates, one can find linear combinations such that those rates do no longer appear on the r.h.s. of the balance equations for H, H_2, O_2 and H_2O. These combinations read

$$
\begin{aligned}
L([H]) + \{L([OH]) + 2L([O]) - L([HO_2])\} &= 2w_1 - 2w_5 + 2w_6 \\
L([H_2]) + \{-L([OH]) - 2L([O]) + L([HO_2])\} &= -3w_1 + w_5 - 3w_6 \\
L([O_2]) + \{L([HO_2])\} &= -w_1 - w_6 \\
L([H_2O]) + \{L([OH]) + L([O])\} &= 2w_1 + 2w_6 .
\end{aligned}
\tag{3.2}
$$

Here, the terms in braces are L-operators of steady state species and are to be neglected. By arranging the r.h.s. such that those rates with equal stoichiometric coefficients are added, one obtains

$$
\begin{aligned}
L([H]) &= 2(w_1 + w_6) - 2w_5 \\
L([H_2]) &= -3(w_1 + w_6) + w_5 \\
L([O_2]) &= -(w_1 + w_6) \\
L([H_2O]) &= 2(w_1 + w_6) .
\end{aligned}
\tag{3.3}
$$

The stoichiometry of these balance equations corresponds to the global mechanism

$$
\begin{array}{lrcl}
(I) & 3H_2 + O_2 &=& 2H + 2H_2O \\
(II) & 2H + M &=& H_2 + M
\end{array}
\tag{3.4}
$$

with the rates

$$
\begin{aligned}
w_I &= w_1 + w_6 \\
w_{II} &= w_5 .
\end{aligned}
\tag{3.5}
$$

In the second of the global reactions the inert body M has been added as a reminder of the third body that appears in reaction 5 of the original scheme in Table I, Chapter 1. This shall illustrate the role of reaction II as a chain breaking global reaction where the only remaining radical, namely H, is being consumed. The role of the first global reaction is that of an overall chain branching step. It also could have been derived by adding reaction 2 and twice reaction 3 to reaction 1 and cancelling the steady state species O and OH. Similarly, the global step II could have been derived by adding reaction 7 to reaction 5 and eliminating HO_2 (and fortuitously also O_2).

Alternatively to choosing w_7 to be eliminated by the steady state equation for HO_2, one could have chosen to eliminate w_6, which is, in fact, about five times faster than w_7 at typical flame temperatures. Employing the same procedure as before, this would result in the alternate global steps

$$\begin{array}{rlll} (I') & 3H_2 + O_2 & = & 2H + 2H_2O \\ (II') & 2H_2 + O_2 & = & 2H_2O \end{array} \tag{3.6}$$

with the rates

$$\begin{aligned} w'_I &= w_1 - w_7 - w_8 \\ w'_{II} &= w_5 \, . \end{aligned} \tag{3.7}$$

After writing the balance equations for this scheme

$$\begin{aligned} L([H]) &= 2(w_1 - w_7 - w_8) \\ L([H_2]) &= -3(w_1 - w_7 - w_8) - 2w_5 \\ L([O_2]) &= -(w_1 - w_7 - w_8) - w_5 \\ L([H_2O]) &= 2(w_1 - w_7 - w_8) + 2w_5 \end{aligned} \tag{3.8}$$

and using the steady state relation for HO_2

$$w_5 - w_6 - w_7 - w_8 = 0 \tag{3.9}$$

one finds that the balance equations are identical with those in eq. (3.3). The balance equations therefore remain independent of the choice of the rates that were eliminated. Different global mechanisms therefore lead to the same solution for a given problem.

The reason that reaction 7 has been chosen in the reduced 4-step mechanism for methane flames, which will be derived below, is essentially tutorial. Since in the methane mechanism many chain breaking steps besides reactions 5 and 7 will play an important role in reducing the H-atom concentration, a chain breaking step was retained to illustrate the general behavior of the global mechanism.

Sometimes, for instance for asymptotic analyses of a flame structure, one may choose to disregard some rates in the remaining global reaction scheme. In a first asymptotic analysis of methane flames, which is reviewed in Chapter 4, only reaction rates w_1 and w_5 were retained in eq. (3.5) as principal rates of the global reactions IV and III corresponding to I and II in eq. (3.4), respectively. The same choice in eq. (3.7) would then lead to different balance equations. Therefore the form of the global mechanism may become important, if not all rates are retained in the final formulation.

4. The reduced four-step mechanism for methane flames

The first step in deriving a reduced mechanism is to define a suitable starting mechanism. This may be viewed as an already reduced form of a much larger "full" mechanism available in the literature. For the specific problem considered, a numerical solution must be obtained using the full mechanism and a sensitivity analysis must be carried out to identify the influence of each individual reaction on the solution. The starting mechanism should contain only those elementary reactions that are necessary to reproduce a characteristic quantity, such as the burning velocity, within about less than five percent accuracy. This simplifies the algebra of the following steps considerably. In addition, the sensitivity analysis helps to choose the fast reactions that are to be eliminated later.

For hydrocarbon flames typically about fifty elementary reactions are necessary to reproduce the burning velocity over the whole range of equivalence ratios and pressures up to 50 atm with reasonable accuracy. For lean-to-stoichiometric methane flames the "skeletal" mechanism of 25 elementary reactions in Table II, Chapter 1, was identified as a sufficiently good representation of the elementary kinetics. This mechanism only contains hydrocarbons of the C_1-chain and is therefore expected to be insufficient for rich methane flames. Here, we will use this mechanism as a starting mechanism for the reduction procedure.

The second step in this procedure is to identify steady state species. Following the first of the two alternate reasonings described following eq. (2.8) above, we will analyse the outcome of a numerical calculation based on the starting mechanism and find out the relative order of magnitude of the intermediate species concentrations.

Differently from a homogeneous system, in flame problems the balance equations are usually formulated in terms of mass fractions Y_i rather than in terms of molar densities. These are related to each other and to the mole fraction X_i by

$$[X_i] = \frac{\rho\,Y_i}{W_i} = \rho\frac{X_i}{\overline{W}}\,. \tag{4.1}$$

Here, ρ is the density, W_i the molecular weight of species i and $\overline{W}$ the mean molecular weight. Defining the quantity

$$\Gamma_i = \frac{Y_i}{W_i} \tag{4.2}$$

one may write the balance equations for the species as

$$L(\Gamma_i) = \sum_{k=1}^{r} \nu_{ik} w_k \qquad i = 1, 2, ..., n \tag{4.3}$$

where the L-operator is now defined by

$$L(\Gamma_i) = \rho\frac{\partial \Gamma_i}{\partial t} + \rho\, v_\alpha \frac{\partial \Gamma_i}{\partial x_\alpha} + \frac{\partial}{\partial x_\alpha}\left(\frac{j_{i\alpha}}{W_i}\right)\,. \tag{4.4}$$

In eq. (4.4) v_α and $j_{i\alpha}$ are the velocity components and the diffusion fluxes, respectively. The quantity Γ_i rather than Y_i was introduced as a dependent variable because then the reaction rates in eq. (4.3) are no longer multiplied by the molecular weight and the convective and diffusive terms are directly comparable between the species equations. For order-of-magnitude estimates of different terms in eq. (4.4), we write the diffusion flux of the i-th species $j_{i\alpha}$ using the binary diffusion coefficient with respect to nitrogen as

$$\frac{j_{i\alpha}}{W_i} = -\rho\, D_{i,N_2}\frac{\partial \Gamma_i}{\partial x_\alpha}\,. \tag{4.5}$$

For a steady-state species, the concentration remains very small and all terms in the operator $L(\Gamma_i)$ on the l.h.s. of eq. (4.3) may be neglected compared to the reaction rates. In the balance equations for steady flames, the L operator contains the convective and

	H_2	CO	H	OH	O	CH_3	CH_2O	HCO	HO_2	CH_3O	H_2O_2
Original	1.96	4.91	0.77	0.77	0.36	0.37	0.27	0.009	0.016	0.0007	0.006
Weighted	5.35	4.91	2.92	0.886	0.42	0.442	0.266	0.009	0.015	0.0007	0.006

Table 1: Maximum percent mole fractions in a premixed stoichiometric methane-air flame.

diffusive term; the assumption therefore requires that both terms are small. Since the binary diffusion coefficients are approximately proportional to $1/\sqrt{W_{i.N_2}}$ where

$$W_{i.N_2} = \frac{2W_i W_{N_2}}{W_i + W_{N_2}} \qquad (4.6)$$

the species concentrations must be weighted with this factor for the order-of-magnitude estimate. The relevant concentrations are the mole fractions X_i and not the mass fractions Y_i, since X_i is proportional to Γ_i. In Table 1, we list the maximum mole fractions of intermediate species obtained from the calculation with a large mechanism for a stoichiometric premixed methane-air flame at one atmosphere. We have also listed the mole fractions weighted with $\sqrt{W_{N_2}/W_{i.N_2}}$. The values fall essentially into two groups: those well below 1% and those well above it. The first group includes the intermediate species OH, O, HO_2, CH_3, CH_2O, CHO, CH_3O and H_2O_2, which in the following are assumed to be in a steady state. The second group contains CO, H_2, and H if the weighted mole fractions are considered. Those species will therefore not be assumed to be in steady state. By weighting the species with the diffusion coefficients, we use a result from the asymptotic analysis of flames (conf. Chapter 4) stating that within the reactive layers, the diffusive terms are dominant compared to the convective terms and are the only ones to balance the reaction terms.

The present choice of retaining H as a non-steady-state species finds a further justification because the first reaction $H + O_2 \rightarrow O + OH$ is the most important one for flame calculations since it is chain branching since H appears as a reactant in both reactions. It competes with $H + O_2 + M \rightarrow HO_2 + M$ as the most important chain breaking reaction. It is important to calculate the H concentration more accurately than those of O and OH.

It is now possible to use the eight steady-state conditions to eliminate at least eight reaction rates from the system. We want to eliminate the fastest reactions that consume each steady-state species and by that construct what will be called the main chain. From a sensitivity analysis it is found that for the oxidation of CH_4 via CH_3, CH_2O and CHO to CO, this main chain is

$$
\begin{array}{llll}
11 & CH_4 + H & \rightarrow & CH_3 + H_2 \\
13 & CH_3 + O & \rightarrow & CH_2O + H \\
14 & CH_2O + H & \rightarrow & HCO + H_2 \\
17 & HCO + M & \rightarrow & CO + H + M .
\end{array}
\qquad (4.7)
$$

We will therefore use the steady-state relations for CH_3, CH_2O and HCO to eliminate the rates w_{13}, w_{14}, and w_{17} from the balance equations. In addition, we will use the steady-state relations for O, OH, and H_2O to eliminate the rates w_2, w_3, and w_7 and by that define the main chain for the chain branching reactions

$$
\begin{array}{llll}
1 & H + O_2 & \rightarrow & OH + O \\
2 & O + H_2 & \rightarrow & H + OH \\
3 & OH + H_2 & \rightarrow & H + H_2O
\end{array}
\qquad (4.8)
$$

as well as for the chain breaking reactions

$$
\begin{array}{llll}
5 & H + O_2 + M & \rightarrow & HO_2 + M \\
7 & H + HO_2 & \rightarrow & H_2 + O_2 \, .
\end{array}
\qquad (4.9)
$$

Finally, the main chain for the conversion of CO to CO_2 consists of the two reactions

$$
\begin{array}{llll}
9 & CO + OH & \rightarrow & CO_2 + H \\
3 & OH + H_2 & \rightarrow & H_2O + H \, .
\end{array}
\qquad (4.10)
$$

By adding the reactions in eqs. (4.7)–(4.10) and cancelling the steady state species where reaction 3 is used twice in eq. (4.8) one obtains the global four step mechanism for methane flames

$$
\begin{array}{llll}
I & CH_4 + 2H + H_2O & \rightleftharpoons & CO + 4H_2 \\
II & CO + H_2O & \rightleftharpoons & CO_2 + H_2 \\
III & H + H + M & \rightleftharpoons & H_2 + M \\
IV & O_2 + 3H_2 & \rightleftharpoons & 2H + 2H_2O \, .
\end{array}
\qquad (4.11)
$$

Differently from the more systematic procedure used above for the H_2-O_2 system the consideration of the main chain by itself does not provide the rates of the global reactions as in eq. (3.5). It only provides as principal rates the rate determining steps, which are the first ones in each of the sequences (4.7)–(4.11), namely w_{11} for I, w_9 for II, w_5 for III and w_1 for IV. These are also the only ones that were not eliminated by the steady state relations. The next step is to add to these appropriate additional reaction rates. To gain some more insight into the properties of the remaining reactions and to avoid linear algebra one may consider each of them as part of an alternate chain that is to be compared to the respective main chain. We will call the remaining reactions side reactions. Beginning with reation 4 one realizes that it is linearly dependent on reactions 2 and 3 since the addition of reactions 2 and 4 leads to reaction 3. Since reactions 2 and 3 were eliminated the rate w_4 will also not appear in the rates of the global reactions. This was already found in eq. (3.5).

The effect of side reaction 6 is determined by comparing it to reaction 7. Subtracting reaction 7 from reaction 6 and adding reaction 3 twice leads to the global step IV

$$
\begin{array}{ll}
6 & H + HO_2 \rightarrow OH + OH \\
7 & -(H + HO_2 \rightarrow H_2 + O_2) \\
3 & +2\,(H_2 + OH \rightarrow H_2O + H) \\
\hline
 & O_2 + 3H_2 = 2H + 2H_2O \, .
\end{array}
$$

Therefore the rate w_6 should be added in w_{IV} as it was already found in eq. (3.5). The effect of reaction 8 is found by subtracting reaction 3

$$
\begin{array}{ll}
8 & OH + HO_2 \rightarrow H_2O + O_2 \\
3 & -(H_2 + OH \rightarrow H_2O + H) \\
\hline
7 & H + HO_2 = H_2 + O_2 \, ,
\end{array}
$$

which leads to reaction 7. Therefore reaction 8 has the same chain breaking effect as reaction 7 and w_8 disappears as w_7 from the global rates.

In a similar way the side reactions of the C_1-hydrocarbon chain may be analyzed. For reactions 10 and 12 the main chain reaction that they should be compared with is reaction 11. Reaction 10 may be obtained by subtracting the global reaction III from reaction 11.

11 $\qquad\qquad CH_4 + H \rightarrow CH_3 + H_2$

III $\qquad\qquad -(H + H + M \rightarrow H_2 + M)$

10 $\qquad\qquad CH_4 + M = CH_3 + H + M$

while subtracting reaction 3 from reaction 12 leads to reaction 11

12 $\qquad\qquad CH_4 + OH \rightarrow CH_3 + H_2O$

3 $\qquad\qquad -(H_2 + OH \rightarrow H_2O + H)$

II $\qquad\qquad CH_4 + H = CH_3 + H_2$.

This indicates that reaction 10 acts as reaction 11 but it has a chain branching effect that is stoichiometrically the inverse of the global reaction III. It should therefore be added in w_I and be substracted in w_{III}. On the other hand, reaction 12 has the similar effect as reaction 11 and should only be added in w_I.

The next side reaction in Table 1 is reaction 15. Subtracting reaction 3 from it shows that it has the same effect as 14 and should therefore not appear. But reaction 16 may be obtained by adding the global reaction III to 17. Therefore it has, when compared to reaction 17, a chain breaking effect and should be added in w_{III}.

Reaction 18 initiates a side chain leading from CH_3 to CH_2O. If one chooses reaction 20 as the fastest intermediate step, which consumes CH_3O in this side chain and adds it to 18, one observes a chain branching effect since two radicals are formed

18 $\qquad\qquad CH_3 + O_2 \rightarrow CH_3O + O$

20 $\qquad\qquad CH_3O + M \rightarrow CH_2O + H + M$

II $\qquad\qquad CH_3 + O_2 = CH_2O + O + H$.

Adding reactions 2 and 3 twice to this one obtains

$\qquad\qquad CH_3 + O_2 \rightarrow CH_2O + O + H$

2 $\qquad\qquad 2(O + H_2 \rightarrow OH + H)$

3 $\qquad\qquad 2(H_2 + OH \rightarrow H_2O + H)$

$\qquad CH_3 + O_2 + 4H_2 + O = CH_2O + 5H + 2H_2O$.

This may be decomposed into the reaction 13 plus global reaction IV minus the global reaction III. This suggests that w_{18} as the principal rate of this side chain should be added to w_{IV} and be substracted from w_{III}. Then, within this side chain, reaction 19 must be compared to reaction 20, which shows that it has a chain breaking effect corresponding to the stoichiometry of the global step III. Its rate therefore must be added to w_{III}.

Another side chain is initiated by reaction 21. Considering reaction 22 as the fastest step to consume H_2O_2 and adding this as well as twice reaction 3 and subtracting twice reaction 7 one obtains

21	$HO_2 + HO_2 \rightarrow H_2O_2 + O_2$
22	$H_2O_2 + M \rightarrow OH + OH + M$
3	$2(H_2 + OH \rightarrow H_2O + H)$
7	$-2(H + H_2O \rightarrow H_2 + O_2)$

$$4H_2 + O_2 = 4H + 2H_2O,$$

which corresponds to reaction IV from which reaction III is substracted. Therefore w_{21} should be added in w_{IV} and substracted in w_{III}. When the effect of reaction 23 is compared to 22 in this side chain, one obtains by adding reaction 23 plus reaction 7 minus 3 times reaction 3

23	$H_2O_2 + OH \rightarrow H_2O + HO_2$
7	$H + HO_2 \rightarrow H_2 + O_2$
3	$-3(H_2 + OH \rightarrow H_2O + H)$

$$H_2O_2 + 2H_2O + 4H = 2OH + 4H_2 + O_2.$$

This is an overall step that may be decomposed into reaction 22 plus the global step III minus the global step IV. This implies that w_{23} should be added in w_{III} and be substracted in w_{IV}.

Finally, reactions 24 and 25 are three body chain breaking reactions whose rates should be added in w_{III}. This is immediately evident for reaction 25 and also by subtraction of reaction 3 from reaction 25. We may therefore summarize the rates of the global reactions as

$$
\begin{aligned}
w_I &= w_{10} + w_{11} + w_{12} \\
w_{II} &= w_9 \\
w_{III} &= w_5 - w_{10} + w_{16} - w_{18} + w_{19} \\
&\quad - w_{21} + w_{23} + w_{24} + w_{25} \\
w_{IV} &= w_1 + w_6 + w_{18} + w_{21} - w_{23}.
\end{aligned}
\tag{4.12}
$$

5. Truncation of steady state relations

The reaction rates must be expressed in terms of the rate constants and the concentrations. Some concentrations are those of steady state species. As shown in eq. (3.1), these may be calculated from their balance equations with the L operator set equal to zero, which will be called steady state relations. Therefore a system of non-linear algebraic equation complements the remaining balance equations for the non-steady state species.

The most important step in reducing mechanisms is a systematic truncation of some steady state relations such that the system of non-linear algebraic equations becomes explicit. In Fig. 2 we have plotted the forward and backward rates of reactions 1–4 and in Fig. 3 those of reactions 9–13. The origin in these figures is at the maximum of production of H. This is considered to be the best approximation for the location of the inner layer, which appears in the asymptotic description (conf. Chapter 4) of premixed methane flames. Downstream of this layer the rates $3f$ and $3b$ are dominant in the

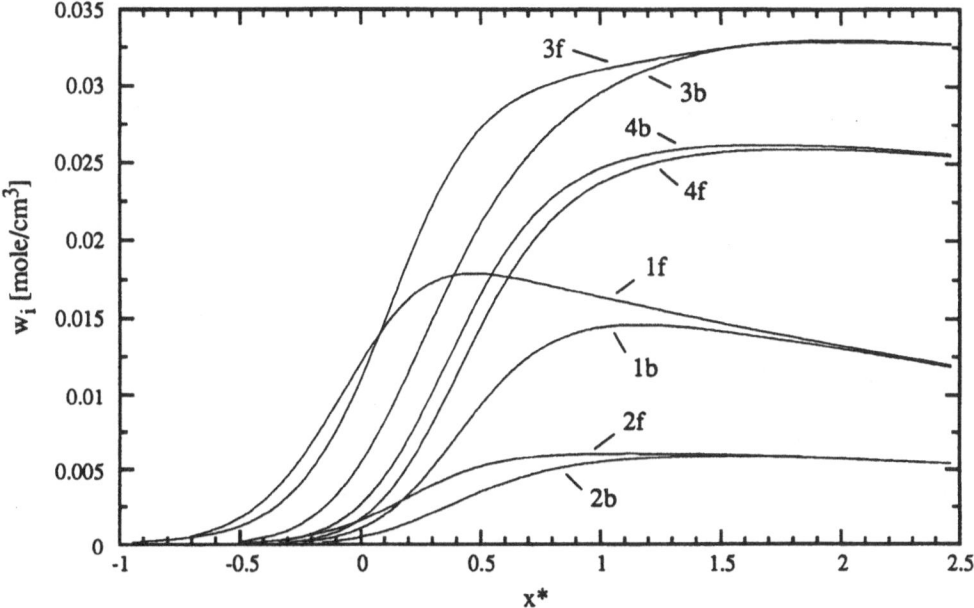

Fig. 2: Reaction rates 1–4 in a stoichiometric methane-air flame at 1 atm as a function of the non-dimensional coordinate $x^* = \rho_u s_L \int_0^x (\lambda/c_p)^{-1} dx$.

steady state equation for OH. Therefore to leading order OH can be calculated from the partial equilibrium of reaction 3

$$[OH] = \frac{k_{3b}[H_2O][H]}{k_{3f}[H_2]} . \tag{5.1}$$

A first order approximation would need to include reaction $1f$ since its rate is large close to the inner layer. To satisfy the transition to equilibrium far downstream, the backward reaction $1b$ also would have been considered, although its rate is small near the inner layer. It may therefore be viewed as a second order term that is retained only for consistency with the downstream equilibrium condition. Since the concentration of O appears in $1b$, an ad-hoc approximation for [O] satisfying the downstream equilibrium is given by partial equilibrium of reaction 4

$$[O] = \frac{k_{4f}[OH]_{eq}^2}{k_{4b}[H_2O]} . \tag{5.2}$$

Here $[OH]_{eq}$ is the partial equilibrium concentration obtained from eq. (5.1). If the steady state relation for [OH] is truncated such that only the forward and backward rates of reactions 1 and 3 appear and the above approximation for [O] is inserted, one obtains

$$[OH] = \frac{k_{3b}[H_2O][H] + k_{1f}[H][O_2]}{k_{3f}[H_2] + k_{1b}k_{4f}k_{3b}^2[H]^2[H_2O]/(k_{4b}k_{3f}^2[H_2]^2)} . \tag{5.3}$$

Since the third reaction does not appear in the steady state relation for O and CH_3, only first and second order terms are to be balanced here.

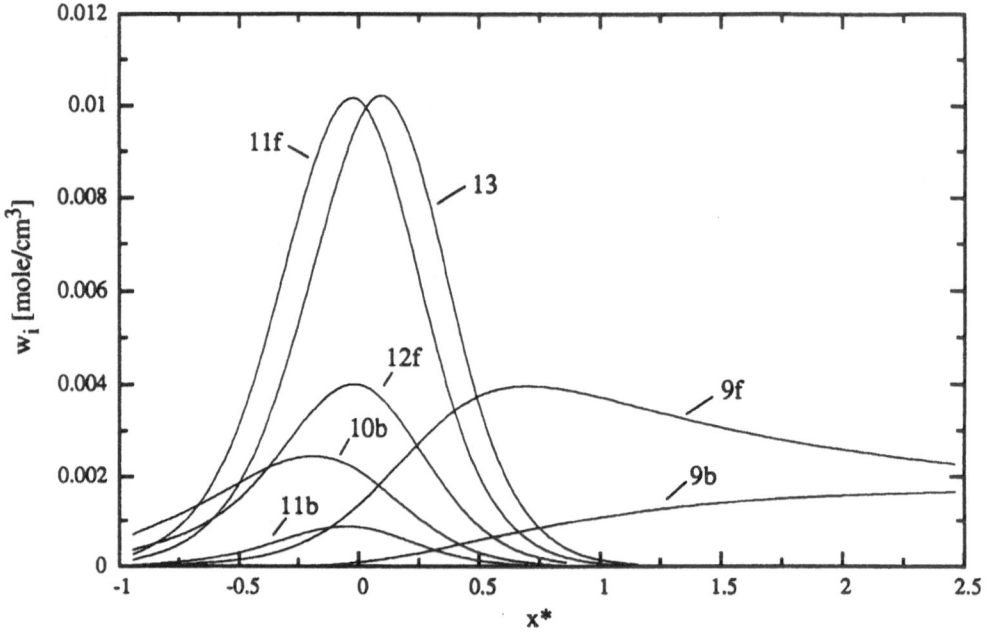

Fig. 3: Reaction rates 9–13 in a stoichiometric methane-air flame at 1 atm as a function of the non-dimensional coordinate $x^* = \rho_u s_L \int_0^x (\lambda/c_p)^{-1} dx$.

O:

$$k_{1f}[H][O_2] + k_{2b}[OH][H] + k_{4f}[OH]^2$$
$$= [O]\{k_{1b}[OH] + k_{2f}[H_2] + k_{4b}[H_2O] + k_{13}[CH_3]\} \quad (5.4)$$

CH_3:

$$\{k_{11f}[H] + k_{12f}[OH]\}[CH_4]$$
$$= [CH_3]\{k_{10b}[H][M] + k_{11b}[H_2] + k_{12b}[H_2O] + k_{13}[O]\}. \quad (5.5)$$

These are again truncated steady state relations based on the comparison of magnitude of the rates in Figs. 2 and 3. Rates of other reactions not shown here are very much smaller.

Since reaction 13 appears in both expressions, $[CH_3]$ must be eliminated and a quadratic equation is obtained for $[O]$

$$[O] = \frac{-b + \sqrt{b^2 - 4ac}}{2a} \quad (5.6)$$

where

$$a = k_{13} B, \quad b = BD + k_{13}(C - A)$$
$$c = -AD \quad (5.7)$$

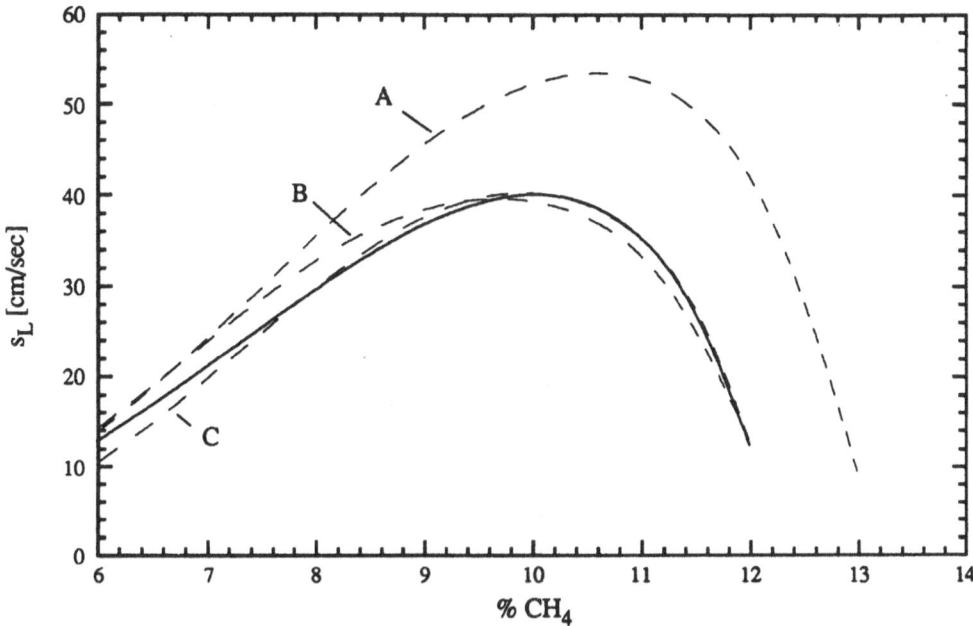

Fig. 4: Burning velocity s_L as a function of the mole fraction of CH_4 in the unburnt gas (starting mechanism: solid line, cases A, B, and C of reduced mechanism formulations: dashed lines).

and

$$A = k_{1f}[H][O_2] + k_{2b}[OH][H] + k_{4f}[OH]^2$$
$$B = k_{1b}[OH] + k_{2f}[H_2] + k_{4b}[H_2O]$$
$$C = \{k_{11f}[H] + k_{12f}[OH]\}[CH_4]$$
$$D = k_{10b}[H][M] + k_{11b}[H_2] + k_{12b}[H_2O].$$

(5.8)

Once solutions to these truncated steady states have been obtained, it is easy to resolve the steady state relations for $[CH_3O], [CH_2O]$ and $[HCO]$ in terms of $[OH], [O]$ and $[CH_3]$

$$[CH_3O] = \frac{k_{18}[CH_3][O_2]}{k_{19}[H] + k_{20}[M]}$$

$$[CH_2O] = \frac{k_{13}[CH_3][O] + (k_{19}[H] + k_{20}[M])[CH_3O]}{k_{14}[H] + k_{15}[OH]}$$

(5.9)

$$[HCO] = \frac{(k_{14}[H] + k_{15}[OH])[CH_2O]}{k_{16}[H] + k_{17}[M]}.$$

The steady state relation for $[HO_2]$ may again be truncated by considering only the rates of reactions 5–8, since the others involving $[HO_2]$ are small. This leads to

$$[HO_2] = \frac{k_5[H][O_2][M]}{(k_6 + k_7)[H] + k_8[OH]}.$$

(5.10)

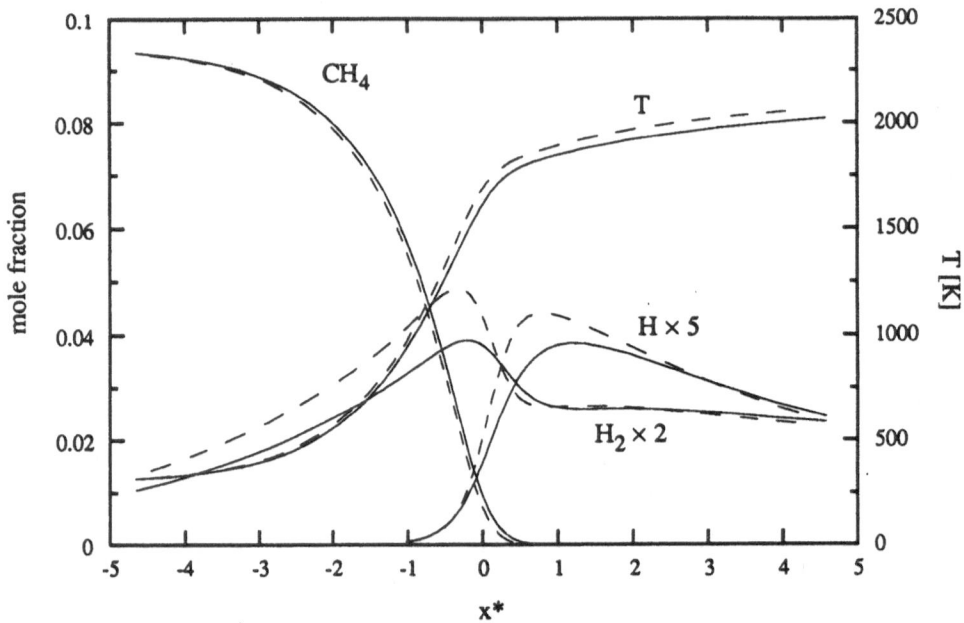

Fig. 5: Mole fractions and temperature for a stoichiometric methane-air flame at 1 atm as a function of the non-dimensional coordinate $x^* = \rho_u s_L \int_0^x (\lambda/c_p)^{-1}\, dx$ (starting mechanism: solid line, reduced mechanism: dashed line).

Finally, the steady state relation for $[H_2O_2]$ leads to

$$[H_2O_2] = \frac{k_{21}[HO_2]^2 + k_{22b}[OH]^2[M] + k_{23b}[H_2O][HO_2]}{k_{22f}[M] + k_{23f}[OH]}.$$ (5.11)

6. Comparison with previous formulations of truncated steady state relations

The explicit algebraic relations derived above have been used in a numerical calculation based on the reduced mechanism for methane flames. They are tested against the solution based on the starting mechanism. Three different formulations of reduced mechanisms were calculated:

A) uses partial equilibrium of reaction 3 for [OH] according to eq. (5.1) and neglects the term involving k_{13} in the steady state relation for [O], such that only reactions of the H_2O_2-system are being balanced there

B) uses partial equilibrium of reaction 3 for [OH] and the quadratic equation (5.6) for [O]

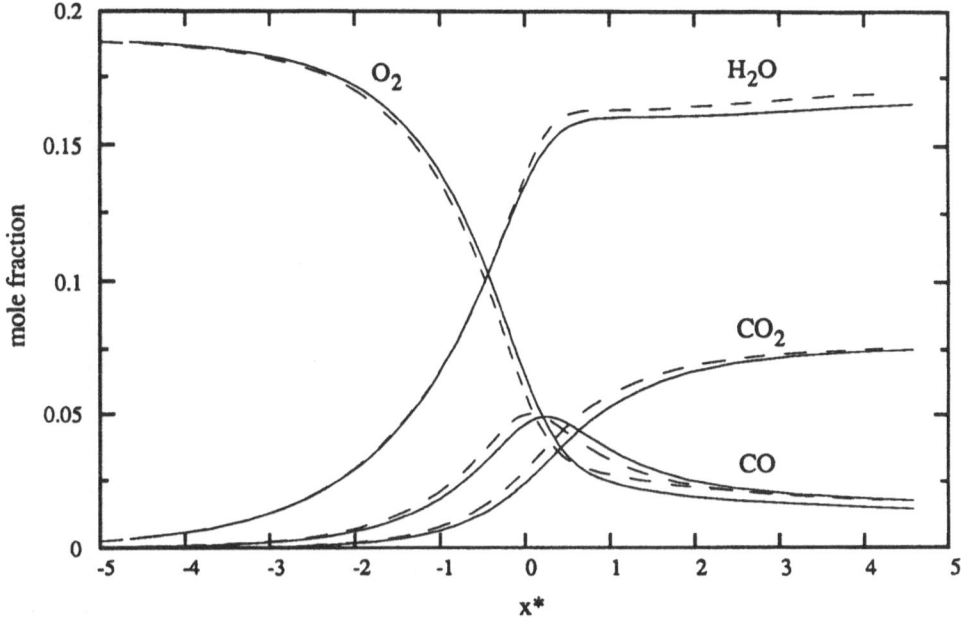

Fig. 6: Mole fractions for a stoichiometric methane-air flame at 1 atm as a function of the non-dimensional coordinate $x^* = \rho_u s_L \int_0^x (\lambda/c_p)^{-1} dx$ (starting mechanism: solid line, reduced mechanism: dashed line).

C) uses the first order correction, eq. (5.3) for [OH] and the quadratic equation (5.5) for [O].

In Fig. 4 the burning velocity at 1 atm is plotted as a function of the mole fraction of CH_4 in the unburnt mixture and the three cases are compared with the starting mechanism. The mechanism in Table 1 has been used for these calculations. While case A gives very high burning velocities, which, at $\phi = 1$, corresponding to $[CH_4]_u = 9.5\%$, is around 50 cm/sec, the cases B and C do well. In case C the burning velocity differs from that of the starting mechanism by less than 1.5 cm/sec over the entire range of equivalence ratios while case B shows maximum derivations of 3 cm/sec or 10% on the lean side.

The mole fractions of the non-steady state species and the temperature for a stoichiometric flame at 1 atm are plotted in Figs. 5 and 6. Here the reduced mechanism formulation based on case C has been used. It is seen, that except for H and H_2 the agreement is quite good, but not as good as for the burning velocity. Even larger deviations are found for the steady state species [O], [OH], and $[CH_3]$ shown in Fig. 7. The larger differences for these species are expected since the steady state assumptions enter their balance equations directly. Finally, a sensitivity analyses of the burning velocity with respect to an increase of the rate of each reaction by 10% is performed for the starting mechanism and the reduced mechanism C for a stoichiometric flame at 1 atm and is shown in Figs. 8 and 9. It is seen that the burning velocity becomes sensitive to the principal rates $1f$, 5, 9 and $11f$. This suggests that by neglecting the diffusion and

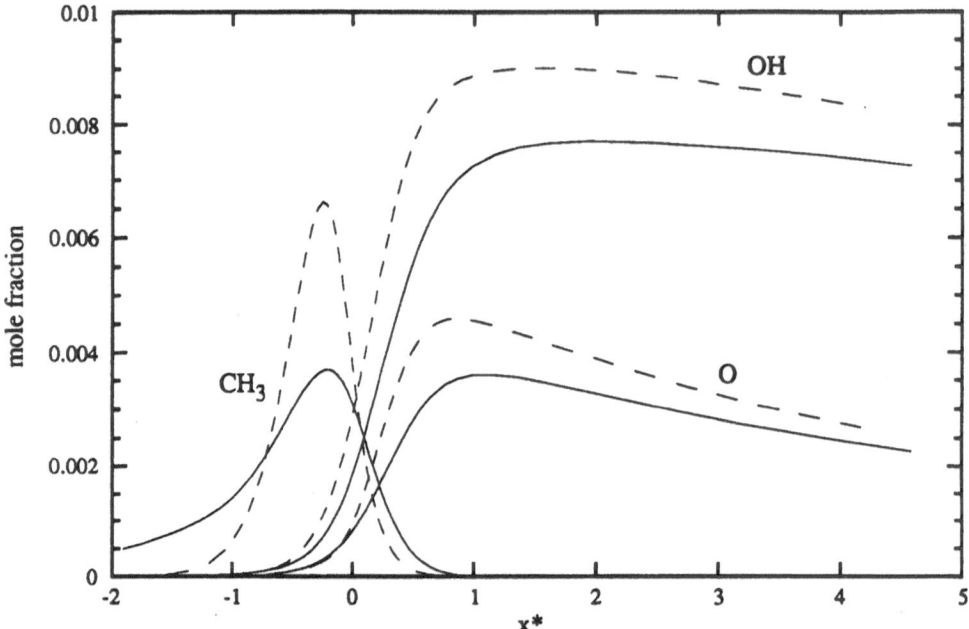

Fig. 7: Mole fractions of steady state species for a stoichiometric methane-air flame at 1 atm as a function of the
non-dimensional coordinate $x^* = \rho_u s_L \int_0^x (\lambda/c_p)^{-1} dx$ (starting mechanism: solid line, reduced mechanism:
dashed line).

convection terms in the steady state equations the chemical interaction focusses on the
principal rates and not on side reactions.

7. Conclusions

Numerical calculations and sensitivity analyses of the chemistry for a specific combustion problem like a methane flame helps to identify those parameters, which are of
primary importance. This knowledge can be used to derive a reduced mechanism with
explicit equations for steady state species. Such a mechanism reduces the chemistry of
methane flames to four global reactions involving seven reacting species. This is considerably less than the 15 reacting species of the starting mechanism or the many more of a
full mechanism. But the validity of the reduced mechanism is restricted to the application for which it has been derived. A mechanism derived for flame applications should
therefore, as an example, not be applied to ignition problems where different elementary
reactions are important. However, the range of application of reduced mechanisms for
flames covers a large range of stoichiometry and pressures.

Acknowledgement

The author is profoundly indebted to F. Mauss, who has provided all the numerical
results and many important observations on the subject.

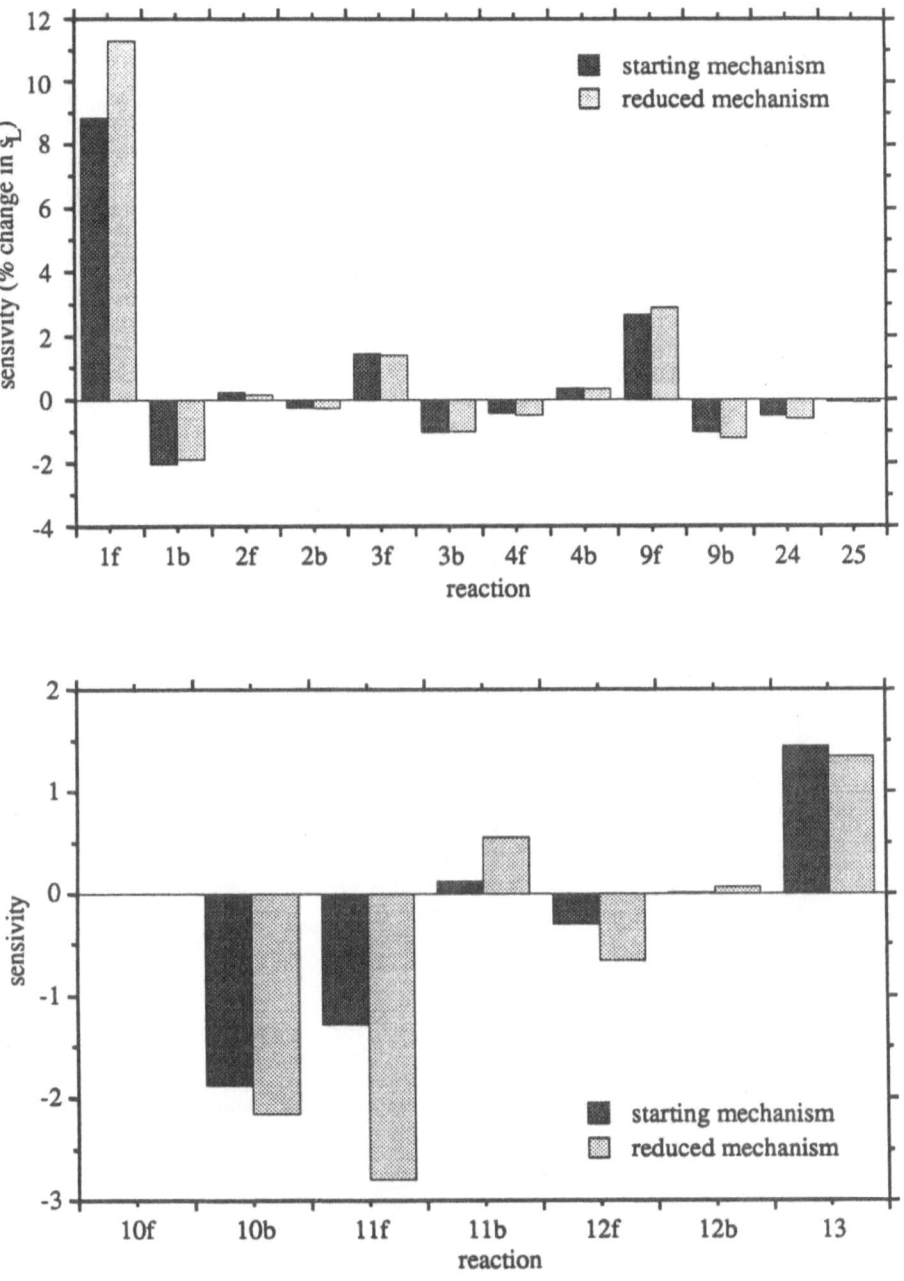

Fig. 8: Sensivity analysis of the burning velocity induced by a change of the reaction rates by 10 %.

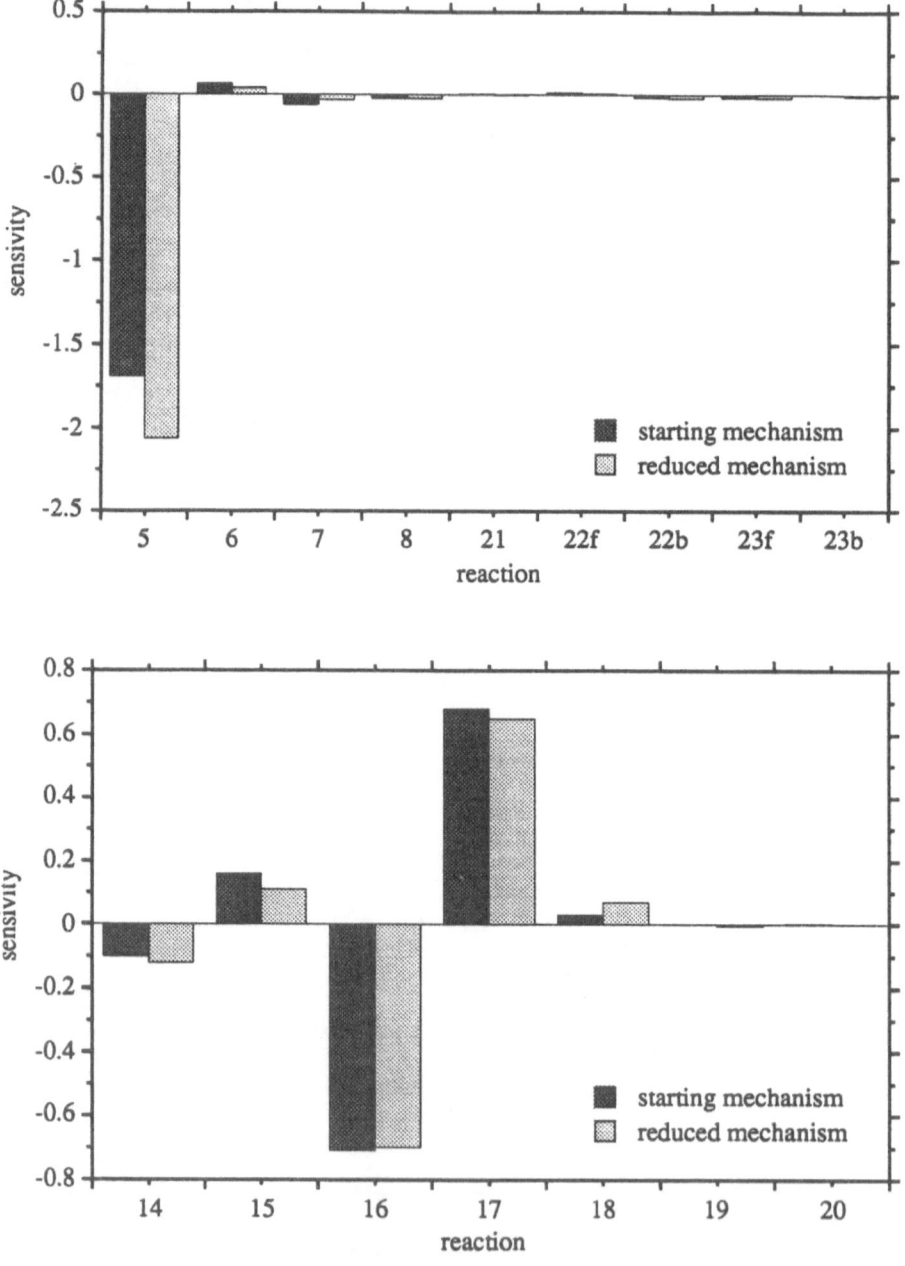

Fig. 9: Sensivity analysis of the burning velocity induced by a change of the reaction rates by 10 %.

References

[1] von Karman, Th., Penner, S.S., *Selected Combustion Problems, Fundamentals and Aeronautical Applications*, AGARD, Butterworths Sci. Publ., London, pp. 5–41, 1954.

[2] Peters, N., "Numerical and asymptotic analysis of systematically reduced reaction schemes for hydrocarbon flames", *Numerical Simulation of Combustion Phenomena, Lecture Notes in Physics* **241**, pp. 90–109, 1985.

[3] Paczko, G., Lefdal, P.M., Peters, N., "Reduced reaction schemes for methane, methanol and propane flames", 21st Symposium (International) on Combustion, The Combustion Institute, pp. 739–748, 1988.

[4] Peters, N., Kee, R. J.,"The computation of stretched laminar methane-air diffusion flames using a reduced four-step mechanism", Comb. and Flame **68**, pp. 17–30, 1987.

CHAPTER 4

OVERVIEW OF ASYMPTOTICS FOR METHANE FLAMES

Forman A. Williams
Department of Applied Mechanics and Engineering Sciences
University of California, San Diego
La Jolla, CA 92093

1. Introduction

The reduced kinetic mechanisms described in the preceding chapter provide the basis needed for the development of asymptotic analyses of flame structures. The purpose of the present chapter is to offer a perspective on the ways in which these reduced mechanisms have led to advancements in the asymptotic analyses. The exposition will proceed sequentially from one-step to four-step mechanisms, with effort made to describe the benefits achieved at each successive level of complication. At each stage in the sequence, premixed flames will be addressed before diffusion flames because of the greater complexity associated with the mixture-fraction variations in the diffusion flames.

2. Asymptotics of One-Step Approximations for Premixed Flames

For many years asymptotic analyses of premixed-flame structures have been based on one-step, Arrhenius approximations for the overall rate of heat release with a nondimensional activation energy treated as a large parameter [1]. If E_a is the overall activation energy and T_u and T_b the unburnt and burnt gas temperatures, respectively, then the large parameter in the analysis is the Zel'dovich number

$$Ze \equiv E_a \left(T_b - T_u\right) / \left(R T_b^2\right) . \tag{2.1}$$

Flame-structure analyses that employ the limit $Ze \to \infty$ are termed activation-energy asymptotics (AEA). The structure in the limit $Ze \to \infty$ is illustrated in Fig. 1; a convective-diffusive preheat zone in which the chemistry is unimportant precedes a reactive-diffusive reaction zone in which convection is negligible at leading order.

The one-step approximation for the methane flame is

$$CH_4 + 2O_2 \to CO_2 + 2H_2O , \tag{2.2}$$

and the resulting burning-velocity formula to leading order in Ze^{-1} is [1]

$$v_u = \left[\frac{2^{2-n}\lambda_b \rho_b^{m+n} B_b T_b^b Y_{O_2 u}^{m+n-1} G(a,m,n)}{\rho_u^2 c_{pb} W_{O_2}^{m+n-1} Ze^{m+n+1} Le_{O_2}^{-m} Le_{CH_4}^{-n}}\right]^{1/2} e^{-E_a/2RT_b} , \tag{2.3}$$

where

$$G(a,m,n) \equiv \int_0^\infty y^m (y+a)^n e^{-y} dy , \tag{2.4}$$

in which

$$a \equiv Ze(\phi^{-1} - 1)/Le_{CH_4 u} , \tag{2.5}$$

the equivalence ratio ϕ having been defined as

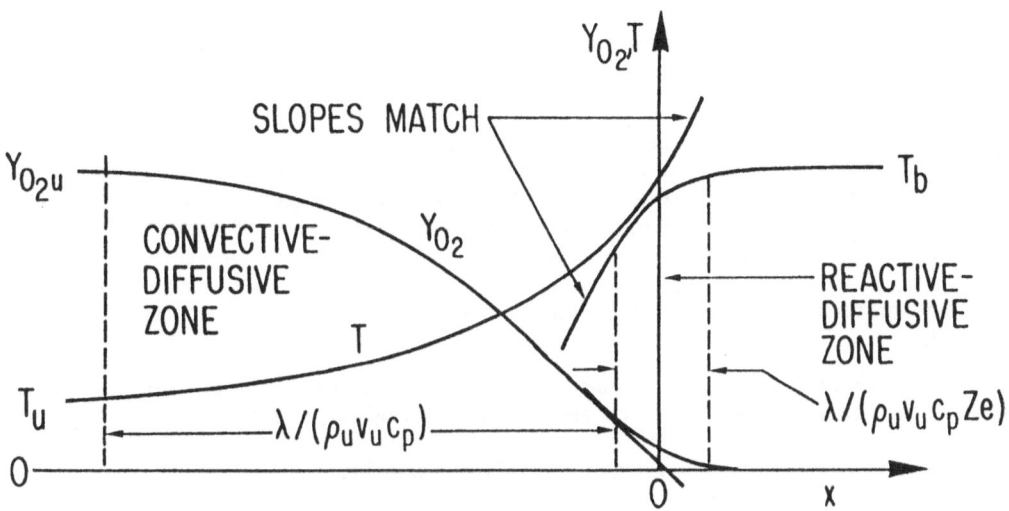

Fig. 1 Asymptotic structure of the premixed flame, according to AEA for one-step chemistry.

$$\phi = W_{\mathrm{CH_4}} Y_{\mathrm{O_2}u} / 2 W_{\mathrm{O_2}} Y_{\mathrm{CH_4}u} . \tag{2.6}$$

Equation (2.3) presumes an empirical overall reaction rate in which the molar rate of consumption of $\mathrm{CH_4}$ is expressed as

$$\omega_{\mathrm{CH_4}} = B T^b e^{-E_a/RT} [\mathrm{CH_4}]^n [\mathrm{O_2}]^m , \tag{2.7}$$

where the concentration $[k]$ of each species k is related to its mass fraction Y_k according to

$$[k] = Y_k \rho / W_k . \tag{2.8}$$

All other symbols are the same as those defined in Chapter 1. The subscripts u and b always will identify unburnt and burnt gas here. In addition to the parameters B, b and E_a of the specific reaction-rate constant, the overall (noninteger) reaction orders m and n with respect to oxygen and methane, respectively, appear in the rate formula (2.7). With this degree of empiricism, B is assigned a dependence on pressure p, typically a power law, $B \sim p^\ell$, to improve agreement with experiment. To provide burning-velocity accuracy commensurate with measurements, the AEA results have been carried to second order [2].

The function G in Eq. (2.4) is readily evaluated by numerical integration or often from tables. For stoichiometric flames we see that $a = 0$, and G becomes

$$G(0, m, n) = \Gamma(m + n + 1) , \tag{2.9}$$

where the Gamma function $\Gamma(m + n + 1)$ is readily available in tabulations [3]. For very lean conditions the limit $a \to \infty$ applies, and G becomes

$$G(a, m, n) = a^n \Gamma(m + 1) . \tag{2.10}$$

A two-term expansion of the resulting burning-velocity formula for very lean flames when $m = 1$ is [1]

$$v_u = \left[\frac{4 \lambda_b \rho_b^{n+1} B_b T_b^b Y_{\mathrm{CH_4}u}^n Y_{\mathrm{O_2}u}}{\rho_u^2 c_{pb} W_{\mathrm{CH_4}}^n Z e^2 / Le_{\mathrm{O_2}}} \right]^{1/2} e^{-E_a/2RT_b}$$

$$\times \left\{ 1 + [1.344 - 3(1 - T_u/T_b)] / Ze \right\} . \tag{2.11}$$

These results are illustrative of expressions for burning velocities derivable through AEA for one-step approximations to the combustion chemistry.

Use of these one-step results requires knowledge of the overall rate parameters B, b, ℓ, m, n and E_a that appear in the formulas. For sufficiently simple flames, such as the ozone decomposition flame with low initial ozone concentrations, these parameters can be expressed directly in terms of rate parameters for elementary reaction steps [4]. However, for the methane-air flame, as with most flames, such a direct relationship does not exist, and the parameters must be viewed as empirical fits, over limited ranges of conditions, to burning-velocity results that fundamentally are more complicated.

Despite the empirical character of the one-step approximation for the methane flame, it remains a useful approximation for describing flame dynamics and flame responses to external perturbations [5]. In many situations premixed flames can be treated as fronts that propagate at velocity v_u into the fresh mixture and that respond to curvature and strain in the flow in a manner determined by the values of response parameters termed Markstein lengths [1,5]. Expressions for parameters in Eq. (2.3) for v_u have been derived for methane flames from asymptotic analyses with reduced kinetic mechanisms, as have corresponding expressions for a Markstein length [6]. For example, it has been shown that in a first approximation for lean or stoichiometric methane flames [7]

$$E_a \approx 4RT_b^2/(T_b - T^0)\,, \tag{2.12}$$

where T^0 is an inner-layer temperature determined by chemical kinetics and remaining relatively fixed (about 1600 K at pressure $p = 1$ atm) as T_b varies through changes in ϕ, for example. This same result has in fact been found to extend to other hydrocarbon-air flames [8]. Since results of this kind help in analyzing flame-front behavior and extinction in various flows, determination of overall rate parameters associated with one-step approximations always will be an important final step in asymptotic analyses that address more complicated chemistry.

However, for methane flames and most other flames, one-step approximations are not satisfying for addressing influences of underlying chemical kinetics. With very few exceptions (one being the production of NO by the Zel'dovich mechanism in the hot burnt gases of lean methane flames, as described in Section 2 of Chapter 3), to describe flame properties of practical and fundamental interest more thorough knowledge is needed of the chemical kinetics occurring within the flame than can be provided by any one-step approximation. This is illustrated in the following discussion of the methane-air diffusion flame. For methane flames, asymptotically correct one-step approximations for flame structures cannot be derived by any logical, systematic procedure having reasonable accuracy. The one-step approximation ever must remain empirical for methane and must be deemed wholly unsatisfactory from the perspective of the chemistry.

3. Asymptotics of One-Step Approximations for Diffusion Flames

Asymptotic analyses of diffusion-flame structures for one-step, Arrhenius chemistry through AEA are more challenging than the corresponding analyses for premixed flames. In the classical solution to this problem [9] four different combustion regimes were identified, (i) an ignition regime, at the highest strain rates a of Eq. (2.21) of Chapter 1, in which there is extensive interpenetration of fuel and oxygen through diffusion, with the reaction widely distributed over the mixture fraction as a small perturbation (the nearly frozen flow identified by the subscript f in Fig. 2), (ii) a partial-burning regime at smaller a, in which there is a thin reaction zone through which fuel and oxygen both leak but which is unstable and hence not anticipated to be observed in real flames, (iii) a premixed-flame regime, in which there is a thin reaction zone through which one reactant leaks but the other does not, thereby causing the inner structure of the reaction zone in this regime to be the same as that for a nonadiabatic premixed flame, and (iv) a diffusion-flame regime at the lowest a's, in which there is a thin reaction zone through which neither reactant leaks at leading order, causing the profiles outside the reaction zone to be the chemical-equilibrium profiles of the Burke-Schumann limit [1] (identified by the subscript e in Fig. 2). The resulting dependence of the peak temperature on a Damköhler number, inversely proportional to the strain rate, is illustrated schematically in Fig. 3. Ignition occurs in the ignition regime, which extends on the lower branch from $a = a_I$ to $a = \infty$ ($1/a = 0$), while extinction occurs either in the diffusion-flame regime, which then extends on the upper branch from $a = 0$ ($1/a = \infty$) to $a = a_E$, or in the premixed-flame regime, which then applies over a range of the upper and middle branches including $a = a_E$; the middle branch is unstable and not expected to be observed.

Formulas for a_E, analogous to Eq. (2.3), are derivable through AEA [1,9]. The approach involves obtaining solutions, influenced by the flow field, in the two outer convective-diffusive zones and matching this to a solution in a stretched variable centered at the stoichiometric point for describing the structure of the reactive-diffusive zone. The scalar dissipation,

$$\chi = 2(dZ/dy)^2\lambda/(\rho c_p)\,, \tag{3.1}$$

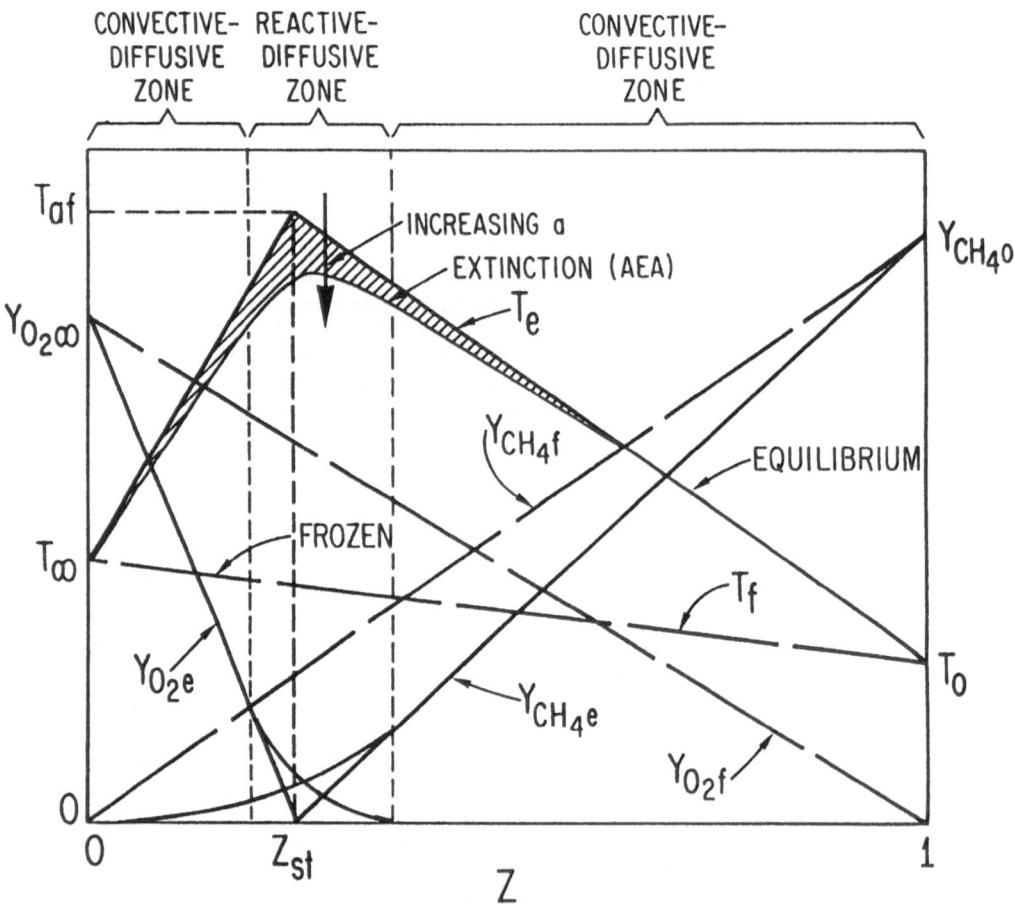

Fig. 2 Diffusion-flame structure in the mixture-fraction coordinate for one-step chemistry.

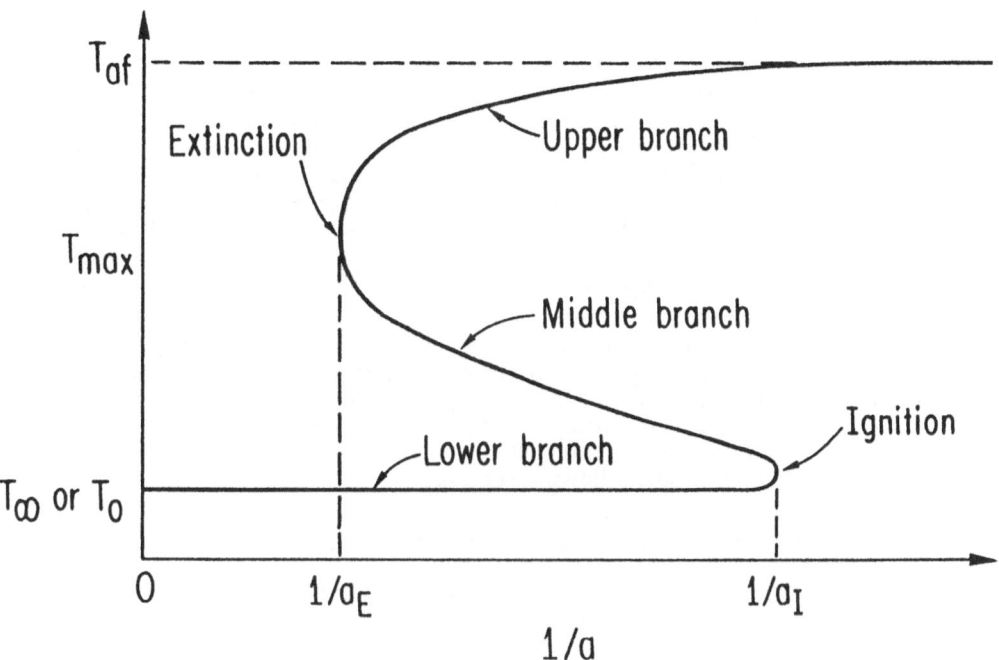

Fig. 3 Schematic illustration of the dependence of the maximum temperature on the reciprocal of the strain rate for diffusion flames, according to AEA for one-step chemistry.

plays a central part in the analysis; the value of this quantity at the stoichiometric mixture fraction, χ_{st}, affects the inner structure most directly and is employed in defining the Damköhler number. The strain rate a influences the flame structure mainly through its influence on χ_{st}, and associated with a_E is an extinction value χ_{stE} [10]. According to AEA, the departures of the profiles from the equilibrium profiles remain small all the way to extinction, as illustrated in Fig. 2.

Just as burning-velocity data can be correlated by the AEA formulas, so can data on a_E [1,11]. However, the one-step AEA predictions have been found to be wrong in rather macroscopic points of detail. For example, for methane-air diffusion flames, since the stoichiometric mixture fraction is $Z_{st} = 0.054$, the AEA analysis predicts that as extinction is approached, there is significant fuel leakage through the reaction zone but negligible oxygen leakage. This may be inferred from Fig. 2, which indicates that because of the shallower slope of the $Y_{CH_4 e}$ line, compared with that of the $Y_{O_2 e}$ line, at finite reaction rates the average value of Y_{CH_4} in the reaction zone exceeds that of Y_{O_2}. This prediction of AEA applies irrespective of whether the extinction occurs in the diffusion-flame or premixed-flame regime (the small Z_{st} favors greater accuracy for the premixed-flame regime than for the diffusion-flame regime near extinction). This AEA prediction is exactly contrary to the results of both experiment and numerical flame-structure computations with full chemistry, which exhibit appreciable O_2 leakage but immeasurably small CH_4 leakage near extinction. The asymptotic structures obtained from two-step approximations are sufficient to correct this erroneous prediction of one-step AEA, as will be seen in the following sections.

4. Asymptotics of Two-Step Approximations for Premixed Flames

Two-step approximations to the chemical kinetics provide the next level of complexity, beyond one-step approximations, in asymptotic descriptions of flame structures. In passing from one-step to two-step descriptions, the range of possible asymptotic structures is greatly increased. For this reason it seems desirable to insist on deriving two-step descriptions from the full chemistry, or at least from a skeletal mechanism (Chapter 1) by systematic procedures. A way to achieve this has been presented in Chapter 2. Before discussing the two-step structure of the methane flame, we shall briefly address flames having simpler kinetic mechanisms, since these illustrate the range of phenomena that may be encountered.

The ozone decomposition flame [6] is an example of a flame that, even with detailed chemistry, fundamentally possesses a two-step mechanism. This may be seen by observing that the only reactive species involved in this flame are O_3, O_2 and O, and in view of the O-atom conservation equation, only two independent differential equations with nonvanishing rate terms can exist for this system. Despite this basic simplicity, a wide range of structures has been found to be possible for this flame. The simplest is the flame structure illustrated in Fig. 1, but with two simultaneous differential equations instead of one describing the chemistry in the reactive-diffusive zone; this has been termed the "merged regime" and applies in practice over a wide range of conditions for this flame [6], corresponding merely to loss of the steady-state approximation for the O-atom in the reaction zone.

However, this is not the only possibility. The two steps may have rate parameters that lead to O-atom recombination occurring slowly, in a broad recombination zone that may maintain a reactive-convective balance, located downstream from the reactive-diffusive zone shown in Fig. 1. Also, the O atom may diffuse into the preheat zone, through a transition zone located at the upstream edge of the reactive-diffusive zone; heat release through recombination in the preheat zone then gives this zone a convective-reactive-diffusive balance. This asymptotic structure was termed a "two-zone" structure [6] and was found never to occur for the actual values of the rate parameters of the ozone flame, although more recent studies indicate that recombination in the preheat zone may be quite important for hydrogen-air flames, for example. This discussion illustrates that a wide variety of different reaction-zone arrangements are possible with two-step chemistry.

A two-step mechanism for the hydrogen-air flame has been derived and discussed in Chapter 3. With this mechanism, too, in advance there are many different possible asymptotic flame structures. A notable occurrence that seems to be emerging from recent asymptotic flame-structure studies is that among all these possibilities, one particular generic two-step asymptotic structure appears to be applicable to a surprisingly wide variety of real flames. This is a structure for which the single reactive-diffusive zone in Fig. 1 becomes replaced by a thin inner reaction zone followed by a thicker reactive-diffusive zone having different chemistry than the inner zone. The inner zone will be called the δ layer and the reaction zone behind it the ϵ layer, where δ and ϵ represent the orders of magnitude of the thicknesses of each layer, measured in the nondimensional coordinate

$$\xi = \rho_u v_u c_p x / \lambda, \tag{4.1}$$

identifiable from Fig. 1 as the natural coordinate for the overall premixed-flame structure. This generic structure has $\delta < \epsilon < 1$ and employs both δ and ϵ as small parameters of expansion in the asymptotic analyses.

For lean H_2–O_2 flames with the two-step mechanism, the δ layer is a layer of transition to partial equilibrium of step 1 of Table II of Chapter 1, and the one-step approximation

$$2H_2 + O_2 \rightarrow 2H_2O, \tag{4.2}$$

derivable by addition from Eq. (3.4) of Chapter 2 by assuming H to be in steady state, applies in the ϵ layer behind, while upstream from the transition layer, in the preheat zone, H-atom recombination initiated by step 5 (Table II, Chapter 1) continues to occur. A similar asymptotic structure may be anticipated for CO–H_2 flames with O_2 as oxidizer [12]. For hydrocarbon flames the different fuel chemistry changes the character of the δ layer and removes the recombination from the preheat zone, in a sense thereby introducing simplification into the asymptotic description. Yet recent studies suggest that the δ layer, ϵ layer approach remains applicable, not only to lean methane and other alkane flames, but also to many other hydrocarbon flames, such as acetylene flames, with only the fuel chemistry in need of individual treatment. The following description of the methane flame therefore should provide a background for discussion of many other flames as well.

Equation (4.11) of Chapter 3 provides a four-step mechanism for methane flames. A systematic derivation of a two-step mechanism from this four-step mechanism may be obtained by putting the H atom in steady state and the water-gas shift in partial equilibrium. The steady state converts steps I, II, III and IV into

$$\left.\begin{array}{rcl} CH_4 + O_2 & \rightleftharpoons & CO + H_2 + H_2O, \\ CO + H_2O & \rightleftharpoons & CO_2 + H_2, \\ 2H_2 + O_2 & \rightleftharpoons & 2H_2O, \end{array}\right\} \tag{4.3}$$

then the partial equilibrium of step II allows the two-step mechanism to be expressed as

$$\left.\begin{array}{rl} \text{I} & CH_4 + O_2 \rightleftharpoons \left(\frac{2}{1+\alpha}\right)(H_2 + \alpha CO) + \left(\frac{2\alpha}{1+\alpha}\right)H_2O + \left(\frac{1-\alpha}{1+\alpha}\right)CO_2, \\ \text{II} & \left(\frac{2}{1+\alpha}\right)(H_2 + \alpha CO) + O_2 \rightleftharpoons \left(\frac{2}{1+\alpha}\right)H_2O + \left(\frac{2\alpha}{1+\alpha}\right)CO_2, \end{array}\right\} \tag{4.4}$$

where α denotes the ratio of CO to H_2 concentrations at water-gas equilibrium. It may be noted in passing that the one-step approximation of Eq. (2.2) is derivable from the two-step approximation in Eq. (4.4) by introducing a steady-state approximation for the intermediate combination $H_2 + \alpha CO$, assuming their concentrations to be small, but this is a poor approximation. The two-step approximation in Eq. (4.4) displays significant qualitative improvement over any one-step approximation.

The asymptotic structure of the premixed flame [7] in the two-step approximation may be illustrated schematically from Fig. 4 by neglecting the layer of CO and H_2 nonequilibrium shown there and discussed later. Step I of Eq. (4.4) occurs in the δ layer and step II in the ϵ layer. The analysis presumes that in the δ layer the ratio of the rate of the branching step 1 ($H + O_2 \rightarrow OH + O$), to that which the fuel-consumption steps 11 and 12 (e.g. $CH_4 + H \rightarrow CH_3 + H_2$) would have if the CH_4 concentration were equal to its value in the fresh mixture, is a small parameter. With the superscript ^{0} identifying conditions at the inner layer, the definition,

$$\delta = (k_{1f}/k_{11f})^0 \left(Y_{O_2}^0 / Y_{CH_4 u} \right) (W_{CH_4}/W_{O_2}) , \qquad (4.5)$$

may be employed, and the assumption $\delta << 1$, expressing the smallness of the ratio of these two reaction rates, results in what is coming to be called rate-ratio asymptotics (RRA), to distinguish the approach from AEA. There is no assumption that the rate of step I (or of step II) possesses a high overall activation energy, so the Ze of Eq. (2.1) plays no role in the analysis. The small expansion parameter for the inner layer is δ, not any Ze^{-1}. Since the concentration of CH_4 in the inner layer is of order δ times its value in the fresh mixture, Eq. (4.5) amounts to identifying a kind of crossover temperature T^0 at which the rate of the chain-branching step 1 equals that of a fuel-consumption step 11. The equality of these two rates in the δ layer balances branching and termination because the hydrocarbon fuel chain has the effect of removing radicals from the system, as may be seen from step I of Eq. (4.11) of Chapter 3. The preheat zone remains inert for hydrocarbon flames because of the very effective radical removal through the fuel chain at the higher fuel concentrations present upstream from the δ layer.

Step I of Eq. (4.4) shows that the fuel consumption in the δ layer produces the intermediates H_2 and CO in addition to some products. The ϵ layer is the layer in which the H_2 and CO are oxidized through step II. Since CH_4 has been depleted before reaching the ϵ layer, steps 10-20 of Table II of Chapter 1 are irrelevant in this zone, and the simpler mechanism associated with steps 1-9 becomes dominant. The main elementary rate determining the overall rate in this layer is found to be step 5 ($H + O_2 + M \rightarrow HO_2 + M$) [7], which has zero activation energy, again making AEA inappropriate. For stoichiometric flames the small parameter ϵ is the reciprocal of the fourth root of a Damköhler number, giving ϵ proportional to $\sqrt{v_u / \sqrt{k_5 p \lambda / c_p}}$ [7]. Energy conservation results in

$$\epsilon = \left(T_b - T^0 \right) / \left(T_b - T_u \right) , \qquad (4.6)$$

within a factor of order unity, so that

$$v_u \sim \left(T_b - T^0 \right)^2 , \qquad (4.7)$$

which leads to Eq. (2.12). Equating burning velocities obtained from analyses of the δ and ϵ layers results in the crossover temperature being determined approximately by

$$k_{11}^0 k_5^0 \left[p / \left(RT^0 \right) \right] = \left(k_1^0 \right)^2 , \qquad (4.8)$$

a competition between the branching step 1 and termination steps 5 and 11. Activation energies of elementary steps certainly do play a role in determining T^0 from Eq. (4.8), but they do not appear in the small expansion parameters for any of the zones.

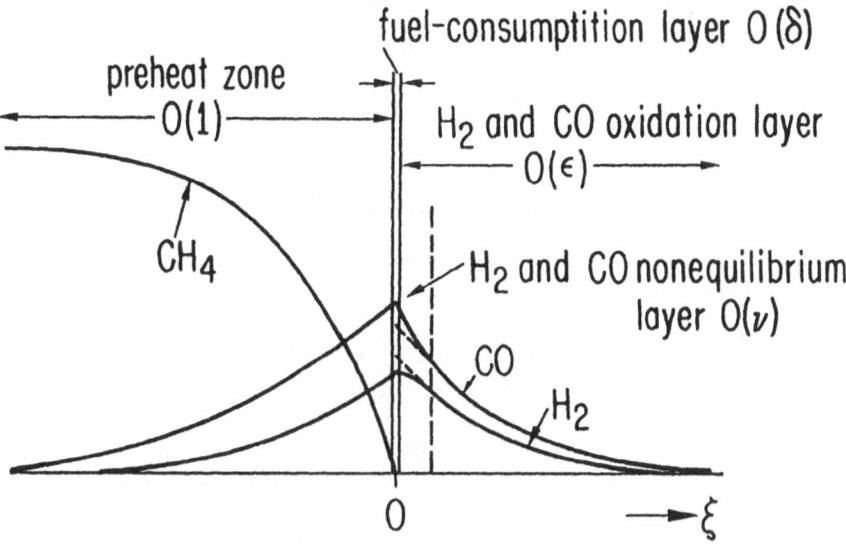

Fig. 4 The asymptotic structure of the premixed methane flame in two-step and three-step approximations.

This outline of the two-step premixed-flame asymptotic structure has been designed to give a general overall view of the zone arrangement and the chemistry of each zone. It will be necessary to consult the literature [7] to find the derivations of the differential equations applicable within each zone from Eqs. (2.1)–(2.14) of Chapter 1 and to see specifically how the matching conditions between zones can be enforced in the asymptotics. The side-reaction additions and steady-state truncations discussed in Chapter 3 play significant roles in both improvement and simplification of the asymptotic analysis. A brief discussion of these influences may be given by considering the rates w_I and w_{II} of the overall steps in the δ and ϵ layers, following the reductions described in Chapter 3.

Exclusion of all but the most essential steps [7] results in

$$\left.\begin{array}{rcl} w_I & = & k_{11f}\,[\mathrm{CH_4}]\,[\mathrm{H}]\ , \\ w_{II} & = & k_5\,[\mathrm{O_2}]\,[\mathrm{H}]\,[M]\ , \end{array}\right\} \tag{4.9}$$

in which truncation replacing the steady states of the hydrogen-oxygen chain by partial equilibria of steps 1-3 gives [7]

$$\left.\begin{array}{rcl} [\mathrm{H}] & = & \sqrt{F}\sqrt{K_1 K_2 K_3^2\,[\mathrm{O_2}]\,[\mathrm{H_2}]^3}\,/\,[\mathrm{H_2O}]\ , \\ F & = & 1 - (k_{11f}/k_{1f})\,[\mathrm{CH_4}]\,/\,[\mathrm{O_2}]\ , \end{array}\right\} \tag{4.10}$$

where K_i denotes the equilibrium constant for step i. In this simplest approximation, the reaction rate vanishes at a finite value of the stretched variable in the δ layer because as T decreases F reaches zero, and the radical concentrations remain zero at lower temperatures ($F < 0$). Relaxing truncation approximations in principal improves accuracy but in practice rapidly complicates algebra, introducing multiple spurious solutions in algebraic equations that must be avoided; truncation remains a nonexact art having a strong bearing on the ease with which asymptotic analyses of flame structures can be completed. On the other hand, inclusion of additional fuel-chemistry steps in the δ layer is rather straightforward since this merely modifies the integrand of an integral that needs to be evaluated; inclusion of the additional steps such as 11b and 12 proves to be important in seeking quantitative accuracy [7]. Although convergence of the integral in approaching the preheat zone can require special consideration, elements of AEA have not been found necessary.

The relative rates of the various steps turn out to maintain the inequality $\delta \ll \epsilon < 1$ as a reasonable approximation over the entire range of stoichiometry and pressure considered in this volume. Steps I and II therefore always occur in different layers; the merged-regime behavior for which $\delta \to \epsilon$ is not encountered. Although asymptotic structures in principle may change from one regime to another as experimental conditions are changed, such transitions have not been found for lean or stoichiometric hydrocarbon flames.

5. Asymptotics of Two-Step Approximations for Diffusion Flames

Just as two-step chemistry opens many possibilities for premixed-flame asymptotic structures, so too does it greatly increase the number of possible asymptotic diffusion-flame structures. The number of possibilities for diffusion flames generally exceeds the number for premixed flames. For example, the two steps could occur in two thin zones widely separated from each other, a situation that might be found in certain hydrogen-halogen flames [12,13]. Attention here is restricted to methane-air flames, although the structures appear likely to be quite prevalent for diffusion flames in which the reactants are hydrocarbons and oxygen.

The asymptotic structure in the two-step approximation for the methane-air diffusion flame is shown schematically in Fig. 5 [15], which should be compared with Fig. 2. The dashed lines in Fig. 5 correspond to the one-step approximation of Fig. 2. Steps I and

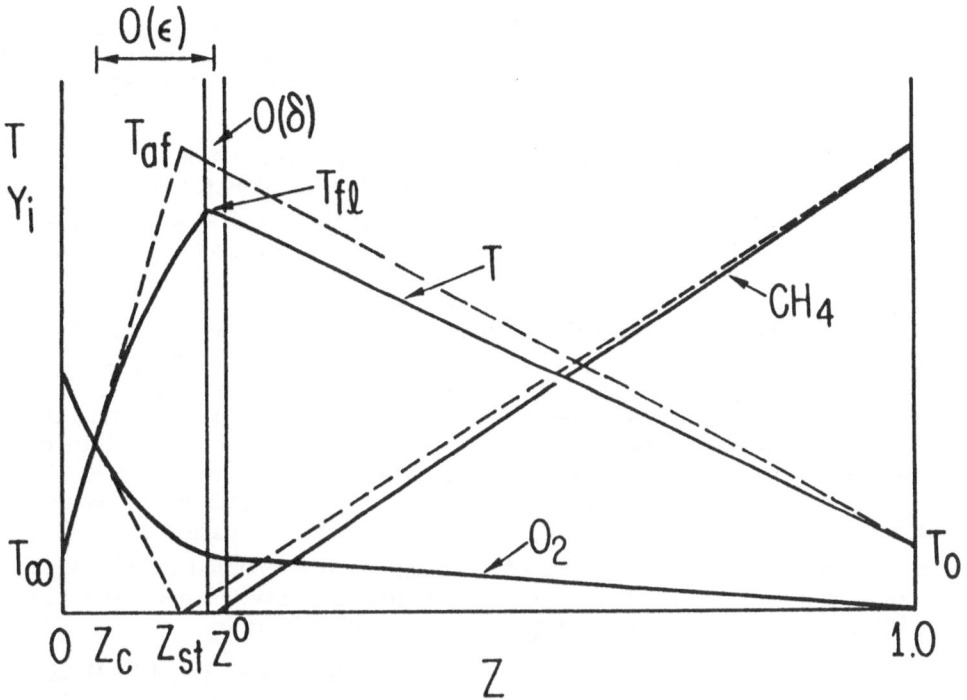

Fig. 5 The asymptotic structure of the methane-air diffusion flame in a two-step
approximation.

II occur in the δ layer and in the ϵ layer, respectively, just as they do in Fig. 4. Also, these layers remain adjacent to each other. The fuel side of the diffusion flame resembles the cold side of the premixed flame, and the oxidizer side of the diffusion flame corresponds to the hot side of the premixed flame, although the temperature decrease and O_2-concentration increase with increasing distance from the flame on the oxidizer side differ from the temperature and O_2 profiles on the downstream side of the adiabatic premixed flame; the diffusion flame looks more like a premixed flame with strong downstream heat loss, although even this is not a precise correspondence because the mixture fraction Z remains fixed throughout the premixed flame but is the principal coordinate for the diffusion flame. Nevertheless, the reaction-zone structures for the two flames are similar in that the chemistry occurring for each is the same.

The peak flame temperature T_{fl} with the two-step approximation lies below the adiabatic flame temperature T_{af} of the one-step approximation, as illustrated in Fig. 5. Here T_{fl} occurs in the δ layer, the fuel-consumption layer located at $Z = Z^0 > Z_{st}$, and the reduction of T_{fl} below T_{af} is caused by the fact that step I produces some H_2 and CO in addition to H_2O, as seen in Eq. (4.4); depending on the value of α, CO_2 may be produced or consumed in this layer. The H_2 and CO are oxidized in the ϵ layer, the thicker of the two reaction zones, located on the oxidizer side of the fuel-consumption layer and always extending from $Z = Z^0 > Z_{st}$ to a value $Z = Z_c < Z_{st}$. The asymptotics involve stretchings about $Z = Z^0$ near Z_{st}, just as in the one-step approximation, there was stretching about $Z = Z_{st}$, but two different stretchings arise in the two different reaction zones near the stoichiometric mixture fraction [15]. The parameter δ can still be defined by Eq. (4.5) if the superscript ^{0} there identifies outer-expansion conditions at $Z = Z^0$ and Y_{CH_4u} the fuel-side mass fraction $Y_{CH_4 0}$. The parameter ϵ again is the reciprocal of the fourth root of a Damköhler number, but now the scalar dissipation of Eq. (3.1) appears in the Damköhler number instead of v_u, making ϵ proportional to $[\chi^0/(k_5 p \rho^0)]^{1/4}$ [15]. Increasing the strain rate to approach extinction thus involves increasing ϵ, which however is found to remain small enough ($\epsilon < 1$) even at extinction.

Equation (4.4) indicates that oxygen is consumed in both steps I and II. However, it has been seen in Eq. (4.9) that step I mainly involves attack of radicals on the fuel, and from Eq. (4.10) it is clear that there is no need for the O_2 concentration to vanish on the fuel side of the δ layer to shut off the chemistry—the decrease in temperature and increase in [CH$_4$] readily give [H] = 0 at a finite position in the δ layer on its fuel side. Therefore, with this two-step mechanism there clearly exists the possibility of O_2 leakage through the δ layer into the inert convective-diffusive zone on the fuel side. On the other hand, complete consumption of CH$_4$ on the oxygen side of this layer at leading order is needed to bring w_I back to zero there. The chemistry of the δ layer thus allows O_2 leakage but not CH$_4$ leakage, and the calculated results verify this, giving O_2 leakage that increases with increasing strain rate (increasing χ^0 or ϵ). The O_2 leakage is illustrated in Fig. 5. This behavior agrees qualitatively with experiment, as discussed previously. It is remarkable that going from a one-step to a two-step description of the chemistry is all that is required to eliminate this erroneous prediction of AEA.

6. Asymptotics of Methane Flames with Three-Step Approximations

Despite these qualitative successes of the two-step approximation for methane-air chemistry, the quantitative predictions of flame structures, burning velocities and diffusion-flame extinction conditions that they produce are not very good. Substantial quantitative improvements are achieved by going to three-step approximations. At least two different three-step approximations can be considered. One is that of Eq. (4.3), which has been applied to both premixed flames [7] and diffusion flames [15]. Another is that in which the H-atom steady state is not imposed but water-gas equilibrium is enforced, which has been employed for diffusion flames [16]. Comparisons of the predictions of these two approaches for the diffusion flames have been made [17,18]. The different three-step approximations result in different flame structures. Let us first consider the possibilities associated with the mechanism of Eq. (4.3).

In comparing Eqs. (4.3) and (4.4) it is evident that their difference lies in the finite rate of the water-gas shift. The rate constants for step 9 then become relevant, rather than just the equilibrium constant for this step. It is possible that the departure from equilibrium is small, in which case a perturbation approach can be developed [7], based on a small parameter ν describing the extent of departure from water-gas equilibrium. For the premixed flame, expansion for small ν leads to identification of a thin layer of water-gas nonequilibrium at the upstream end of the ϵ layer, as illustrated in Fig. 4. For the diffusion flame, correspondingly in Fig. 5 there would be a ν layer within the ϵ layer adjacent to the δ layer on its oxidizer side. Numerical evaluations have shown that in fact ν is not small compared with ϵ [7]. If ν were large, then the three-step chemistry would have the water-gas step nearly frozen downstream from the δ layer in the premixed flame, and it would then occur in a thick convective-reactive zone downstream without affecting the burning velocity. In fact, ϵ and ν are roughly equal, resulting in a kind of merged-layer structure for the ϵ layer, requiring two independent rates to be considered at the same time in this layer. This most realistic limit for the premixed flame is addressed by Seshadri and Göttgens in a later chapter in this volume [19].

The agreements achievable between asymptotic predictions and experiment with the three-step chemistry of Eq. (4.3) are illustrated in Figs. 6 and 7 for the burning velocities of premixed flames and for the peak temperatures of diffusion flames, respectively. The agreements are seen to be quite good, even though not based on the latest kinetic data of the present volume. Further study is needed in testing accuracies of these predictions.

The alternative three-step approximation, in which water-gas equilibrium is maintained but the H-atom of the hydrogen-oxygen chain is not put in steady state, leads to a qualitatively different diffusion-flame structure [13,16,17]. In this approximation, O_2 no longer is consumed in the overall fuel-consumption step. Instead, the fuel is consumed by radicals, and O_2 is consumed only in the oxidation of H_2 and CO in the ϵ layer, in an overall step that also produces radicals. The ϵ layer then becomes an oxygen-consumption, radical-production layer having a different character than discussed previously. The third step, dominated by three-body rates, represents radical consumption by recombination and overall typically removes H_2 and CO as well as radicals, producing H_2O and CO_2. This third step occurs broadly throughout the flame on the oxygen side of the fuel-consumption layer. The consumption of fuel by radicals occurs in what amounts to a diffusion flame within the diffusion flame, so that the δ-layer structure now is quite different as well. In this limit AEA methods become important for the δ layer, and they also play a role in describing freezing of the reverse of the oxygen-consumption step in the ϵ layer [16,17].

These different three-step descriptions have been studied because neither water-gas equilibrium nor H-atom steady state are too good approximations to the four-step mechanism of Chapter 3. The success of the four-step mechanism in numerical integrations indicates that its asymptotic flame-structure predictions are well worth developing. The studies of the two-step and three-step mechanisms have served to explore potential asymptotic structures to which the four-step chemistry might be applied. It is therefore now straightforward, although complicated, to address asymptotic structures for four-step mechanisms.

7. Asymptotics of Methane Flames with Four-Step Approximations

The four-step mechanism of Chapter 3 has now been applied to derive asymptotic structures of methane flames [20]. A limiting case that approaches the structure of Fig. 4 is one in which all radicals nearly maintain steady states, but there is a thin radical-consumption layer (not affecting the burning velocity) in the upstream part of the δ layer

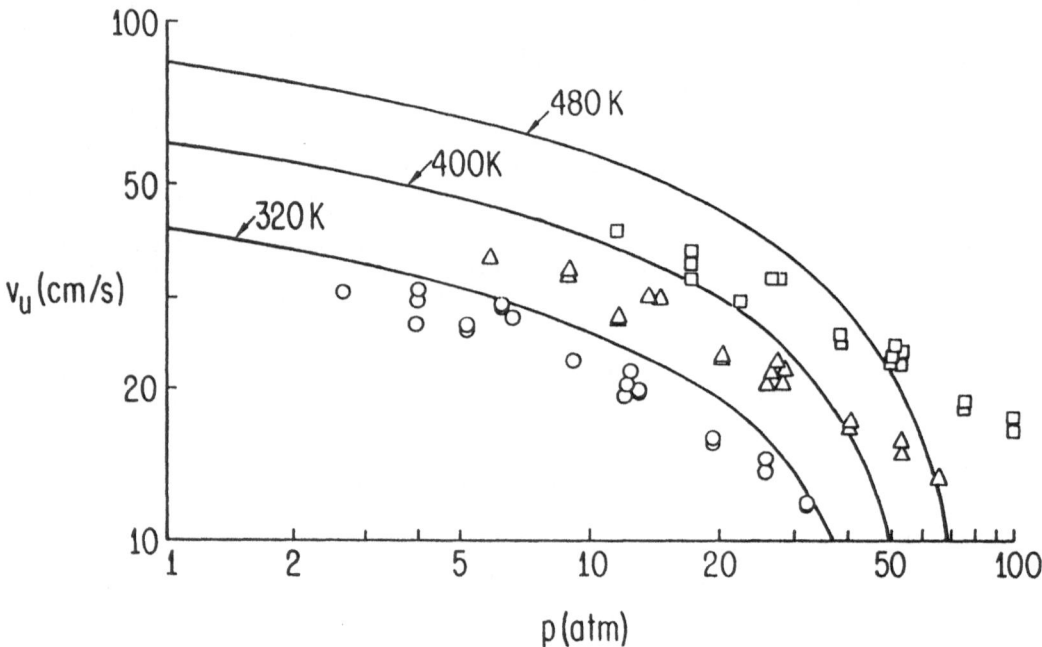

Fig. 6 Burning velocities of stoichiometric methane-air flames at different initial temperatures, calculated from asymptotic analysis with a three-step mechanism (curves), compared with experiment (points) [7].

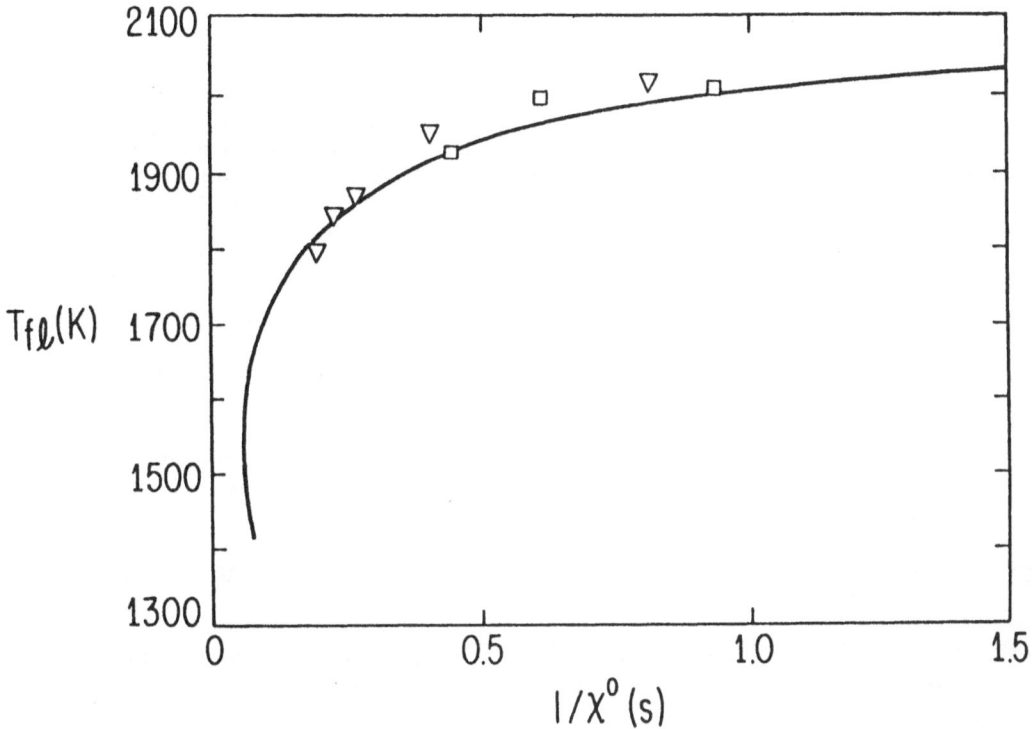

Fig. 7 Dependence of the maximum temperature on the effective inverse scalar
dissipation for methane-air diffusion flames, according to asymptotic analysis
with a three-step mechanism (curve), numerical integration with a four-step
mechanism (triangles), and experiment (squares) [14].

in which H-atom departure from a steady state must be considered [7,20]. Another limit is one in which the fuel and radicals are consumed in a diffusion-flame layer, of the type just described, within the premixed flame [20]. Estimates indicate that the true situation lies between these two limiting cases [20]. Although a large number of parameters turn out to be involved in the analysis, agreements for burning velocities are quite good, and diffusion-flame structures can readily be calculated in the same way. Further exploitation of the four-step mechanism for describing asymptotic structures of methane flames therefore seems warranted.

8. Conclusions

In the past five years remarkable advances have been made in defining asymptotic structures of methane flames. The structure derived for two-step chemistry, although inaccurate numerically, greatly improves our understanding of what can be expected from these flames. The basis therefore now exists for employing reduced mechanisms, including the rather accurate four-step mechanisms, for deriving asymptotic structures of methane flames. The following chapters in Part II report some recent progress in this direction.

References

[1] Williams, F. A., Combustion Theory, 2nd Ed., Addison-Wesley Publishing Co., Menlo Park, CA, (1985), Ch. 3,5,9.

[2] Chelliah, H.K. and Williams, F.A., "Asymptotic Analysis of Two-Reactant Flames with Variable Properties and Stefan-Maxwell Transport," *Comb. Sci. and Tech.*, **51**, (1987), p. 129.

[3] Abramowitz, M. and Stegun, I.A., Handbook of Mathematical Functions, Dover, New York, (1965), p. 255.

[4] Rogg, B., Liñán, A. and Williams, F.A., "Deflagration Regimes of Laminar Flames Modeled after the Ozone Decomposition Flame," *Comb. and Flame*, **65**, (1986), p. 79.

[5] Clavin, P., "Dynamic Behavior of Premixed Flame Fronts in Laminar and Turbulent Flows," *Prog. Energy Comb. Science*, **11**, (1985), p. 1.

[6] Rogg, B. and Peters, N., "The Asymptotic Structure of Weakly Strained Stoichiometric Methane-Air Flames," *Comb. and Flame*, (1990), to appear.

[7] Peters, N. and Williams, F.A., "The Asymptotic Structure of Stoichiometric Methane-Air Flames," *Comb. and Flame*, **68**, (1987), p. 185.

[8] Kennel, C., Göttgens, J. and Peters, N., "The Basic Structure of Lean Propane Flames," Twenty-Third Symposium (International) on Combustion, The Combustion Institute, Pittsburgh, (1990), to appear.

[9] Liñán, A., "The Asymptotic Structure of Counterflow Diffusion Flames for Large Activation Energies," *Acta Astronautica*, **1**, (1974), p. 1007.

[10] Peters, N., "Laminar Diffusion Flamelet Models in Non-Premixed Turbulent Combustion," *Prog. Energy Combust. Sci.*, **10**, (1984), p. 319.

[11] Williams, F.A., "A Review of Flame Extinction," *Fire Safety Journal*, **3**, (1981), p. 163.

[12] Rogg, B. and Williams, F.A., "Structures of Wet CO Flames with Full and Reduced Kinetic Mechanisms," Twenty-Second Symposium (International) on Combustion, The Combustion Institute, Pittsburgh, (1989), p. 1441.

[13] Williams, F.A., "Influences of Detailed Chemistry on Asymptotic Approximations for Flame Structure," Mathematical Modeling in Combustion and Related Topics, C.-M. Brauner and C. Schmidt-Lainé, Eds., Martinus Nijhoff Publishers, Dordrecht, The Netherlands, (1988), p. 315.

[14] Williams, F.A., "Asymptotic Methods for Flames with Detailed Chemistry," Mathematical Modeling in Combustion Science, J.D. Buckmaster and T. Takeno, Eds., Springer-Verlag, New York, (1988), p. 44.

[15] Seshadri, K. and Peters, N., "Asymptotic Structure and Extinction of Methane-Air Diffusion Flames," Comb. and Flame, 73, (1988), p. 23.

[16] Treviño, C. and Williams, F.A., "Asymptotic Analysis of the Structure and Extinction of Methane-Air Diffusion Flames," in Dynamics of Reactive Systems Part I: Flames, A.L. Kuhl, J.R. Bowen, J.-L. Leyer and A. Borison, Eds., Progress in Astronautics and Aeronautics, Vol. 113, AIAA, Washington, DC, ISBN 0-930403-46-0, (1988).

[17] Chelliah, H.K. and Williams, F.A., "Aspects of the Structure and Extinction of Diffusion Flames in Methane-Oxygen-Nitrogen Systems," Comb. and Flame, 80, (1990), p. 17.

[18] Chelliah, H.K., Treviño, C. and Williams, F.A., this volume.

[19] Seshadri, K. and Göttgens, J., this volume.

[20] Seshadri, K. and Peters, N., "The Inner Structure of Methane-Air Flames," Comb. and Flame, 81, (1990), p. 96.

ON REDUCED MECHANISMS FOR
METHANE-AIR COMBUSTION

R.W.Bilger, M.B.Esler and S.H.Starner
Department of Mechanical Engineering
The University of Sydney
NSW 2006, Australia

1. Introduction

In an earlier paper [1] we present the reduction of a 58-step C_1 mechanism of methane to a four-step reduced mechanism. A feature of that work is the treatment of the steady-state approximations for O-atom and CH_3. These lead to a quadratic expression for O-atom concentration which gives better estimates of radicals and minor species on the fuel side of the inner part of the reaction zone. Here we obtain a similar four step mechanism reduced from the somewhat different starting mechanism agreed for the Sydney/Yale/UCSD workshop [2]. Numerical solutions are used to compare this reduced mechanism with the skeletal version of the starting mechanism and with other reduced mechanisms [3-7]. The effect of the improved modeling of O-atom and CH_3 is examined. Williams and his co-workers [6,7] give considerable theoretical attention to the competition between the chain-branching and HO_2 forming reactions of H with O_2; the significance of such chain termination via the HO_2 route is also examined.

Conserved scalars are widely used in theoretical analyses to link the concentrations of reactive species and the temperature. There is a difficulty associated with the non-unity of Lewis numbers in the definition of such conserved scalars. The numerical solutions with simple transport that are given here are used to address this question. Reactive scalars are used to measure the progress of the chemical reaction, particularly in premixed flames. A four-step mechanism requires appropriate definition of four separate progress variables: this is also considered. Finally, comments are made on the nature of controlling mechanisms in the flames and a possible alternative approach to the analysis of the flame structure.

2. Reduced Mechanisms

The skeletal mechanism originally agreed for the Sydney/Yale/UCSD Workshop is shown in Table I. This differs from the final form given in Chapter 1 [2] in that there is no correction applied to the Lindemann form for Reactions 10f and 10b. Consequently the flame speeds are a little higher than obtained by Smooke [2]. The reduced mechanism obtained by the procedures in [1] is

$$CH_4 + 2H + H_2O \rightarrow CO + 4H_2, \qquad (I)$$

$$CO + H_2O \rightleftharpoons CO_2 + H_2, \qquad (II)$$

TABLE I

Skeletal Methane-Air Reaction Mechanism
Rate Coefficients in the Form $k_f = AT^\beta \exp(-E_0/RT)$.
Units are moles, cubic centimeters, seconds, Kelvins and calories/mole.

	REACTION	A	β	E
1f.	$H + O_2 \rightarrow OH + O$	2.000E+14	0.000	16800.
1b.	$OH + O \rightarrow H + O_2$	1.575E+13	0.000	690.
2f.	$O + H_2 \rightarrow OH + H$	1.800E+10	1.000	8826.
2b.	$OH + H \rightarrow O + H_2$	8.000E+09	1.000	6760.
3f.	$H_2 + OH \rightarrow H_2O + H$	1.170E+09	1.300	3626.
3b.	$H_2O + H \rightarrow H_2 + OH$	5.090E+09	1.300	18588.
4f.	$OH + OH \rightarrow O + H_2O$	6.000E+08	1.300	0.
4b.	$O + H_2O \rightarrow OH + OH$	5.900E+09	1.300	17029.
5.	$H + O_2 + M \rightarrow HO_2 + M^a$	2.300E+18	-0.800	0.
6.	$H + HO_2 \rightarrow OH + OH$	1.500E+14	0.000	1004.
7.	$H + HO_2 \rightarrow H_2 + O_2$	2.500E+13	0.000	700.
8.	$OH + HO_2 \rightarrow H_2O + O_2$	2.000E+13	0.000	1000.
9f.	$CO + OH \rightarrow CO_2 + H$	1.510E+07	1.300	-758.
9b.	$CO_2 + H \rightarrow CO + OH$	1.570E+09	1.300	22337.
10f.	$CH_4 + (M) \rightarrow CH_3 + H + (M)^b$	6.300E+14	0.000	104000.
10b.	$CH_3 + H + (M) \rightarrow CH_4 + (M)^b$	5.200E+12	0.000	-1310.
11f.	$CH_4 + H \rightarrow CH_3 + H_2$	2.200E+04	3.000	8750.
11b.	$CH_3 + H_2 \rightarrow CH_4 + H$	9.570E+02	3.000	8750.
12f.	$CH_4 + OH \rightarrow CH_3 + H_2O$	1.600E+06	2.100	2460.
12b.	$CH_3 + H_2O \rightarrow CH_4 + OH$	3.020E+05	2.100	17422.
13.	$CH_3 + O \rightarrow CH_2O + H$	6.800E+13	0.000	0.
14.	$CH_2O + H \rightarrow HCO + H_2$	2.500E+13	0.000	3991.
15.	$CH_2O + OH \rightarrow HCO + H_2O$	3.000E+13	0.000	1195.
16.	$HCO + H \rightarrow CO + H_2$	4.000E+13	0.000	0.
17.	$HCO + M \rightarrow CO + H + M$	1.600E+14	0.000	14700.
18.	$CH_3 + O_2 \rightarrow CH_3O + O$	7.000E+12	0.000	25652.
19.	$CH_3O + H \rightarrow CH_2O + H_2$	2.000E+13	0.000	0.
20.	$CH_3O + M \rightarrow CH_2O + H + M$	2.400E+13	0.000	28812.
21.	$HO_2 + HO_2 \rightarrow H_2O_2 + O_2$	2.000E+12	0.000	0.
22f.	$H_2O_2 + M \rightarrow OH + OH + M$	1.300E+17	0.000	45500.
22b.	$OH + OH + M \rightarrow H_2O_2 + M$	9.860E+14	0.000	-5070.
23f.	$H_2O_2 + OH \rightarrow H_2O + HO_2$	1.000E+13	0.000	1800.
23b.	$H_2O + HO_2 \rightarrow H_2O_2 + OH$	2.860E+13	0.000	32790.
24.	$OH + H + M \rightarrow H_2O + M^a$	2.200E+22	-2.000	0.
25.	$H + H + M \rightarrow H_2 + M^a$	1.800E+18	-1.000	0.

[a] Third body efficiencies: $CH_4 = 6.5, H_2O = 6.5, CO_2 = 1.5, H_2 = 1.0, CO = 0.75, O_2 = 0.4, N_2 = 0.4$ All other species $= 1.0$

[b] Lindemann form, $k = k_\infty/(1 + \alpha/[M])$ where $\alpha = .0063 \exp(-18000/RT)$.

$$2H_2 + O_2 \rightarrow 2H_2O, \tag{III}$$

$$3H_2 + O_2 \rightleftharpoons 2H_2O + 2H. \tag{IV}$$

The global reaction rates are:

$$\omega_I = \omega_{13} + \omega_{18}, \tag{2.1a}$$
$$\approx \omega_{13}; \tag{2.1b}$$
$$\omega_{II} = \omega_9; \tag{2.2}$$
$$\omega_{III} = \omega_5 - \omega_{10} + \omega_{16} - \omega_{20} - \omega_{22} + \omega_{24} + \omega_{25}, \tag{2.3a}$$
$$\approx \omega_5 + \omega_{10b}; \tag{2.3b}$$
$$\omega_{IV} = \omega_1 - \omega_7 - \omega_8 + \omega_{10} - \omega_{16} + \omega_{18} + \omega_{20} - \omega_{24} - \omega_{25}, \tag{2.4a}$$
$$\approx \omega_1 - \omega_{10b}. \tag{2.4b}$$

Here the use of a numerical subscript without the f or b modifier (denoting forward or backward rate) signifies the net rate. The contributions of Reactions 21 to 23 are omitted as they give negligible contributions to the major species balances. Reaction 21 does, however, make a contribution to the estimation of HO_2 needed for calculating ω_7 and ω_8. The approximations given in Equations (2.1b), (2.3b) and (2.4b) are found to be quite adequate for diffusion flames at strain rates in excess of $50s^{-1}$. At lower strain rates the forward direction of Reaction 10 and Reaction 18 become significant.

Assumption of Reaction 3 being in partial equilibrium yields

$$\Gamma_{OH} = \Gamma_H \Gamma_{H_2O}/(K_3 \Gamma_{H_2}). \tag{2.5}$$

The modified partial equilibrium of [1] included a contribution from Reaction 5 in Eqn.(2.5). This is abandoned here since, although it gives slightly better peak H-atom concentrations, it gives anomalously high values of H_2 concentration on the air side of the counterflow diffusion flames. Other intermediate species are obtained by assuming them to be in steady state:

$$\Gamma_O = \frac{1}{2}\left[(B^2 + 4C)^{1/2} - B\right], \tag{2.6a}$$

where

$$B = \left(k_{10b}^* \Gamma_H + k_{11b}\Gamma_{H_2} + k_{12b}\Gamma_{H_2O} + k_{18}\Gamma_{O_2}\right) \Big/ k_{13} +$$

$$\left[\Gamma_{CH_4}\left(k_{10f}^*/\rho + k_{11f}\Gamma_H + k_{12f}\Gamma_{OH}\right) - k_{1f}\Gamma_H\Gamma_{O_2} - k_{2b}\Gamma_{OH}\Gamma_H - k_{4f}\Gamma_{OH}^2\right]$$

$$\Big/ \left[k_{1b}\Gamma_{OH} + k_{2f}\Gamma_{H_2} + k_{4b}\Gamma_{H_2O}\right], \tag{2.6b}$$

$$\approx k_{11b}\Gamma_{H_2}\Big/ k_{13} + \left(k_{11f}\Gamma_{CH_4}\Gamma_H - k_{1f}\Gamma_H\Gamma_{O_2} - k_{4f}\Gamma_{OH}^2 \right)$$

$$\Big/ \left(k_{1b}\Gamma_{OH} + k_{4b}\Gamma_{H_2O} \right), \tag{2.6c}$$

$$C = \left[\left(k_{1f}\Gamma_H\Gamma_{O_2} + k_{2b}\Gamma_{OH}\Gamma_H + k_{4f}\Gamma_{OH}^2 \right)\left(k_{10b}^*\Gamma_H + k_{11b}\Gamma_{H_2} + k_{12b}\Gamma_{H_2O} \right) \right.$$

$$+ k_{18}\Gamma_{O_2}\Gamma_{CH_4}\left(k_{10f}^*/\rho + k_{11f}\Gamma_H + k_{12f}\Gamma_{OH} \right) \Big] \Big/ \Big[\left(k_{1b}\Gamma_{OH} \right.$$

$$\left. + k_{2f}\Gamma_{H_2} + k_{4b}\Gamma_{H_2O} \right) k_{13} \Big], \tag{2.6d}$$

$$\approx \left[k_{11b}\Gamma_{H_2}\left(k_{1f}\Gamma_H\Gamma_{O_2} + k_{4f}\Gamma_{OH}^2 \right) + k_{11f}k_{18}\Gamma_H\Gamma_{O_2}\Gamma_{CH_4} \right]$$

$$\Big/ \left[k_{13}\left(k_{1b}\Gamma_{OH} + k_{4b}\Gamma_{H_2O} \right) \right], \tag{2.6e}$$

$$\Gamma_{CH_3} = \Gamma_{CH_4}\left(k_{10f}^*/\rho + k_{11f}\Gamma_H + k_{12f}\Gamma_{OH} \right) \Big/ \left(k_{10b}^*\Gamma_H + k_{11b}\Gamma_{H_2} \right.$$

$$\left. + k_{12b}\Gamma_{H_2O} + k_{13}\Gamma_O + k_{18}\Gamma_{O_2} \right), \tag{2.7a}$$

$$\approx K_{11}\Gamma_{CH_4}\Gamma_H \Big/ \Gamma_{H_2}\left(1 + k_{13}\Gamma_O \Big/ k_{11b}\Gamma_{H_2} \right), \tag{2.7b}$$

$$\Gamma_{HO_2} = \frac{1}{2}\left[\left(D^2 + 2\rho k_5\Gamma_H\Gamma_{O_2}\Gamma_M \Big/ k_{21} \right)^{1/2} - D \right], \tag{2.8a}$$

$$D = \left(k_6\Gamma_H + k_7\Gamma_H + k_8\Gamma_{OH} + k_{23b}\Gamma_{H_2O} \right) \Big/ 2k_{21}. \tag{2.8b}$$

In the above $\Gamma_i \equiv Y_i/W_i$ is the "specific abundance" of species i, equal to its mass fraction, Y_i, divided by its molecular mass, W_i. Rate constants k_k are as given and defined in Table I; for reaction 10, k_k^* are the k_∞ values divided by the Lindemann factor $(1 + \alpha RT/P)$.

Seshadri and Peters [5] and others give the rate of Reaction I as the net rate of Reactions 10, 11 and 12. Equations (2.1a) and (2.7a) are exactly equivalent to this. If the simpler form of (2.7b) for the CH_3 concentration is used Eqn. (2.1a) gives better results than the net rates of Reactions 10, 11 and 12. The preference for the mole-reducing oxidation of H_2 as step III of the reduced mechanism over the recombination of H atoms has already been discussed [1] but it will be justified in the next section.

3. Numerical Solutions

Numerical solutions were carried out for counterflow diffusion flames in the forward stagnation region of a porous cylinder (the Tsuji burner [8]) and for freely propagating planar premixed flames using modified versions of codes [9] provided by Professor M. D. Smooke. The solutions were carried out for "full" transport, using the Curtiss-Hirschfelder approximations [10] for multicomponent transport, or for "simple" transport using Fick's Law with constant Lewis numbers as correlated for this workshop by Smooke [2] and shown in Table II. For the four-step mechanism a special subroutine was written that returns the net species reaction rates for the given temperature and major species concentrations. The codes were modified to run on a NEC micro-computer with a Definicon fast processing board.

TABLE II

Simplified Transport Assumptions [2]

Thermal properties:

$$\lambda/c_p = 2.58 \times 10^{-4}(T/298)^{0.7} gcm^{-1}s^{-1}$$

Lewis numbers

Species	Value
CH_4	0.97
O_2	1.11
H_2O	0.83
CO_2	1.39
H	0.18
O	0.70
OH	0.73
HO_2	1.10
H_2	0.30
CO	1.10
H_2O_2	1.12
HCO	1.27
CH_2O	1.28
CH_3	1.00
CH_3O	1.30
N_2	1.00

Results for the counterflow diffusion flames are shown in Table III and Figure 1. The Seshadri and Peters [5] reduced four-step mechanism is very similar to that presented here but uses a truncated steady state for O-atom which does not include Reaction 13.

TABLE III

Numerical Solutions for Counterflow
Diffusion Flame at $a=300s^{-1}$

Flame Parameters	Sk. Mech. Full Transp.	Sk. Mech. Modified HO_2	4-step Simple Transp.	Simple 4-Step	Seshadri and Peters[5]
Peak T (K)	1884	1887	1953	1943	1924
Peak H (mol%)	0.327	0.340	0.370	0.379	0.454
Peak H_2 (mol%)	1.82	1.86	1.89	1.81	2.57
Peak CO (mol%)	4.12	4.11	4.44	4.31	4.99
Peak CO_2 (mol%)	6.12	6.18	5.79	5.83	5.49
Peak H_2O (mol%)	16.25	16.25	15.69	15.56	15.65
Fuel Rate Peak:					
$\xi_o - \xi_s$	0.0130	0.0130	0.0038	0.0041	0.0070
T (K)	1883	1885	1918	1910	1914
Peak H (mol%)	0.145	0.149	0.243	0.243	0.211
Peak H_2 (mol%)	1.63	1.66	1.32	1.31	1.96
Peak O_2 (mol%)	1.40	1.33	2.03	2.95	1.70
H Atom Peak:					
$\xi_s - \xi_p$	0.0040	0.0040	0.0059	0.0063	0.0049
T (K)	1794	1795	1800	1785	1800
Peak H_2 (mol%)	0.865	0.875	0.668	0.645	0.796
Peak CO (mol%)	2.77	2.73	2.50	2.44	2.58
Peak O_2 (mol%)	3.28	3.25	3.90	4.04	3.64
Leakage O_2 (mol%)	0.985	0.925	0.812	0.988	1.09

It is seen that in general the predictions using the four-step mechanisms are very good. Peak temperatures and temperatures on the rich side are somewhat high, but on the lean side agreement is excellent. Fuel concentrations are somewhat overpredicted on the rich side. This is a consequence of the peak fuel reaction rate occurring at a mixture fraction, ξ_o, which is closer to the stoichiometric mixture fraction, ξ_s, as shown in Table III. The mixture fraction is defined

$$\xi \equiv (\beta - \beta_{ox})/(\beta_{Fu} - \beta_{ox}), \tag{3.1}$$

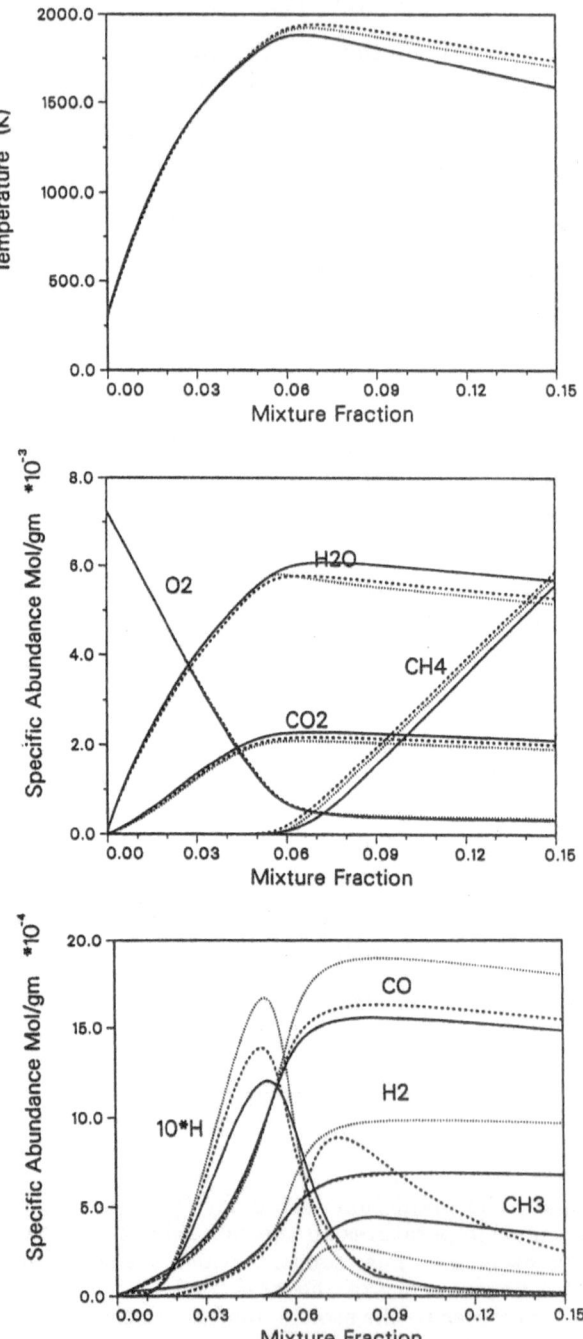

Figure 1. Comparison of calculated flame structure for counterflow diffusion flames at a=300 s^{-1}: ———skeletal mechanism with full transport; - - - - simplified version of four-step mechanism presented here using simplified transport, ...Seshadri and Peters [5] four-step mechanism with simplified transport.

with

$$\beta \equiv 2Z_C/W_C + \frac{1}{2}Z_H/W_H - Z_O/W_O, \qquad (3.2)$$

where Z_l and W_l are the mass fractions and atomic masses, respectively, of the elements carbon, hydrogen and oxygen. The subscripts fu and ox refer to the fuel and oxidant streams, respectively. For methane/air the stoichiometric mixture fraction, $\xi_s = 0.055$. The cause of this shift in the fuel reaction zone is not clear at this time.

The oxygen concentrations, including the important leakage of oxygen through the reaction zone to the rich side of the flame, are predicted very well. In Table III the "leakage" oxygen concentration is that determined for $\xi = 0.10$. The four-step mechanisms underpredict CO_2 and H_2O on the rich side, a consequence of the shift in the fuel reaction zone. It seems anomalous that high temperatures are associated on the rich side with low product concentrations CO_2 and H_2O and higher amounts of unreacted fuel. It is likely to be the effect of differential diffusion of enthalpy and species being predicted differently for the simplified transport used with the four-step calculations. At this time of writing, solutions for simplified transport with the skeletal mechanism are not available to clarify this point. It is apparent that there are also differences in the oxygen element balances on the rich side. The correlation found by Smooke [2] for the thermal conductivity and shown in Table II is fitted primarily for premixed flames and fits the diffusion flames well on the lean side only, with values being lower on the rich side. There is no such systematic trend in the fitting of the Lewis numbers of the various species. We return to questions of differential diffusion in the section following on Conserved and Reactive Scalars.

Significant differences appear in the predictions of the four-step mechanisms for CO, H_2 and H. The full steady state for O-atom that is used here gives better results. It is apparent that the quadratic form of Eqn. (2.6a) is necessary for giving good predictions of CO, H_2 and H on the rich side. Peak H-atom concentrations are also improved. Figure 1(c) shows the resulting predictions for CH_3. It seems that the diffusional effects that cause the marked overshoot in CH_3 predictions when the full steady state is used are more than offset by excluding Reaction 13 from the O-atom balance.

Table IV and Figs. 2 and 3 show results of calculations made for premixed flames. In the figures distance through the flame is in terms of the normalized distance $\hat{x}$, where

$$\hat{x} \equiv m(x - x^\circ)c_p/\lambda. \qquad (3.3)$$

Here m is the mass flux through the flame and λ/c_p is evaluated from Smooke's correlation of Table II using the unburnt temperature. The maximum fuel consumption rate is used to evaluate the arbitrary origin x° and reference temperature T° given in Table IV.

Flame speed predictions for the four-step mechanisms are in general 10 to 20% higher than for the skeletal mechanism. Calculations made with simplified transport and the skeletal mechanism indicated that the discrepancy is not due to any deficiency in the transport model. Of the four-step reduced mechanisms the full version presented here appears to be better than either its simplified version (Eqns. 2.6(c), (e) and 2.7(b)) or the Seshadri and Peters [5] version. The subroutine used to calculate the four-step chemical production rates arbitrarily sets all rates to zero below 750K. This was done to avoid arithmetic problems in calculations for the counterflow diffusion flame: the

TABLE IV

Numerical Solutions for Premixed Flames

Flame Parameters	Sk. Mech. Full Transp.	Sk. Mech. Modified HO_2	4-step Simple Transp.	Simple 4-Step	Seshadri and Peters[5]
1 atm, Stoich.:					
Flame speed (cm/s)	42.0	45.8	45.6	48.0	52.6
Peak H (mol%)	0.854	0.979	0.871	1.07	1.09
T (K)	1587	1589	1689	1654	1561
5 atm, Stoich.:					
Flame speed (cm/s)	20.0	-	19.9	23.4	22.8
Peak H (mol%)	0.241	-	0.250	0.315	0.340
T (K)	1764	-	1882	1850	1813
1 atm, Eq. Rat 0.7:					
Flame speed (cm/s)	19.3	-	23.1	23.5	22.7
Peak H (mol%)	0.176	-	0.220	0.245	0.237
T (K)	1415	-	1460	1420	1436

steady-state approximations on which the production rates are based become invalid at low temperatures in this flame. Examination of the CH_3 and H-atom profiles in Figs. 2 and 3 indicate that this may not be a valid procedure in premixed flames. Some of the discrepancy in flame speed may arise from this source. In general the four-step reduced mechanisms give very good predictions of the flame structure on the basis of comparison with the skeletal mechanism. The full version presented here gives better predictions of H-atom, CO and H_2 but overpredicts the temperature at which fuel consumption peaks. Results obtained at five atmospheres pressure, stoichiometric and for one atmosphere, 0.7 equivalence ratio show similar good agreement for the flame structure.

Figure 4 and 5 show comparisons for net species reaction rates for diffusion flames and premixed flames, respectively. For the diffusion flames the molar reaction rate is divided by the local mass density. For the premixed flames this ratio is non dimensionalized by dividing by the factor

$$\rho_u S_u^2 \Gamma_{O_2,u} \left(c_p/\lambda \right)_u$$

where S_u is the flame speed and the other quantities are evaluated at the unburnt condition. For stoichiometric mixtures at one atmosphere and 298K this factor has a value of 55.9 mol/gm- s for a flame speed of 42 cm/s. Peak reaction rates for the

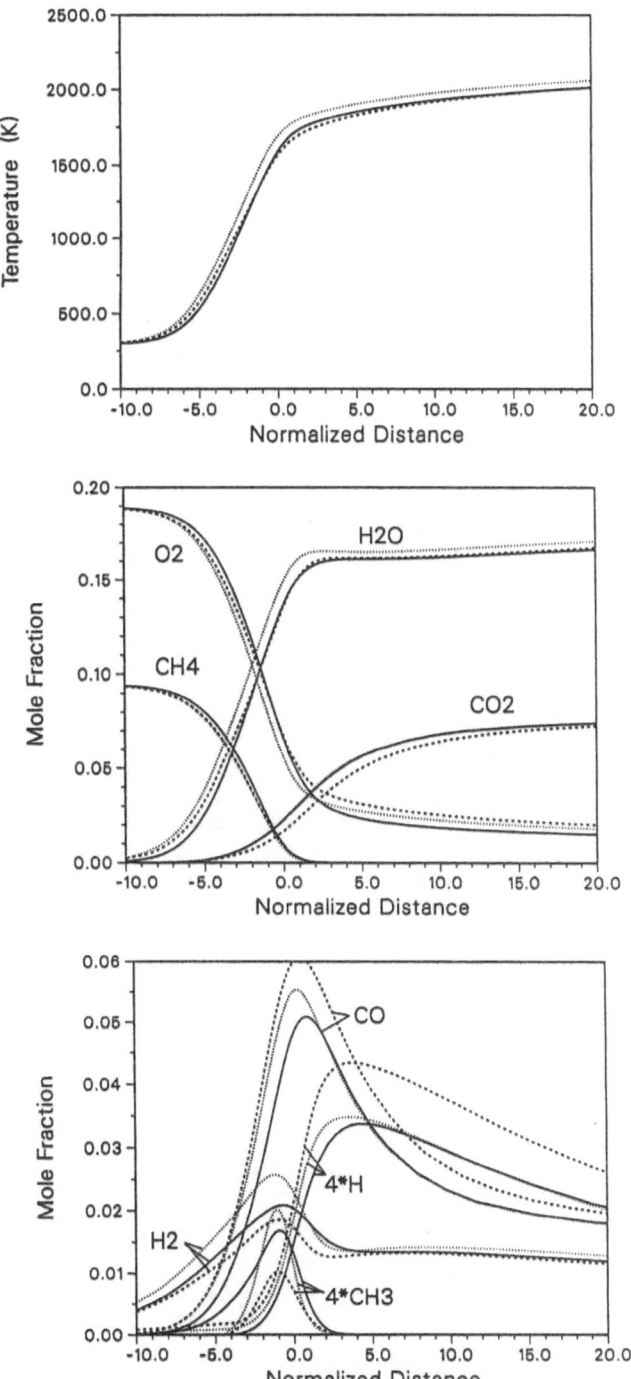

Figure 2. Comparison of calculated flame structure for premixed flames at one atmosphere, stoichiometric: ———skeletal mechanism with full transport; ....full four-step mechanism presented here with simplified transport; - - - - - Seshadri and Peters [5] four-step mechanism with simplified transport.

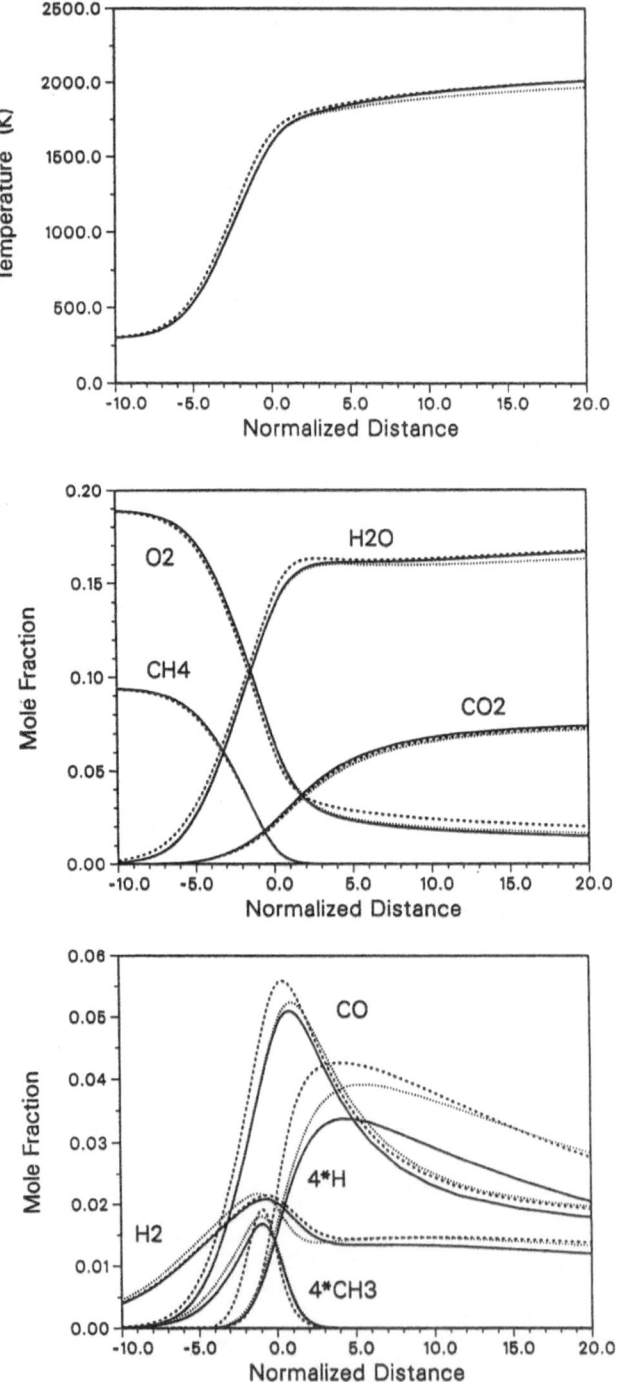

Figure 3. Further comparisons of calculated flame structure for premixed flames at one atmosphere, stoichiometric: ——skeletal mechanism with full transport; - - - - - simplified version of four-step mechanism presented here using simplified transport; ....skeletal mechanism with modified HO_2 chemistry.

reactants and products in the diffusion flames are a little lower than in the premixed flame but would be much the same near extinction which occurs around 400 s^{-1} for both the skeletal and four-step mechanisms. The reaction rates for the four-step mechanisms show higher peaks than for the skeletal mechanisms, particularly for the intermediates H_2 and CO and for H-atom. In diffusion flames the peaks are also shifted toward the lean side. For H_2 and H the increased peakiness of the net reaction rates, which are equal to the net transport, may be due to the neglect of transport for OH, CH_3 and the other steady-state species. Figures 4 and 5 show these for the skeletal mechanisms. For diffusion flames the CH_3 peaks are about 70% and OH about 35% of the peaks for H-atom. For the premixed flame the CH_3 and OH peaks are both around 30% of the H-atom peaks, indicating that the steady-state assumption is better for these species in premixed flames.

The numerical solutions indicate that the four-step reduced mechanisms considered here may be validly used over the wide range of flame conditions considered. Further validation efforts should include study of rich premixed flames and diffusion flames at high pressure. The incorporation of the quadratic form of Eqn. (2.6) appears to be warranted, particularly for diffusion flames. The simplified version of Eqns. 2.6(c) and 2.7(b) appears to be valid for analytical purposes. It can be noted in passing that reaction of O with H_2O in Reaction (4b) dominates over reaction with H_2 in Reaction (2f), particularly in diffusion flames where H_2O concentrations are high in the parts of the reaction zone where the "shuffle" Reactions (1) to (4) are out of partial equilibrium.

4. Recombination and Mole Reduction

Peters and Williams and their co-workers [3-7] prefer to write the third step of the reduced mechanism as

$$2H \rightarrow H_2 \qquad\qquad (IIIP)$$

The rate for this reaction is still given by Eqn. (2.3) but for the fourth step which remains the same as IV for purposes of stoichiometry the rate becomes the sum of ω_{III} and ω_{IV}:

$$\omega_{IVP} = \omega_1 + \omega_6 + \omega_{18} + \omega_{22}, \qquad\qquad (4.1)$$

where the steady state for HO_2 has been used. The production rates for the species remain the same and the choice of form for the summarizing four steps is from that viewpoint completely arbitrary. The choice of the form of the steps can, however, help to clarify our thinking about the structure of the flame. We argue here that the mole reduction form of our step III is more in keeping with the significant processes in the flame than the recombination form IIIP.

A major consideration in our argument is that Reaction (6) dominates over Reactions (7) and (8) and the H_2O_2 chain so that HO_2 is essentially a chain carrier. Reaction (5), which is the dominant term in ω_{III} is thus essentially chain carrying and does not in the main result in the recombination of radicals. Chelliah and Williams [6] and Trevino and Williams [7] make much of the chain terminating aspects of the HO_2 chain and investigate the question of competition between the chain branching of the forward rate of Reaction

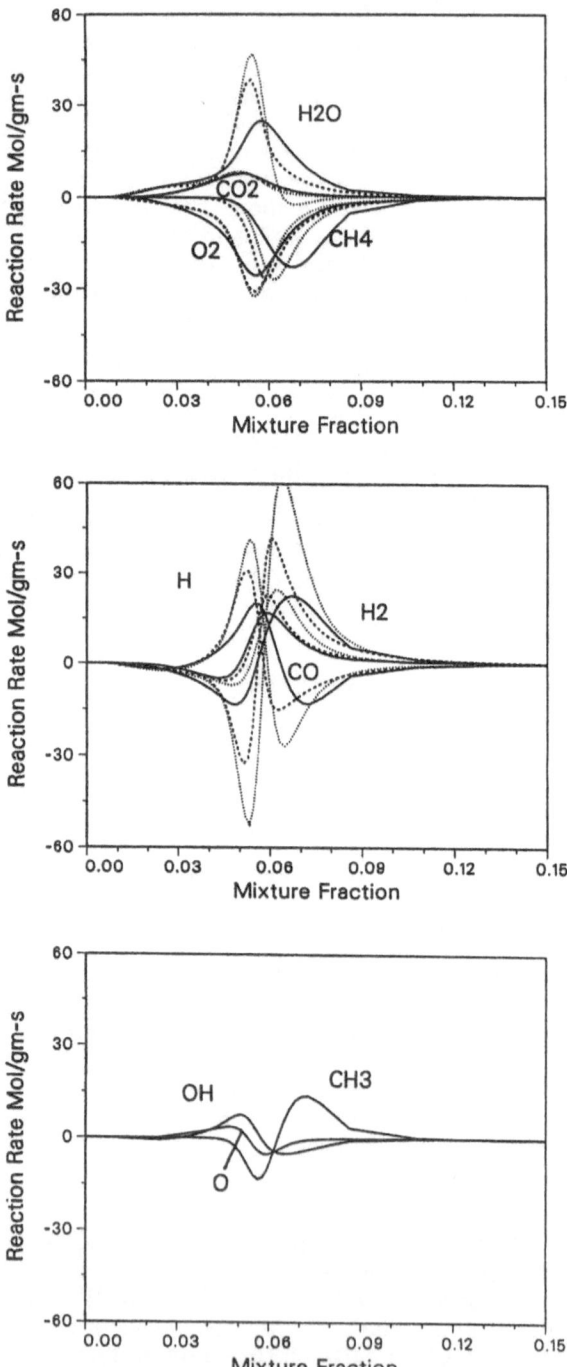

Figure 4. Comparison of calculated reaction rates for counterflow diffusion flames at a = $300s^{-1}$: ————skeletal mechanism with full transport; - - - - simplified version of four-step mechanism presented here using simplified transport;.........Seshadri and Peters [5] four-step mechanism with simplified transport.

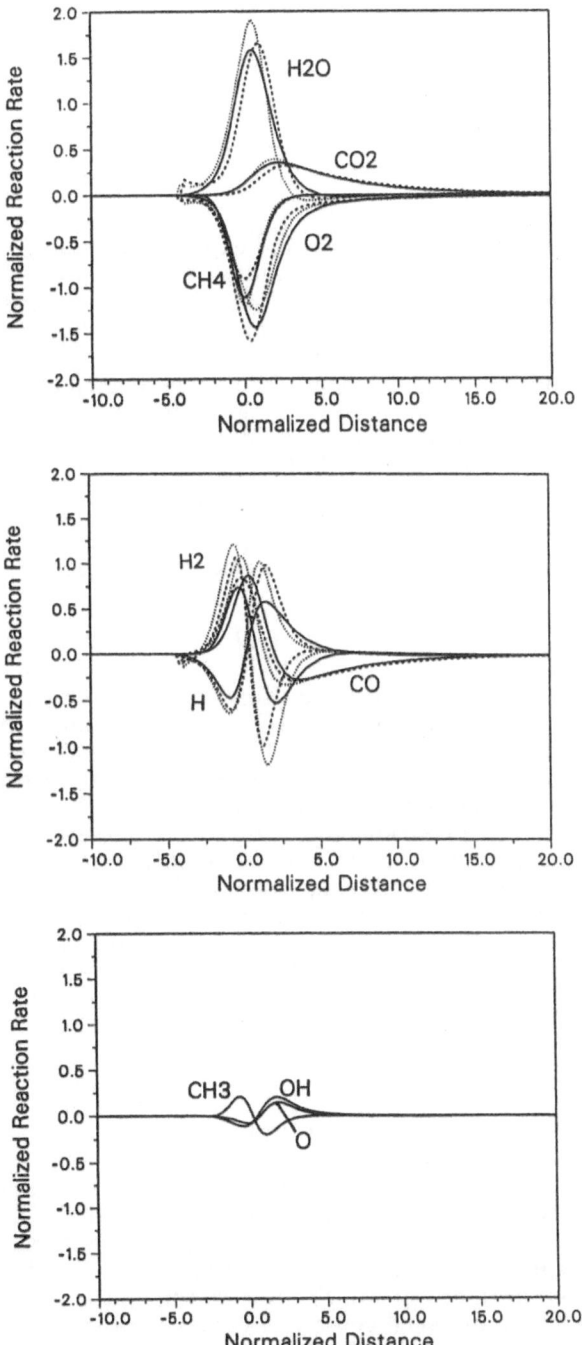

Figure 5. Comparison of calculated reaction rates for premixed flames at one atmosphere, stoichiometric: ———skeletal mechanism with full transport; ....full four-step mechanism presented here with simplified transport; - - - - Seshadri and Peters [5] four-step mechanism with simplified transport.

$$H + O_2 \rightarrow OH + O \hspace{7cm} R1f$$

and chain terminating arising from Reaction 5

$$H + O_2 + M \rightarrow HO_2 + M \hspace{6cm} R5$$

This competition is important in autoignition and explosions leading to the well-known second limit for H_2 - O_2 explosions. Such competition is not important in most flames, however. This can be seen from Tables III and IV and Fig. 2 where results are presented for calculations in which the chain terminating routes for HO_2 have been switched off. In diffusion flames the effects on flame structure are insignificant and they are little more so in premixed flames. In flames below 1000K, where such competition between branching and terminating would be significant, the radical pool of H, OH and O has concentrations above those given by partial equilibrium, due to diffusion from radical rich regions of the flame; and Reaction 1 is in fact going in *reverse*.

The principal role of Reaction 5, known from sensitivity analyses to be a highly significant reaction, is in mole reduction. Total specific abundance, $\Sigma \Gamma_i$, or the total number of moles per unit mass is conserved by the common bimolecular form of reaction

$$A + B \rightleftharpoons C + D$$

but is increased in pyrolysis reactions such as Reactions 10f and 17 and can only be reduced by reactions such as Reactions 5 and 10b in which the sum of the stoichiometric coefficients decreases. For the overall reaction of methane with O_2 to form CO_2 there is no change in mole number. All the CH_4 is oxidized, however, by routes which yield HCO. And since Reaction 17 dominates over Reaction 16 and other mole invariant routes for HCO consumption the conversion of CH_4 to CO and H_2 results in a close to unity mole increase as is expressed in Reaction I. The only route for restoring the net zero balance of moles is via the mole reducing step III. The total conversion by step III must thus be equal to that of step I. The adoption of IIIP for the third step indicates that the total conversion by step IV must be twice that of step I, and since $\omega_6 \approx \omega_5$ half of it is coming from that source. The use of IIIP, then, results in third and fourth steps which do not clearly delineate the important and separate roles of Reactions 1 and 5, the two most important reactions in combustion.

5. Conserved and Reactive Scalars

Conserved scalars [11] or Shvab-Zeldovich variables [12] are important in the analysis of both premixed and diffusion flames. The mixture fraction, ξ, defined in Eqn. (3.2) is only one of many variables that can be defined which will have a zero net reaction rate. The mixture fraction has found widespread use as a transformed variable for describing the structure of laminar and turbulent diffusion flames. In laminar flame analysis of both premixed and diffusion types, use is made of such variables to relate the concentrations of reactive species and their gradients [3-7]. The role of differential molecular diffusion is significant in flames, however, and such combined variables can have significant source terms arising from their transport terms. In flame structure studies [eg. 3-7] it is often considered that diffusion dominates over convection in the reaction zone and species concentrations are divided by the Lewis number before summing to form the conserved scalar. The numerical solutions for the flames that have been obtained in these studies may be used to address the question of how best to define such scalars.

Figure 6 shows various conserved scalars computed for the premixed and diffusion flames. The normalized element mass fractions $\hat{Z}_m$ (ZHN, ZON, ZCN and ZNN in Figure 6) are defined

$$Z_m \equiv \Sigma \mu_{mi} \Gamma_i, \tag{5.1a}$$

$$\hat{Z}_m = 1 + \left(Z_m - Z_m^t(\xi) \right) \bigg/ Z_m^t(\xi_s), \tag{5.1b}$$

$$Z_m^t(\xi) = Z_{m.ox} + \xi(Z_{m,fu} - Z_{m,ox}), \tag{5.1c}$$

where μ_{mi} is the number of atoms of element m in species i and ξ is the mixture fraction defined by Eqns. (3.1), (3.2), with ξ_s its value at stoichiometric. The choice of this mixture fraction for purposes of normalization is arbitrary, but it does locate the stoichiometric point properly. As an example of conserved scalars employing Lewis number scaling we show the Lewis number scaled element mass fractions $\hat{Z}_m^L$ (ZLH, ZLO, ZLC in Fig. 6) with

$$Z_m^L \equiv \Sigma \mu_{mi} \Gamma_i / Le_i, \tag{5.2}$$

and the normalization as in Eqns. (5.1b) and (5.1c); the Lewis number of species i being denoted by Le_i. The normalization is such that departures from unity represent the enrichment or depletion of the quantity relative to the amount of that quantity present at a stoichiometric mixture.

Also shown in the standardized enthalpy, h, where

$$h \equiv \Sigma \Gamma_i \left(h_{f,i} + \int_{T_1}^{T} \hat{C}_{p,i} dT \right). \tag{5.3}$$

Here h_{fi} and $\hat{C}_{p,i}$ are the enthalpy of formation at the reference temperature T_1 and the molar specific heat for species i , respectively. The standardized enthalpy is also normalized as in Eqns. (5.1b) and (5.1c) except that is scaled by the sensible enthalpy rise at stoichimetric. It is denoted in Fig. 6 as "Enthalpy." For the premixed flame the variation in ξ is also shown (denoted by fmix in Fig. 6) in normalized form

$$\hat{\xi} = 1 + (\xi - \xi_u)/\xi_s, \tag{5.4}$$

where the subscript u denotes the unburnt condition. For the other conserved scalars in the premixed flames the "theoretical" (superscript t) values used in Eqns. (5.1b) and (5.1c) are evaluated at the unburnt condition.

It can be seen from Fig. 6 that scaling by the Lewis number makes things worse, on the whole. The only exception is for oxygen element in diffusion flames where Lewis number scaling gives a significant improvement. Smooke [2] finds for the temperature version of the energy equation that the conduction term never dominates the convection term in premixed flames. It can be presumed that the diffusion terms in species balances also never dominate the convection terms and so the basis for Lewis number scaling is questionable.

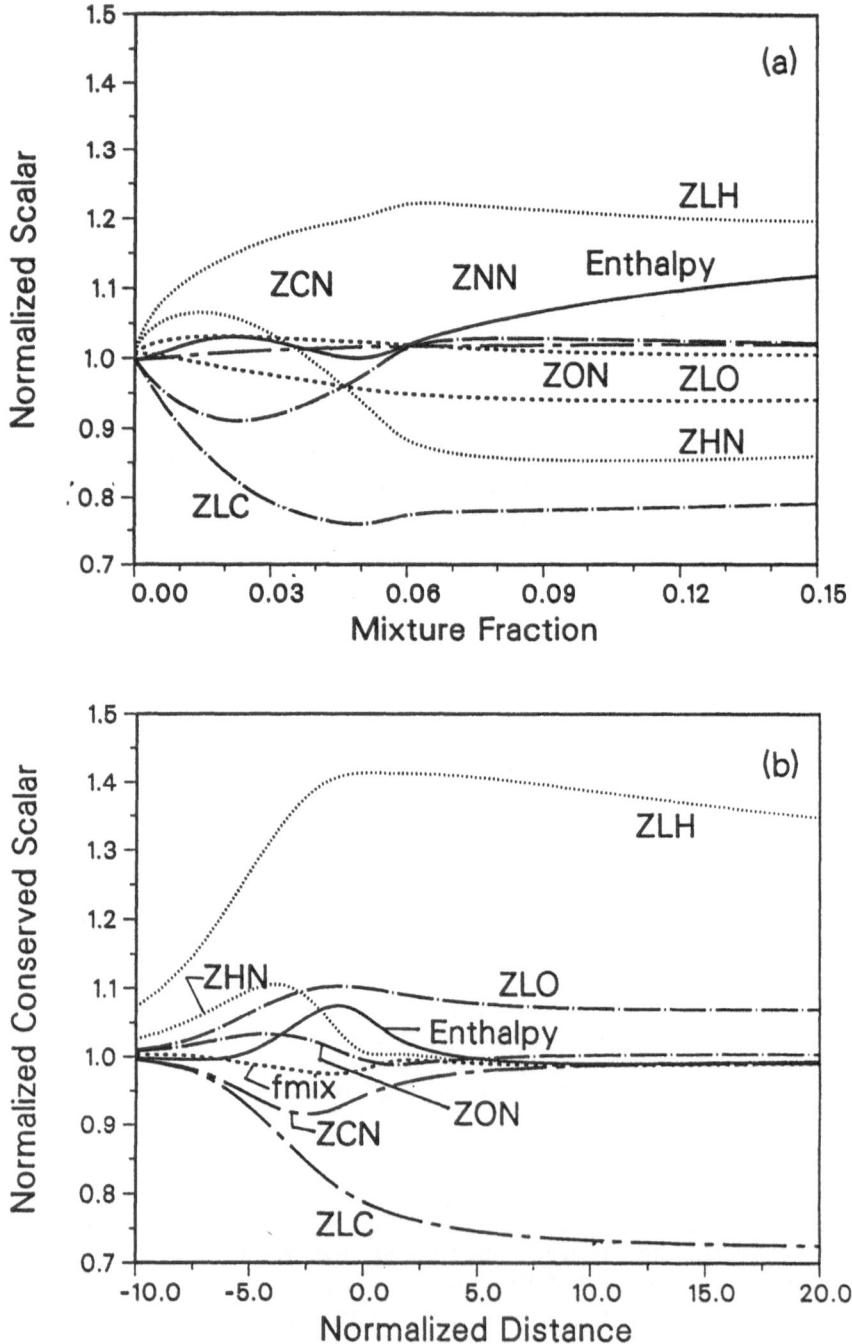

Figure 6. Variation of selected conserved scalars in flames: (a) counterflow diffusion flame at a $= 300s^{-1}$ with simplified version of four-step mechanism presented here and simplified transport; (b) premixed flame at 1 atm and stoichiometric with full version of four-step mechanism presented here and simplified transport. The conserved scalars for elements (ZHN, ZON, ZCN, ZNN), elements scaled by species Lewis numbers (ZLN, ZLO, ZLC), standardized enthalpy and mixture fraction are defined and normalized in Eqns. (5.1) to (5.4) of the text.

In simple theories of turbulent premixed combustion the reaction is assumed to be one-step and measured by a reaction progress variable normalized to have a value of zero in the unburnt mixture and unity in the fully burnt mixture. It is evident from the profiles in Figs. 2 and 3 and from the four-step modeling of flames that it may be appropriate to define progress variables for each of the four steps. Steps I and III are irreversible reactions and scalar variables can be defined that clearly evaluate the degree of completion of these reactions. Steps II and IV, on the other hand are reversible reactions and may be more appropriately described by partial equilibrium concepts where the question of their degree of completion is being considered.

In view of the foregoing discussion on the effects of differential diffusion on conserved scalars, it seems appropriate to define progress variables in terms of the sensible enthalpy, h^s, and of species such as CH_4, O_2, and CO, whose Lewis numbers are near unity. For simplicity we assume that these have a diffusivity equal to the thermal diffusivity $\lambda/(\rho c_p)$ so that they obey the balance equations

$$\mathcal{L}\{h^s\} = \left(\omega_I q_I + \omega_{II} q_{II} + \omega_{III} q_{III} + \omega_{IV} q_{IV}\right)\hat{Q}_c, \tag{5.5}$$

$$\mathcal{L}\{\Gamma_{CH_4}\} = -\omega_I, \tag{5.6a}$$

$$\mathcal{L}\{\Gamma_{O_2}\} = \omega - \omega_{IV}, \tag{5.6b}$$

$$\mathcal{L}\{\Gamma_{CO}\} = \omega_I - \omega_{II}, \tag{5.6c}$$

where

$$\mathcal{L}\{\} \equiv \rho\frac{\partial}{\partial t} + \rho u_j \frac{\partial}{\partial x_j} - \frac{\partial}{\partial x_j}\left(\frac{\lambda}{c_p}\frac{\partial}{\partial x_j}\right). \tag{5.7}$$

Also $\hat{Q}_c$ is the heat release per mole of fuel consumed with proportions $q_I, q_{II}, q_{III}, q_{IV}$ being associated with steps I to IV respectively. It is of course obvious that Γ_{CH_4} is an appropriate scalar by which to measure the progress of step I. For step III we find

$$\varsigma_{III} \equiv \left\{\frac{h^s}{\hat{Q}_c} + (q_I + q_{II})\Gamma_{CH_4} + q_{II}\Gamma_{CO} + q_{IV}\Gamma_{O_2}\right\}\bigg/(q_{III} - q_{IV}), \tag{5.8}$$

to be the variable which obeys the desired balance equation

$$\mathcal{L}\{\varsigma_{III}\} = \omega_{III}. \tag{5.9}$$

It is noted that the source term for ς_{III} is independent of the rates of reactions I, II and IV so the ς_{III} marks the progress of step III alone.

At 298K $\hat{Q}_c = 191.76$ kcal/mol and $q_I = 0.286, q_{II} = 0.051, q_{III} = 0.603$ and $q_{IV} = 0.060$. Figure 7 shows normalized values $\hat{\varsigma}_I$, defined from Γ_{CH_4}, $\hat{\varsigma}_{II}$ defined from Γ_{CO_2}, and $\hat{\varsigma}_{III}$ together with $\hat{h}^s$ where the normalizing has been done for the premixed flame by the form

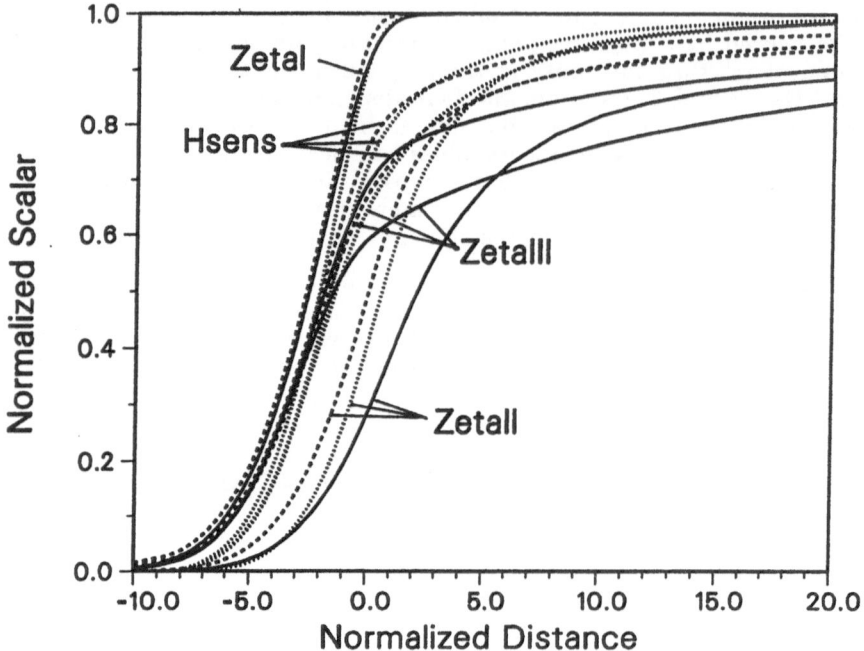

Figure 7. Variation of reactive scalars in premixed flames with full version of four-step mechanism presented here: ———1 atm, stoichiometric, - - - - - 5 atm, stoichiometric;............1 atm, 0.7 equivalence ratio. The variables are normalized values of the sensible enthalpy, the fuel and CO_2 concentration which give progress variables for steps I of the four-step mechanism, and ς_{III} defined in Eqn. (5.8) which measures the progress of step III of the four-step mechanism.

$$\hat{\varsigma} \equiv (\varsigma - \varsigma_u)/(\varsigma_b - \varsigma_u), \qquad (5.10)$$

with subscripts b and u denoting fully burnt and unburnt, respectively. It is seen that $\hat{\varsigma}_{III}$ emphasizes the slow completion of the mole reduction reactions. It also has a larger slope thickness than $\hat{\varsigma}_I$ or the sensible enthalpy. Table V lists the slope thicknesses derived. They are scaled by the length scale used to normalize the distance through the flame in Eqn. (3.3). It is also seen that the slope thickness does not fully characterize the ς_{III} profiles. It has a very long tail. It may be that in turbulent premixed flames the turbulence can affect the main reaction zone by interaction with this tail region.

Figure 8 shows features of the flame structure plotted against $\hat{\varsigma}_{III}$. It is seen that the fuel reaction is complete when $\hat{\varsigma}_{III}$ is about 0.65 and that step I occurs over a relatively narrow range of $\hat{\varsigma}_{III}$ as does step IV. Step III, however, occurs over a broad range,

TABLE V

Slope Thicknesses* for Premixed Flames

Variable	$\phi=1.0$ 1atm	$\phi=1.0$ 5atm	$\phi=0.7$ 1atm
Sensible Enthalpy	7.6	6.9	5.9
CH_4	5.4	5.6	4.7
CO_2	9.5	6.6	6.4
ς_m	9.2	7.9	7.1

*Nondimensionalised by $(\lambda/c_p m)_u$ as in Eqn (11)

particularly the component of ω_{III}, coming from Reaction 5. The component coming from Reaction 10b appears overall to be much more significant than Reaction 5. Noting from Eqn. (5.9) that it is only ω_{III} which is the reaction term for ς_{III} , it is ω_{III} which is responsible for determining the slope thickness derived from ς_{III} as well as the long tail. It appears that Reaction 10b is mainly responsible for the determination of the slope thickness and Reaction 5 for the long tail. Table VI shows the flame speed sensitivity coefficients to the rate coefficients, S_S, at one atmosphere, stoichiometric for the skeletal mechanism. We have

$$S_S \equiv d(ln S_u)/d(ln A_k),\tag{5.11}$$

where A_k is the pre-exponential factor for the kth reaction as listed in Table I. It is seen that the flame speed is more sensitive to the rate of Reaction 5 than the rate of Reaction 10b, even though the latter appears to dominate. It is apparent that the flame speed is quite sensitive to what happens in the tail. The slope thickness may thus not be an adequate characterization of the length scale associated with mole reduction.

The possibility that ς_{III} may be a useful variable for Crocco transformation [14] has been investigated. In a coordinate system affixed to isopleths of ς_{III}, and with ς_{III} the stretched coordinate in the normal direction, derivatives in the other two coordinate directions lying in the isopleth plane are likely to be small and make minor contributions to species and energy balances in premixed systems. The transformation yields

$$\begin{aligned}\omega_i &= \mathcal{L}\{\Gamma_i\} \\ &= \rho\left(\frac{\partial\Gamma_i}{\partial t}\right)_{\varsigma_{III}} + \frac{\partial\Gamma_i}{\partial\varsigma_{III}}\mathcal{L}\{\varsigma_{III}\} - \frac{\lambda}{c_p Le_i}\frac{\partial\varsigma_{III}}{\partial x_j}\frac{\partial\varsigma_{III}}{\partial x_j}\frac{\partial^2\Gamma_i}{\partial\varsigma_{III}^2} \\ &= \rho\left(\frac{\partial\Gamma_i}{\partial t}\right)_{\varsigma_{III}} + \frac{\partial\Gamma_i}{\partial\varsigma_{III}}\omega_{III} - \frac{\lambda}{c_p Le_i}\frac{\partial\varsigma_{III}}{\partial x_j}\frac{\partial\varsigma_{III}}{\partial x_j}\frac{\partial^2\Gamma_i}{\partial\varsigma_{III}^2}.\end{aligned}\tag{5.12}$$

TABLE VI

Flame Speed Sensitivity Coefficients*
for Selected Reactions

Reaction Number	Sensitivity Coefficient, S_*
1f	0.82
1b	-0.24
5	-0.17
9f	0.25
9b	-0.11
10b	-0.13
11f	-0.11
13	0.10

*Skeletal mechanism at one atmosphere, stoichiometric.

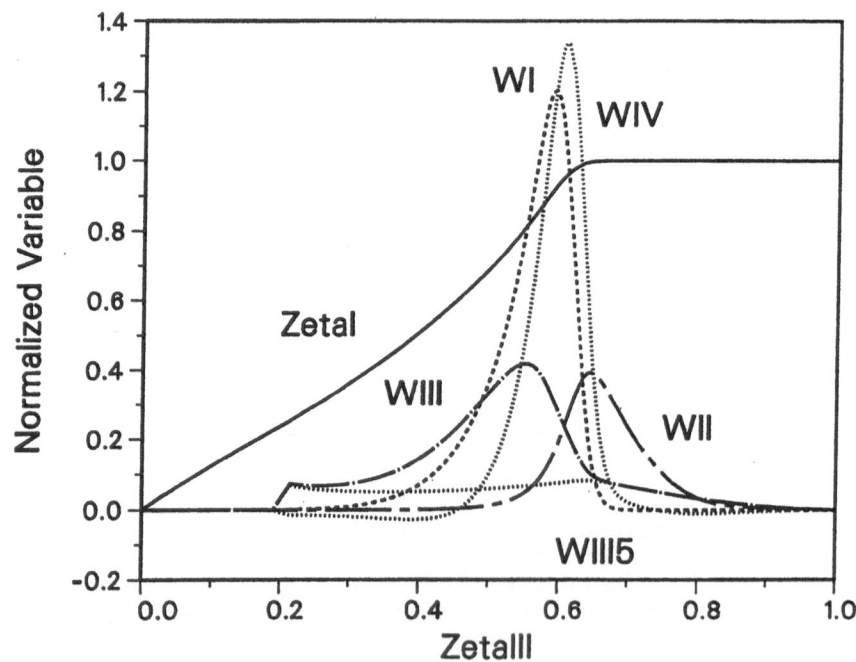

Figure 8. Structure of premixed flame in terms of transformed coordinate $\hat{\varsigma}_{III}$. Zeta I is the normalized progress variable for step I and WI, WII, WIII and WIV are the normalized reaction rates for steps I to IV of the four-step "full" mechanism presented here. WIII 5 is the contribution of elementary Reaction 5 to WIII. Results from numerical solutions at 1 atmosphere, stoichiometric.

The unsteady term is zero in steady laminar flow and the transformation has been tested by evaluating the other two terms on the right hand side (RHS) using the numerical solutions for the full four-step mechanism at one atmosphere, stoichiometric. The results are shown in Fig. 9. It is seen that the agreement is excellent for the fuel reaction and quite good even for the low Lewis number species H. It would be of great interest if the last term on the RHS of Eqn. (5.12) dominated the transformation. Unfortunately it does not. The second term is by far the dominant one and this largely arises from the contributions of Reaction 10b. This limits the usefulness of this transformation. It would be interesting to investigate methanol flames where the formation of methyl radicals in unimportant and Reaction 5 should dominate ω_{III}. The transformation in terms of a reactive scalar based on CO_2 and involving reaction step II is an alternative but has the same domination by the second term in Eqn. (5.12).

6. Flame Structure and Analysis

Table VI indicates the dominant role played by the chain-branching Reaction 1 in the determination of flame speed. It is also dominant [2] in the determination of extinction values of stretch in diffusion flames. The pioneering analysis of the structure of these flames by Peters and Williams and their co-workers [3-7] focuses on fuel consumption and centers the inner layer of the reaction zone at the peak of the fuel consumption rate.

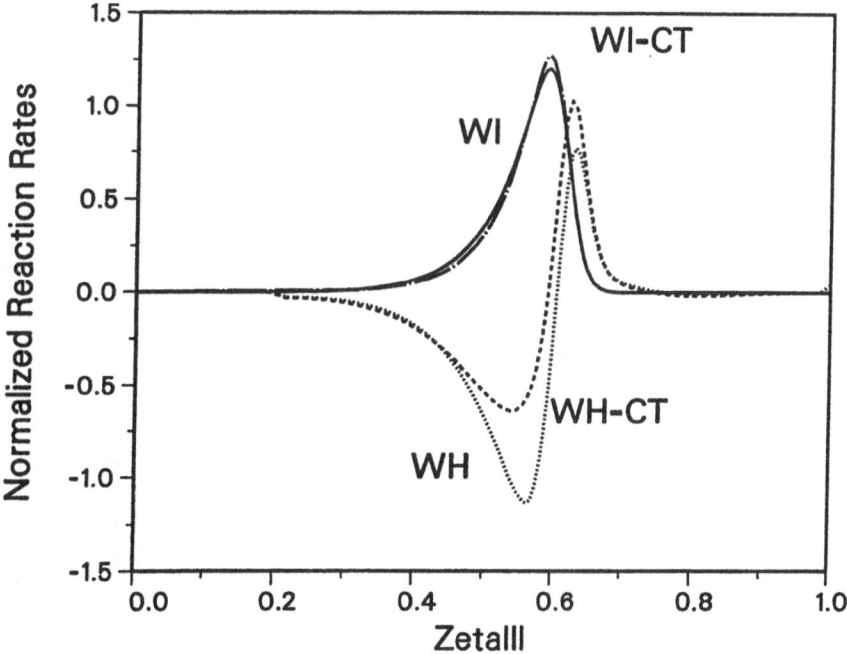

Figure 9. Crocco transformation for CH_4 and H-atom reaction rates in premixed flame at 1 atmosphere, stoichiometric using four-step "full" mechanism. WI is the negative of the normalized full reaction rate and WI - CT is that obtained from the Crocco transformation of Eqn. (5.12). WH is the normalized H-atom reaction rate and WH - CT that obtained from the Crocco transformation.

The resulting formulation does not show directly the dominant influence of k_{1f} on the flame speed and extinction that sensitivity analysis implies. It may be more appropriate to focus on the production of H-atoms by the chain-branching step IV and center the inner layer where this is a maximum. Both diffusion and local consumption by the fuel are significant in the H-atom balance at this point. The production of H-atoms (which are here a surrogate for OH and O as well) is basically controlled by the competition between the forward and backward rate of Reaction 1, with the backward rate being abetted by Reaction 10b. The production of H-atoms by step IV will be a maximum at some value of H-atom concentration between zero, where production is zero, and its partial equilibrium value, where production is also zero. Achievement of this optimum H-atom concentration at a point in the flame where H_2 and O_2 concentrations are such as to also maximize H-atom production could be the mechanism controlling flame speed of premixed flames and extinction of diffusion flames.

In the region of the flame where ω_{IV} peaks, B in Eqn. (2.6) is negative and $B^2 \gg 4C$ so that we have

$$
\begin{aligned}
\Gamma_O &= -\frac{B}{2}\left(1 + \frac{4C}{B^2}\right)^{1/2} - \frac{B}{2} \\
&\approx -B \\
&\approx \left(k_{1f}\Gamma_H\Gamma_{O_2} + k_{4f}\Gamma_{OH}^2 - k_{11f}\Gamma_{CH_4}\Gamma_H\right)\Big/\left(k_{1b}\Gamma_{OH} + k_{4b}\Gamma_{H_2O}\right) \\
&\quad - k_{11b}\Gamma_{H_2}\Big/k_{13}.
\end{aligned}
$$
(6.1)

Also

$$
\Gamma_{CH_3} \approx k_{11f}\Gamma_{CH_4}\Gamma_H\Big/\left(k_{13}\Gamma_O\right).
$$
(6.2)

Substitution of Eqns. (2.5), (6.1) and (6.2) into Eqn. (2.4b) yields

$$
\begin{aligned}
\omega_{IV} &= \rho^2 k_{1f}\Gamma_H\left\{\Gamma_{O_2} - K_4\Gamma_{H_2O}^2\Gamma_H^2\Big/\left(K_3^3 K_1\Gamma_{H_2}^3\right)\right\}\Big/\left\{1 + k_{1b}\Gamma_H\Big/\left(k_{4b}K_3\Gamma_{H_2}\right)\right\} \\
&\quad - \rho^2 k_{1b}\Gamma_H\Gamma_{H_2O}\left\{k_{11b}\Gamma_{H_2}\Big/k_{13} + k_{11f}\Gamma_{CH_4}\Gamma_H\Big/\left(k_{1b}\Gamma_{OH} + k_{4b}\Gamma_{H_2O}\right)\right\} \\
&\quad - \rho^2 k_{10b}^*\Gamma_{CH_4}\Gamma_H^2\Big/\left(k_{13}\Gamma_O\right).
\end{aligned}
$$
(6.3)

For simplicity of presentation the terms involving Γ_{CH_4} in Eqn. (6.3) have not had Γ_O and Γ_{OH} expressed in terms of Γ_H. It can be seen that for a given mixture fraction and values of $\Gamma_{CH_4}, \Gamma_{CO_2}$ and ς_{III} (representing progress of the other steps I to III, respectively) there will be a value of Γ_H which maximizes ω_{IV}. In the original asymptotic analyses [3,4] Γ_H is placed in steady state within the critical layer while in another approach [6,7] all the negative terms in Eqn. (6.3) are neglected. It may be better to assume that Γ_H at the ω_{IV} peak is given by the value which maximizes ω_{IV} and use this in solving equations for Γ_{CH_4} and Γ_H around this peak. The total production of H-atom on the fuel side of the H-atom peak can then be set equal to the fuel consumption required of step I.

7. Conclusions

It can be concluded that formally reduced four-step mechanisms give excellent models for the complex kinetics of methane-air combustion in a wide range of premixed and diffusion flames. Four-step mechanism which involve full steady-state assumptions for O-atom and CH_3 appear to give predictions closer to their parent skeletal mechanism. A simplified version of the mechanism in which many of the less important terms are dropped works well for diffusion flames, and with the inclusion of one or two terms (yet to be found) should also give good results for premixed flames.

Inclusion of the Lewis number in definitions of conserved scalars does not appear to be justified. It appears to be best to formulate conserved scalars in terms of the sensible enthalpy and species concentrations for species with Lewis numbers close to unity. A reactive scalar is defined which marks the progress of step III of the mechanism. It indicates that the thickness of the flame associate with mole reduction kinetics is significantly higher than that associated with the consumption of the fuel. Use of this variable as a Crocco variable has been demonstrated but the resulting transformation of the balance equations for species and temperature do not appear to be very useful.

It is suggested that analysis of the inner layer may be more appropriately cast in terms of the H-atoms being at a concentration which maximizes their production at the peak of their production.

Acknowledgements

This work has been supported by a grant from the Australian Research Council and a University Research Grant from the University of Sydney. Travel assistance from the Department of Science, Canberra under bilateral agreements with the USA, Germany and Mexico is also gratefully acknowledged. Professor M. D. Smooke provided copies of his flame codes and much assistance in getting them running properly. The work was completed by the first author while on sabbatical at Sandia and their generous support through the Department of Energy, Office of Basic Energy Sciences is gratefully acknowledged. The help of Drs. J.-Y. Chen at Sandia and Dr. A. R. Green of the Londonderry Occupational Safety Centre, Londonderry, N.S.W. are also gratefully acknowledged.

References

[1] Bilger, R. W., Starner, S. H. and Kee, R. J., "On Reduced Mechanisms for Methane-Air Combustion in Non-Premixed Flames, *Comb. Flame* **80**, (1990), p. 135.

[2] Smooke, M. D., "Formulation of the Premixed and Nonpremixed Test Problems," Chapter 1, and "Premixed and Nonpremixed Test Problem Results," Chapter 2, This volume.

[3] Peters, N. and Williams, F.A., "The Asymptotic Structure of Stoichiometric Methane Air Flames," *Comb. Flame* **68**, (1978), p. 185.

[4] Seshadri, K. and Peters, N., "Asymptotic Structure and Extinction of Methane-Air Diffusion Flames," *Comb. Flame* **73**, (1988), p. 23.

[5] Seshadri, K. and Peters, N. "The Inner Structure of Methane-Air Flames," *Comb. Flame* **81**, (1990), p. 96.

[6] Chelliah, H. K. and Williams, F. A., "Aspects of the Structure and Extinction of Diffusion Flames in Methane-Oxygen-Nitrogen Systems," *Comb. Flame* **80**, (1990), p. 17.

[7] Trevino, C. and Williams, F. A., "Asymptotic Analysis of the Structure and Extinction of Methane-Air Diffusion Flames," *Progress in Astro. Aero* **113**, (1988), p. 129.

[8] Tsuji, H. and Yamaoka, I., "Structure Analysis of Counterflow Diffusion Flames in the Forward Stagnation Region of a Porous Cylinder," Thirteenth Symposium (International) on Combustion, The Combustion Institute, Pittsburgh, PA, 1971, p. 723.

[9] Smooke, M. D, Miller, J. A. and Kee, R. J. "Solution of Premixed and Counterflow Diffusion Flame Problems by Adaptive Boundary Value Methods ," <u>Numerical Boundary Value ODEs</u>, U. M. Ascher and R. D. Russell, Eds., Birkhäuser Boston, 1985, p. 303.

[10] Curtiss, C. F. and Hirschfelder, J. O., "Transport Properties of Multicomponent Gases," *J. Chem Phys.* **17**, (1949), p. 550.

[11] Bilger, R. W., "Turbulent Nonpremixed Combustion," in <u>Turbulent Reacting Flows</u>, P. A. Libby and F. A. Williams, Ed., Springer, 1980, p. 65.

[12] Williams, F. A., <u>Combustion Theory</u>, 2nd Ed., Benjamin/Cummings, Menlo Park, CA, 1985.

[13] Libby, P. A., "Theory of Normal Premixed Turbulent Flames Revisited," *Prog. Energy. Comb. Sci* **11**, (1985), p. 83.

[14] Williams, F. A., "Crocco Variables in Combustion," <u>Recent Advances in the Aerospace Sciences,</u> C. Casci, Ed., Plenum, 1985, p. 415.

CHAPTER 6
STRUCTURE OF THE OXIDATION LAYER
FOR STOICHIOMETRIC AND LEAN METHANE-AIR FLAMES

K. Seshadri and J. Göttgens
Institut für Technische Mechanik
RWTH-Aachen, D-5100 Aachen
Federal Republic of Germany

1. Introduction

Reduced chemical kinetic mechanisms have been successfully employed previously [1,2] to describe the structure of laminar, premixed methane-air flames. The reduced, four-step, chemical-kinetic mechanisms used in these analyses was deduced systematically from a detailed chemical-kinetic mechanism by following procedures described in detail elsewhere [3]. The essential common feature in these analyses was that the basic structure of the flame was presumed to consist of distinct layers which includes an inner layer of thickness of $O(\delta)$ and an oxidation layer of thickness of $O(\varepsilon)$ with the presumed ordering $\delta \ll \varepsilon \ll 1$. All of the hydrocarbon chemistry was presumed to occur in the inner layer which is defined as the merged structure of the fuel and radical consumption layers. In this layer the fuel is attacked by radicals to form CO and H_2, and the ratio of the thickness of the fuel consumption layer to the thickness of the radical consumption layer was denoted previously by the parameter ω [2]. In the oxidation layer the CO and H_2 formed in the inner layer are oxidized to CO_2 and H_2O. In the previous analyses [1,2] the water gas shift reaction was assumed to be in equilibrium everywhere in the oxidation layer except in a thin region of thickness of $O(\nu)$ which is embedded between the inner layer and the oxidation layer with $\delta \ll \nu \ll \varepsilon$.

The analysis of Peters and Williams [1] considered the limit $\omega \to \infty$, wherein the thickness of the radical consumption layer is much smaller than the thickness of the fuel consumption layer. In this limit the concentration of the H radicals are in steady state everywhere in the reaction zone except in the radical consumption layer, which is embedded in the fuel consumption layer. Hence, the reduced four-step chemical-kinetic mechanism effectively reduces to a three-step mechanism. Asymptotic analysis was performed by Peters and Williams for stoichiometric flames [1] using only the principal elementary reactions to characterize the overall rates of the reduced mechanism. When only the terms of leading order are retained in the analytical expression for the burning velocity v_u, the predicted value of this quantity at p equal to 1 atm was 48 cm/s, which is slightly higher than the measured value of v_u at these conditions. Also, the predicted value of v_u decreased too rapidly with increasing values of p [1]. The results of the asymptotic analysis [1] were improved by including a number of additional chemical reactions in the overall rates of the reduced mechanism through numerical evaluation of an integral involving an iteration for determination of the integrand.

Seshadri and Peters [2] subsequently refined the analysis of Peters and Williams [1] by considering a four-step mechanism and including the effects of additional reactions by algebraic parameters to avoid the iteration of Ref. 1, which would be inapplicable with the four-step mechanism. The elementary chemical kinetic mechanism employed in Ref. 2 to deduce the reduced four-step mechanism, and the rate parameters for these reactions were identical to that used here, with the exception of reaction 10b. Analytical solutions for the burning velocity eigenvalue were obtained [2] in the limit $\omega \to 0$, and $\omega \to \infty$, and by use of numerical integration an approximation for the eigenvalue was postulated as a function of ω, which includes these limiting expressions. Using the results of the analysis the burning velocity was calculated for stoichiometric methane-air flames for values of p between 1 atm and 80 atm. At $p = 1$ atm the calculated burning velocity was 48 cm/s which is higher than the measured value at these conditions. However, the predicted decrease in the value of v_u with increasing values of p was in agreement with the measurements.

The results of the previous asymptotic analyses show the values of ε and δ to decrease with increasing values of p [1,2], and the presumed ordering $\delta \ll \varepsilon$ was retained over the entire range of values of p, considered in these analyses. Hence, the inner layer and oxidation layer do not appear to merge. However, the presumed ordering $\nu \ll \varepsilon$ does not appear to be generally valid [1,2]. In fact, the numerical value of ν appears to be larger than the value of ε. This observation implies that the effects of non-equilibrium of the water gas shift reaction must be considered everywhere in the oxidation layer, and this is the motivation for the present study.

The analysis reported here attempts to improve the description of the structure of the oxidation layer, and the previous asymptotic description of the inner layer is retained [2]. Fuel-lean flames are also considered in the analysis. One of the major goals of this work and previous work [1,2] is to provide guidance for the numerical approximation of laminar flame data in the construction of flamelet libraries.

2. Structure of the Post-Flame Zone

Downstream of the flame, in the post-flame zone the reaction products are in chemical equilibrium, and the temperature is equal to the adiabatic flame temperature. The equilibrium concentrations of the products are determined as described elsewhere [4] by assuming that only the species O_2, H_2, CO, CO_2 and H_2O are present and are in chemical equilibrium, and that the enthalpy and the element mass fractions are equal to those in the unburnt gas. For given values of ϕ, p and the initial temperature, T_u, the adiabatic flame temperature, T_b, and the mass fraction of species i in the post-flame zone, Y_{ib}, can be calculated. The equivalence ratio is related to the mixture fraction Z by

$$\phi = \frac{Z}{Z_{st}} \frac{(1 - Z_{st})}{(1 - Z)}, \tag{2.1}$$

where $Z_{st} = 0.055$ is the stoichiometric mixture fraction. Although the asymptotic analysis is performed for non zero concentrations of O, OH, H and HO_2 in the post-flame zone, these species were neglected in the calculation of T_b because their concentrations are negligibly small in this zone. In Fig. 1 results of thermo-chemical calculations are plotted showing the conditions in the post-flame zone as functions of ϕ.

Shown in Fig. 1a, in addition to T_b, is the adiabatic flame temperature T_c for complete combustion with $Y_{H_2} = Y_{CO} = 0$, which corresponds to the overall reaction $CH_4 + 2O_2 \to CO_2 + 2H_2O$. In this limit the products would contain unburnt O_2 for fuel-lean flames. If the heat of combustion for this overall reaction per mole of CH_4 consumed is denoted by $(-\Delta H)$, then

$$\int_{T_u}^{T_c} c_{p,P} dT = (-\Delta H) Y_{Fu}/W_F \quad \text{for } \phi \leq 1,$$

where $c_{p,P}$ is the specific heat at constant pressure of the product mixture (N_2, CO_2, H_2O and O_2 or CH_4). Y_{Fu} denotes the mass fraction of the fuel in the initial reactant stream, and W_i is the molecular weight of species i. Figure 1 shows that there is a region

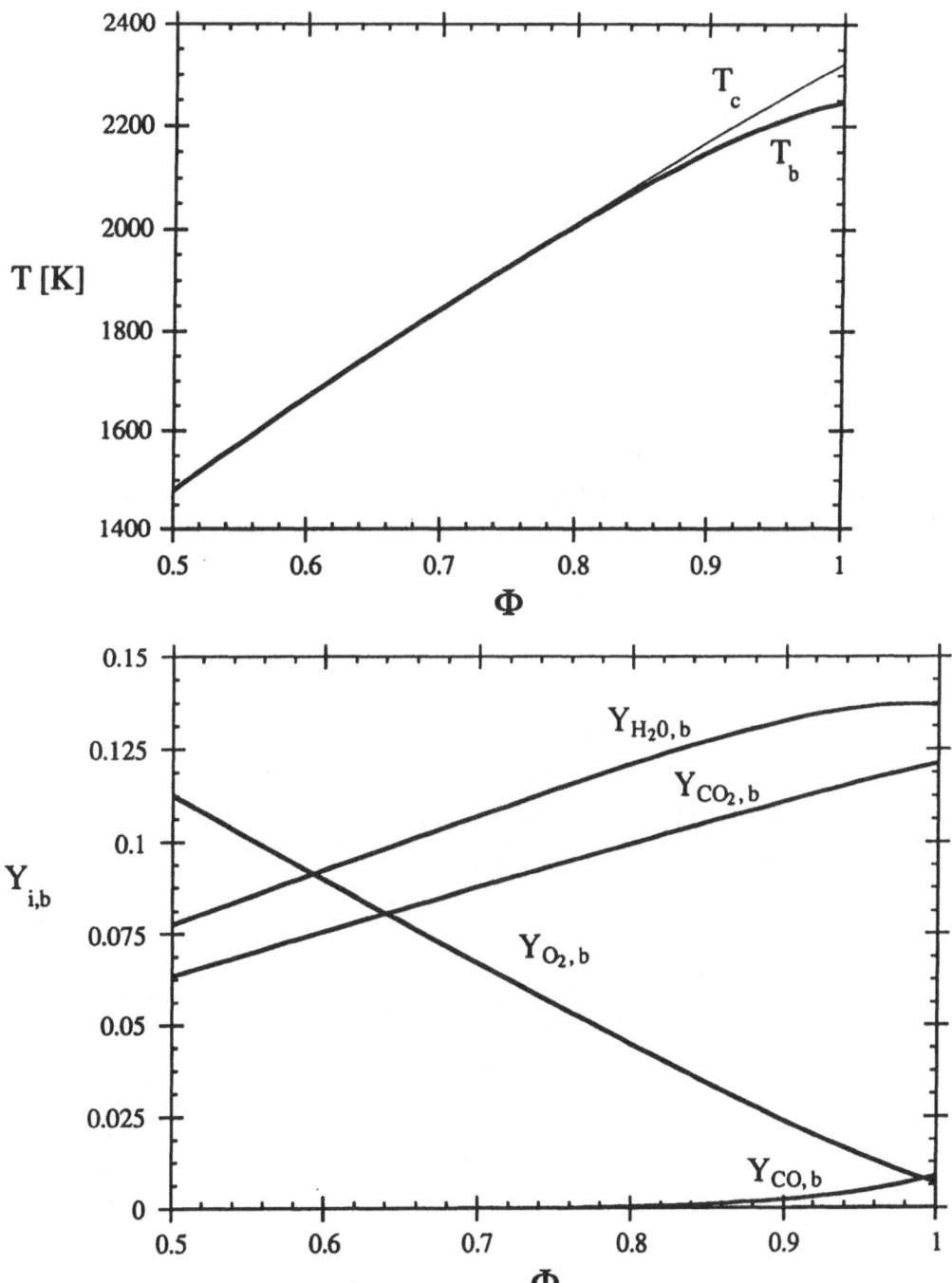

Figure 1. Results of thermochemical calculations showing the conditions in the burnt gas zone of the flame at $p = 1$ atm and $T_u = 300$ K for (a) temperatures and (b) mass fractions.

in which the products contain O_2, H_2 and CO simultaneously. Expansions later will be performed about T_c rather than T_b.

3. Reduced Chemical-Kinetic Mechanism

A chemical-kinetic mechanism describing the oxidation of methane is shown in Table I. Employing the procedure described in Ref. 4, a four-step mechanism can be deduced from this mechanism by assuming that there exists a dynamic steady state for the concentrations of the species O, OH, HO_2, H_2O_2, CH_3, CH_2O, CH_3O and HCO. The four-step mechanism can be written as

I $\qquad\qquad\qquad\qquad CH_4 + 2H + H_2O \rightleftharpoons CO + 4H_2,$

II $\qquad\qquad\qquad\qquad CO + H_2O \rightleftharpoons CO_2 + H_2,$

III $\qquad\qquad\qquad\qquad H + H + M \rightleftharpoons H_2 + M,$

IV $\qquad\qquad\qquad\qquad O_2 + 3H_2 \rightleftharpoons 2H_2O + 2H.$

The reaction rates $w_k, k = $ I, II, III, IV for the overall reactions I–IV can be related to the reaction rates of the elementary reactions $w_n, n = 1, 2, ..., 25$ shown in Table 1 and are

$$w_I = w_{10f} - w_{10b} + w_{11f} - w_{11b} + w_{12f} - w_{12b},$$

$$w_{II} = w_{9f} - w_{9b},$$

$$w_{III} = w_{5f} - w_{5b} - w_{10f} + w_{10b} + w_{16} - w_{18} + w_{19} \qquad\qquad (3.1)$$

$$- w_{22f} + w_{22b} + w_{24f} - w_{24b} + w_{25f} - w_{25b},$$

$$w_{IV} = w_{1f} - w_{1b} + w_6 + w_{18} + w_{22f} - w_{22b},$$

where the subscripts f and b identify forward and backward rates, respectively. The values of the reaction rates w_n shown in Eq. (3.1) are proportional to the product of the concentration of the reactants and the rate constant k_n of the elementary reaction. Results of numerical calculations have shown that reactions 10f, 19, 21, 22f, 22b, 23f and 23b have only a minor influence on the burning velocity of premixed flames; therefore, as in Refs. 2 and 3 they are neglected in the analysis reported here. Differently from the previous asymptotic analysis [1,2] the backward steps of reactions 5, 24 and 25 are included in the analysis, hence the overall rate of reaction III shown in Eq. (3.1) is modified to include the rates of these backward reactions. Although reactions 5b, 24b and 25b have negligible influence on the value of v_u, they are included in the analysis to facilitate matching of the oxidation layer with the post-flame zone. In addition to the steady-state assumptions, as in previous analyses [1,2], the elementary reaction 3 shown in Table 1 is assumed to be in equilibrium yielding the algebraic relation $C_{OH} = \gamma C_H$, where C_i is the molar concentration of species i and

$$\gamma \equiv \frac{C_{H_2O}}{K_3 C_{H_2}} \qquad\qquad (3.2)$$

in which $K_3 = 0.23 \exp(7530/T)$ is the equilibrium constant of the elementary reaction 3.

TABLE I

Skeletal Methane-Air Reaction Mechanism

Rate Coefficients in the Form $k_f = AT^\beta \exp(-E_0/RT)$.

Units are moles, cubic centimeters, seconds, Kelvins and calories/mole.

	REACTION	A	β	E
1f.	$H + O_2 \rightarrow OH + O$	2.000E+14	0.000	16800.
1b.	$OH + O \rightarrow H + O_2$	1.575E+13	0.000	690.
2f.	$O + H_2 \rightarrow OH + H$	1.800E+10	1.000	8826.
2b.	$OH + H \rightarrow O + H_2$	8.000E+09	1.000	6760.
3f.	$H_2 + OH \rightarrow H_2O + H$	1.170E+09	1.300	3626.
3b.	$H_2O + H \rightarrow H_2 + OH$	5.090E+09	1.300	18588.
4f.	$OH + OH \rightarrow O + H_2O$	6.000E+08	1.300	0.
4b.	$O + H_2O \rightarrow OH + OH$	5.900E+09	1.300	17029.
5.	$H + O_2 + M \rightarrow HO_2 + M^a$	2.300E+18	-0.800	0.
6.	$H + HO_2 \rightarrow OH + OH$	1.500E+14	0.000	1004.
7.	$H + HO_2 \rightarrow H_2 + O_2$	2.500E+13	0.000	700.
8.	$OH + HO_2 \rightarrow H_2O + O_2$	2.000E+13	0.000	1000.
9f.	$CO + OH \rightarrow CO_2 + H$	1.510E+07	1.300	-758.
9b.	$CO_2 + H \rightarrow CO + OH$	1.570E+09	1.300	22337.
10f.	$CH_4 + (M) \rightarrow CH_3 + H + (M)^b$	6.300E+14	0.000	104000.
10b.	$CH_3 + H + (M) \rightarrow CH_4 + (M)^b$	5.200E+12	0.000	-1310.
11f.	$CH_4 + H \rightarrow CH_3 + H_2$	2.200E+04	3.000	8750.
11b.	$CH_3 + H_2 \rightarrow CH_4 + H$	9.570E+02	3.000	8750.
12f.	$CH_4 + OH \rightarrow CH_3 + H_2O$	1.600E+06	2.100	2460.
12b.	$CH_3 + H_2O \rightarrow CH_4 + OH$	3.020E+05	2.100	17422.
13.	$CH_3 + O \rightarrow CH_2O + H$	6.800E+13	0.000	0.
14.	$CH_2O + H \rightarrow HCO + H_2$	2.500E+13	0.000	3991.
15.	$CH_2O + OH \rightarrow HCO + H_2O$	3.000E+13	0.000	1195.
16.	$HCO + H \rightarrow CO + H_2$	4.000E+13	0.000	0.
17.	$HCO + M \rightarrow CO + H + M$	1.600E+14	0.000	14700.
18.	$CH_3 + O_2 \rightarrow CH_3O + O$	7.000E+12	0.000	25652.
19.	$CH_3O + H \rightarrow CH_2O + H_2$	2.000E+13	0.000	0.
20.	$CH_3O + M \rightarrow CH_2O + H + M$	2.400E+13	0.000	28812.
21.	$HO_2 + HO_2 \rightarrow H_2O_2 + O_2$	2.000E+12	0.000	0.
22f.	$H_2O_2 + M \rightarrow OH + OH + M$	1.300E+17	0.000	45500.
22b.	$OH + OH + M \rightarrow H_2O_2 + M$	9.860E+14	0.000	-5070.
23f.	$H_2O_2 + OH \rightarrow H_2O + HO_2$	1.000E+13	0.000	1800.
23b.	$H_2O + HO_2 \rightarrow H_2O_2 + OH$	2.860E+13	0.000	32790.
24.	$OH + H + M \rightarrow H_2O + M^a$	2.200E+22	-2.000	0.
25.	$H + H + M \rightarrow H_2 + M^a$	1.800E+18	-1.000	0.

^{a} Third body efficiencies: $CH_4 = 6.5, H_2O = 6.5, CO_2 = 1.5, H_2 = 1.0, CO = 0.75, O_2 = 0.4, N_2 = 0.4$ All other species $= 1.0$

^{b} Lindemann form, $k = k_\infty/(1 + k_{fall}/[M])$ where $k_{fall} = .0063 \exp(-18000/RT)$.

4. The Conservation Equations for a Steady Premixed Flame

For a steady, planar, adiabatic deflagration at low Mach number the equation of motion implies that the pressure is essentially constant. The equation for mass conservation can be written as

$$\rho\, v = \rho_u\, v_u \tag{4.1}$$

where ρ is the density and v the gas velocity. The index u denotes conditions in the unburnt gas. Lewis numbers for species i are defined as $L_i = \lambda/(\rho\, c_p\, D_i)$, where λ is the thermal conductivity and c_p is the mean specific heat; the diffusion coefficient D_i is taken to be that of species i with respect to nitrogen, and the binary-diffusion approximation is employed. The values of the Lewis numbers for all species are assumed to be constant. Using the notation of the previous analysis [1,2], the non-dimensionalized species and energy balance equations can be written as

$$
\begin{aligned}
\mathcal{L}_F(X_F) &= -\omega_I\,,\\[4pt]
\mathcal{L}_H(X_H) &= -2\omega_I - 2\omega_{III} + 2\omega_{IV}\,,\\[4pt]
\mathcal{L}_{H_2}(X_{H_2}) &= 4\omega_I + \omega_{II} + \omega_{III} - 3\omega_{IV}\,,\\[4pt]
\mathcal{L}_{H_2O}(X_{H_2O}) &= -\omega_I - \omega_{II} + 2\omega_{IV}\,,\\[4pt]
\mathcal{L}_{O_2}(X_{O_2}) &= -\omega_{IV}\,,\\[4pt]
\mathcal{L}_{CO}(X_{CO}) &= \omega_I - \omega_{II}\,,\\[4pt]
\mathcal{L}_{CO_2}(X_{CO_2}) &= \omega_{II}\,,\\[4pt]
\mathcal{L}(\tau) &= Q_I\omega_I + Q_{II}\omega_{II} + Q_{III}\omega_{III} + Q_{IV}\omega_{IV}\,,
\end{aligned}
\tag{4.2}
$$

The operators are defined as $\mathcal{L}_i \equiv d/dx - (1/L_i)d^2/dx^2$, and $\mathcal{L} = d/dx - d^2/dx^2$. The non-dimensional independent variable x is related to the spatial coordinate x' as

$$x = \int_0^{x'} (\rho\, v\, c_p/\lambda)\, dx'\,,$$

and the quantities X_i and τ are related to the mass fraction of species i, Y_i and the gas temperature, T as

$$X_i = \frac{Y_i W_F}{Y_{Fu} W_i}\,, \qquad \tau = \frac{T - T_u}{T_c - T_u}\,, \tag{4.3}$$

where the subscript F denotes the fuel. In the analysis the average molecular weight $\overline{W}$ is assumed to be a constant equal to 27.62 kg/kmol; hence X_i is the mole fraction of species i divided by the initial mole fraction of the fuel. The non-dimensionalized reaction rates ω_k and the non-dimensionalized heats of reaction Q_k of the reduced four-step mechanism are defined as

$$\omega_k = \frac{\lambda W_F\, w_k}{c_p\, Y_{Fu}\, (\rho_u\, v_u)^2}\,, \qquad Q_k = \frac{Y_{Fu}(-\Delta H_k)}{c_p(T_c - T_u)\, W_F}\,. \tag{4.4}$$

The non-dimensionalization of Eq. (4.4) will also be applied to the rates of the elementary steps. Since assuming steady states and negligible concentrations for CO, H_2 and H enables the overall reaction $CH_4 + 2O_2 \rightarrow CO_2 + 2H_2O$ to be deduced from the four-step mechanism by adding twice reaction IV to the sum of reactions I, II, and III, it follows from the definition of Q_k given in Eq. (4.4) that $Q_I + Q_{II} + Q_{III} + 2Q_{IV} = 1$.

A schematic illustration of the presumed structure of the premixed flame is shown in Fig. 2. It consists of a chemically inert preheat zone, followed by the inner layer, the H_2-CO oxidation layer and the equilibrium post-flame zone. The structure of the inner layer is similar to that shown in Refs. 2 and 3. In the inner layer all the hydrocarbon chemistry occurs, resulting in the formation of H_2 and CO as well as some H_2O and CO_2. The oxidation layer is governed by the overall reactions II–IV and H_2 and CO are oxidized to form H_2O and CO_2. The concentration of fuel is zero in the oxidation layer, and the H-radicals are in steady state. In the post-flame zone downstream of the $H_2 - CO$ oxidation layer, H_2 and CO are in partial equilibrium according to reaction II.

5. Asymptotic Analysis of the Inner Layer

Since the asymptotic analysis of the inner layer is identical to that of Ref. 2, only the results of the previous analysis will be shown here. Also following the development in Ref. 2 the backward rates of reactions 5, 24 and 25 will be neglected in this layer. The thickness of the inner layer is presumed to be of order δ, where

$$\delta = \left[\frac{k_{1f}k_{13}''X_{O_2}}{k_{11f}'k_{13}'L_F}\right]^0 , \tag{5.1}$$

in which the superscript 0 implies that these quantities are evaluated at the origin, $x = 0$, which is taken to coincide with the location of the inner layer, and at this point $T = T^0$ and $X_i = X_i^0$. Other quantities in Eq. 8 are

$$k_{13}'' \equiv k_{13}' + [(k_{2f}'k_{10b} + \gamma k_{1b}k_{11b}')/k_{1f}](C_{H_2}/C_{O_2}) ,$$

$$k_{11f}' \equiv k_{11f} + \gamma k_{12f} ,$$

$$k_{13}' \equiv k_{13} + \gamma k_{18}/K_1 , \tag{5.2}$$

$$k_{2f}' \equiv k_{2f} + \gamma K_3 k_{4b} ,$$

$$k_{11b}' \equiv k_{11b} + \gamma K_3 k_{12b} .$$

For near stoichiometric flames the value of $X_{O_2}^0$ is of order ε, where ε is a measure of the thickness of the H_2-CO oxidation layer. In addition, the ratio k_{1f}/k_{11f}' is small, and k_{13}''/k_{13}' is roughly of order unity; hence δ will be presumed to be smaller than ε, and the ordering $\delta \ll \varepsilon$ used in the previous analyses [1–2] is retained here. Asymptotic analysis show this ordering to be valid also for highly fuel lean flames.

Following the analysis in Ref. 2 a quantity L which contains the burning velocity is defined as

$$L \equiv A\delta^2(k_{11f}'k_{13}'/k_{13}'')^0 L_F R^0 , \tag{5.3}$$

where

$$A \equiv \frac{Y_{Fu}}{v_u^2 W_F}\left[\frac{\lambda}{c_p}\right]^0\left[\frac{T_u}{T^0}\right]^2 ,$$

$$R \equiv \left\{\frac{k_{1f}k_{2f}'X_{O_2}X_{H_2}}{\gamma k_{1b}[\gamma k_{2b}' + (\gamma k_{24f} + k_{25f})C_M]}\right\}^{1/2} , \tag{5.4}$$

$$k_{2b}' \equiv k_{2b} + \gamma k_{4f} .$$

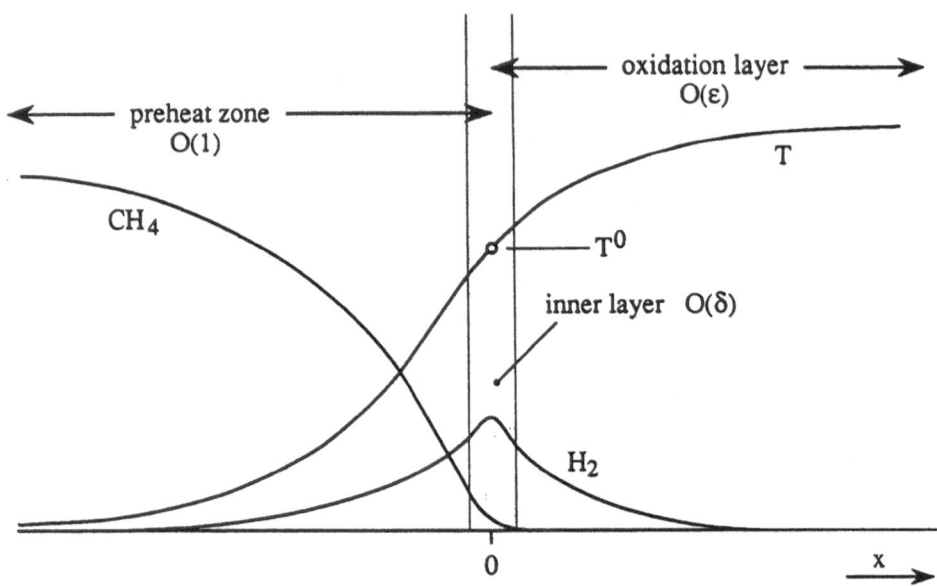

Figure 2. A schematic illustration of the overall flame structure.

The concentration of the third body C_M appearing in Eq. (5.4) can be written in terms of the Chaperon efficiency η_i of species $i(i = 1, ...N)$ as

$$C_M = [p\overline{W}/(\hat{R}T)] \sum_{i=1}^{N} \eta_i Y_i/W_i,$$

where the gas constant is $\hat{R} = 82.05$ atm cm^3/(mol K), and p is in atmospheres. In the previous analysis [2] two limiting structures for the inner layer were identified depending on the value of a parameter ω defined as

$$\omega = 2\delta L_H/R^0. \tag{5.5}$$

The quantity ω represents the ratio of the thickness of the fuel-consumption layer (of order δ), where the reaction I is presumed to occur, to the thickness of the radical-consumption layer of the previous analysis [1].

A schematic illustration of the presumed structure of the inner layer in the limit $\omega \to 0$, and $L/\omega = 0(1)$ is shown in Fig. 3a [2]. The fuel-consumption reaction I will occur in a thin layer of thickness of order $\delta\omega^{-2/3}$, which for convenience is presumed to be located at $\varsigma = x/\delta = 0$. The structure of this layer resembles that of a diffusion flame into which fuel diffuses from one side and H-radicals from the other. Outside this diffusion layer in the region of positive x there exists a radical non-equilibrium layer of thickness of order $\delta\omega^{-1}$, where the concentration of fuel is zero. From analyzing the structure of the diffusion-flame layer shown in Fig. 3a it can be shown [2] that the quantity L is given by the expression

$$L_0 = 2\omega(1 + 2\kappa + 4\theta/3 + 8\Psi/15 + 2\sigma), \tag{5.6}$$

where

$$\kappa \equiv \left[\frac{k_{5f}\,C_M(k_7+\gamma k_8)}{k_{1f}(k_6+k_7+\gamma k_8)}\right]^0, \qquad \theta \equiv \left[\frac{\gamma k_{1b}\kappa}{k'_{2f}\,X_{H_2}} + \frac{(\gamma\,k_{24f}+k_{25f})C_M}{k_{1f}\,X_{O_2}}\right]^0 R^0$$

$$\Psi \equiv \left[\frac{\gamma k_{1b}R}{k'_{2f}\,X_{H_2}}\right]^0, \qquad \sigma \equiv \left[\frac{k'_{2f}k_{10b}X_{H_2} - 2\gamma k_{1b}k_{18}X_{O_2}}{k_{1f}\,k'_{13}\,X_{O_2}}\right]^0 .$$

(5.7)

A schematic illustration of the presumed structure of the inner layer in the limit $\omega \to \infty$ with $L = 0(1)$ is shown in Fig. 3b. Here, the fuel-consumption reaction I occurs in a relatively broad layer, and embedded in this layer is a thin layer of thickness of order $\delta\,\omega^{-1/3}$, where reaction IV is not equilibrium. Asymptotic analysis of the structure of this layer shows that [2]

$$L_\infty^{-1} = \frac{8}{15}\left[\left\{1 + \frac{5}{2}\kappa + \frac{15}{16}\theta + \frac{5}{8}\Psi + \frac{5}{8}\chi + 2\sigma + \frac{15}{4}\mu\left[\frac{3-4\beta^2+\beta^4}{\beta^4}\ln(1+\beta)\right.\right.\right.$$
$$\left.\left.\left. - \frac{2\beta^4 - 4\beta^3 - 9\beta^2 + 3\beta + 6}{2\beta^3(1+\beta)}\right]\right\}^{-1} - \mu(\ln\mu + \ln E_\mu + \tilde{C} - 1)\right].$$

(5.8)

The various parameters appearing in Eq. (5.8) are defined as

$$\beta \equiv \left[\frac{\gamma\,k'_{2b}\,k_{13}\,R}{k_{1f}\,k'_{13}\,X_{O_2}}\right]^0 ,$$

$$\mu \equiv \left[\frac{k'_{2f}\,k'_{11b}\,X_{H_2}^2}{k_{1f}\,k''_{13}\,X_{O_2}\,R}\right]^0 \left[1 + \frac{k_{18}\,X_{O_2}}{k'_{11b}\,X_{H_2}}\right]^0 ,$$

$$\chi \equiv \left[\frac{k_{16}\,R\,Y_{Fu}\,\overline{W}}{k_{17}\,W_F}\right]^0 ,$$

$$E_\mu \equiv \frac{(E_{11b} + n_{11b}\,\hat{R}T_r + E_{2f} + n_{2f}\,\hat{R}T_r - E_{1f})\,\delta(T_c - T_u)\tau^0}{\hat{R}T^{0\,2}} ,$$

(5.9)

where E_n and n_n refer to the activation energy and the temperature exponent of the frequency factor of the elementary chemical reaction shown in Table 1, T_r being the reference temperature, set equal to 1600 K [1,2], and $\tilde{C} = 0.5772$ is Euler's constant.

An ad-hoc approximation to determine L for all values of ω has been proposed and tested previously [2] and is given by the expression

$$L = L_\infty \left[1 - \left(1 + \frac{L_0}{0.18 L_\infty}\right)^{-0.18}\right] .$$

(5.10)

Equation (5.10) will be used with the equations derived from analyzing the structure of the H_2-CO oxidation layer to calculate the burning velocity of the flame.

6. Analysis of the H_2-CO Oxidation Layer

The burning velocity v_u can be calculated from Eqs. (5.1-5.10) if $T^0, X_{H_2}^0, X_{O_2}^0$ and $X_{H_2O}^0$ are known. To determine these quantities the structure of the H_2-CO oxidation layer downstream from the inner layer must be analyzed. In this layer simplifications to the conservation equations arise because convection can be neglected in the first approximation and $X_F = 0$; hence $\omega_I = 0$. The H radicals are presumed to be in steady state in this layer, hence from Eq. (4.2) $\omega_{III} = \omega_{IV}$. It can then be shown [2] that for

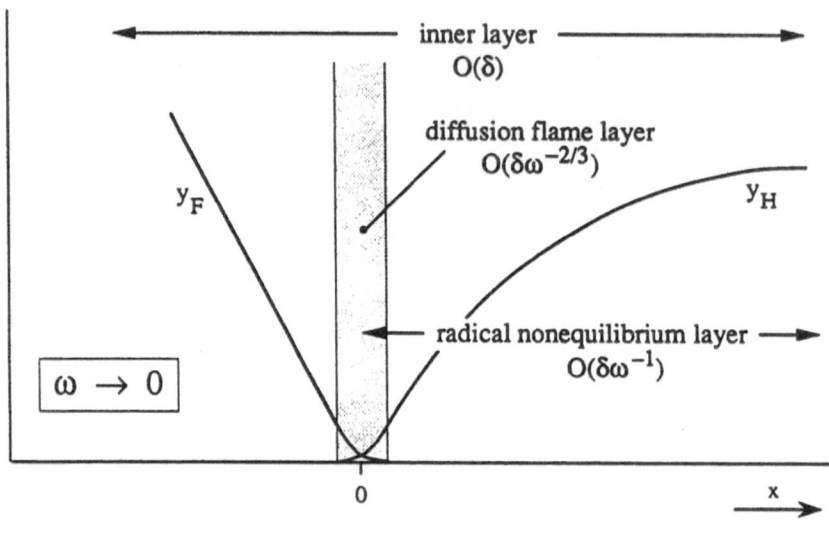

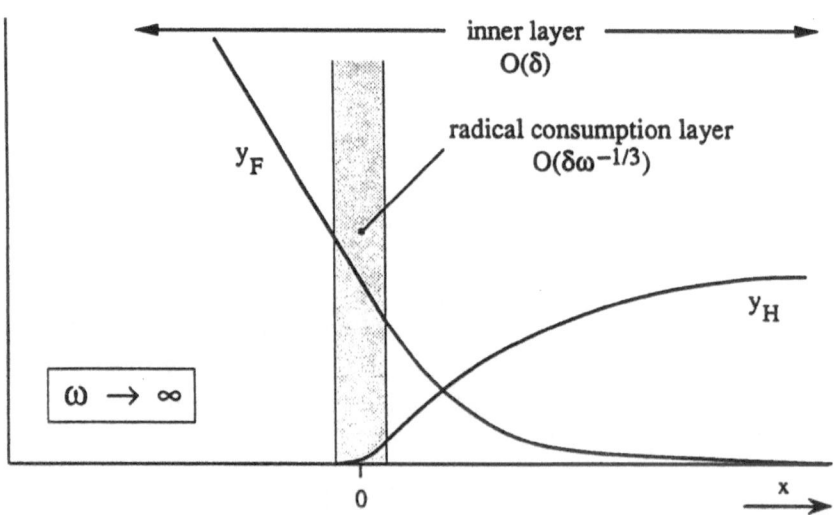

Figure 3. A schematic illustration of the inner layer in the limit (a) $\omega \to 0, L/\omega = O(1)$ and (b) $\omega \to \infty, L = O(1)$.

small θ and κ and neglecting the backward rates of reactions 5, 24 and 25, X_H is given by the expression

$$X_H = R(1 - \theta/2 - \kappa/2). \tag{6.1}$$

With $\omega_{III} = \omega_{IV}$ and the neglect of the convective terms in Eqs. (4.2) the expressions

$$\frac{d^2 x_{CO}}{d x^2} = \omega_{II}$$

$$\frac{d^2}{d x^2}[x_{H_2} + x_{CO}] = 2\omega_{III}$$

$$\frac{d^2}{d x^2}[x_{H_2} + x_{CO} - 2x_{O_2}] = 0$$

$$\frac{d^2}{d x^2}[x_{H_2} + x_{H_2O}] = 0 \tag{6.2}$$

$$\frac{d^2}{d x^2}[x_{CO} + x_{CO_2}] = 0$$

$$\frac{d^2}{d x^2}[(Q_{III} + Q_{IV})(x_{H_2} + x_{CO})/2 + Q_{II} x_{CO} + \tau] = 0,$$

where for convenience the definition

$$x_i = X_i/L_i \tag{6.3}$$

is introduced. From thermo-chemical tables the numerical values of $Q_{III} + Q_{IV}$ and Q_{II} are 0.6232 and 0.0363 respectively.

For stoichiometric and near stoichiometric flames the values of X_{H_2b}, X_{COb} and X_{O_2b} are of order ε. Hence, the following expansions are introduced

$$2qx = \varepsilon \eta, \quad qx_{H_2} = \varepsilon(b + 0.5 z_{H_2}), \quad qx_{CO} = \varepsilon(ba + 0.5z_{CO})$$
$$2qx_{O_2} = \varepsilon(a + z_{O_2}), \quad \tau = \tau_b - \varepsilon t, \quad x_i = x_{ib} - \varepsilon z_i, \quad i = H_2O, CO_2, \tag{6.4}$$

where

$$\alpha = \frac{K_3 X_{CO_2} L_{H_2}}{K_9 X_{H_2O} L_{CO}} \tag{6.5}$$

in which $K_9 = 0.0096 \exp(11623/T)$ is the equilibrium constant of the elementary reaction 9. Other quantities appearing in Eq. (6.4) are to be evaluated in the post-flame zone with $b \equiv qX_{H_2b}/(\varepsilon L_{H_2})$, $a \equiv 2qX_{O_2b}/(\varepsilon L_{O_2})$ and $\tau_b \equiv (T_b - T_u)/(T_c - T_u)$ and for comparison with the previous analysis [2] the quantity $q = 0.33$. In the analysis ε is presumed to be a small quantity, and η, z_{H_2}, z_{CO}, z_{O_2}, q, a, b, α, z_i, τ_b and t are of order unity. Introducing the expansions (6.4) into the coupling relations Eqs. (6.2)$_3$, (6.2)$_4$, (6.2)$_5$ and (6.2)$_6$, and integrating using the matching conditions that the values and gradients of the quantities z_{H_2}, z_{CO}, z_i ($i = H_2O, CO_2$) and t vanish at $\eta \to \infty$, the following relations are obtained to the leading order

$$z_{O_2} = z, \quad qz_{H_2O} = z - 0.5z_{CO}, \quad qz_{CO_2} = 0.5z_{CO},$$
$$2qt = (Q_{III} + Q_{IV}) z + Q_{II} z_{CO}, \tag{6.6}$$

where the quantity z is defined as

$$z \equiv 0.5z_{H_2} + 0.5z_{CO}. \tag{6.7}$$

Under the assumption that $(\gamma k_{24f} + k_{25f}) C_M/(\gamma k'_{2b})$ is small, an expression for the R of Eq. (5.4) may be obtained from Eqs. (3.2) and (6.4) and substituted into Eq. (6.1) to show that X_H can be expressed as

$$X_H = \frac{\varepsilon^2 \, K_1^{1/2} \, K_2^{1/2} \, K_3 \, L_{H_2}^{3/2} \, L_{O_2}^{1/2} \, (z + b - 0.5 z_{CO})^{3/2} (z + a)^{1/2}}{2^{1/2} \, X_{H_2O} q^2} \left[1 - \frac{\theta}{2} - \frac{\kappa}{2} \right], \qquad (6.8)$$

where $K_1 = 12.7 \exp(-8108/T)$ and $K_2 = 2.25 \exp(-1040/T)$ are the equilibrium constants of elementary reactions 1 and 2, respectively. Using the expansion for X_{O_2} shown in Eq. (6.4) the steady state concentration of HO_2 can be written as

$$X_{HO_2} = \frac{\varepsilon \, k_{5f} \, C_M \, L_{O_2}(z + a)}{2q(k_{6f} + k_{7f} + \gamma \, k_{8f})}, \qquad (6.9)$$

where the formation of HO_2 via the backward rates of reactions 6, 7, and 8 and the destruction of HO_2 via the backward rates of reaction 5 are presumed to be small. The source terms ω_{III} appearing in Eq. $(6.2)_2$ can be written from Eqs. (3.1) and (4.4) in terms of the expansions shown in Eqs. (6.4) as

$$\omega_{III} = 2q \, \varepsilon^3 \, D_{III} \, G_{III} \{ (z + a)^{3/2} \, (z + b - 0.5 z_{CO})^{3/2} - K_{III}^0 \, K_{III}' \, (z + a)$$
$$+ G_{III}' \, S^0 [(z + a) \, (z + b - 0.5 z_{CO})^2 - K_{24}^0 \, K_{24}'] \}, \qquad (6.10)$$

where

$$D_{III} = \frac{A(k_{5f} \, C_M \, K_1^{1/2} \, K_2^{1/2} \, K_3)^0 \, (L_{H_2} \, L_{O_2})^{3/2} \, (1 - \kappa/2 - \theta/2)}{2^{5/2} \, X_{H_2O}^0 \, q^4}$$

$$S = \frac{(2 K_1 \, K_2 \, L_{H_2})^{1/2} \, (k_{24f} + k_{25f}/\gamma)(1 - \kappa/2 - \theta/2)}{k_{5f} \, L_{O_2}^{1/2}}$$

$$K_{III} = \frac{2^{1/2} \, k_{5b} \, X_M \, X_{H_2O} q^2}{\varepsilon^2 \, (k_{6f} + k_{7f} + \gamma k_{8f}) K_1^{1/2} \, K_2^{1/2} \, K_3 \, (1 - \kappa/2 - \theta/2) L_{H_2}^{3/2} \, L_{O_2}^{1/2}} \qquad (6.11)$$

$$K_{24} = \frac{2(\gamma \, K_3 \, k_{24b} + k_{25b}) \, X_M \, X_{H_2O}^2 q^3}{\varepsilon^3 \, (\gamma k_{24f} + k_{25f}) C_M \, K_1 \, K_2 \, K_3^2 \, (1 - \kappa/2 - \theta/2)^2 L_{O_2} \, L_{H_2}^2}$$

$$G_{III} = \frac{k_{5f} \, C_M \, K_1^{1/2} \, K_2^{1/2} \, K_3 \, (\lambda/c_p) \, X_{H_2O}^0}{(k_{5f} \, C_M \, K_1^{1/2} \, K_2^{1/2} \, K_3)^0 \, (\lambda/c_p)^0 \, X_{H_2O}} \left(\frac{T^0}{T} \right)^2$$

$$G_{III}' = S/S^0, \quad K_{III}' = K_{III}/K_{III}^0, \quad K_{24}' = K_{24}/K_{24}^0.$$

In the inner layer $G_{III} = G_{III}' = 1$, and following previous analyses [1,2] the quantity ε will be presumed to be

$$\varepsilon = D_{III}^{-1/4}. \qquad (6.12)$$

Introduction of the expansions Eqs. (6.4) into Eq. $(6.2)_2$ followed by use of Eqs. (6.6), (6.7), (6.10), (6.11) and (6.12) with $G_{III} = G_{III}' = K_{III}' = K_{24}' = 1$, results in the leading-order problem

$$\frac{d^2 z}{d\eta^2} = (z + a)^{3/2} \, (z + b - 0.5 z_{CO})^{3/2} - K_{III}^0(z + a)$$
$$+ S^0 [(z + a) \, (z + b - 0.5 z_{CO})^2 - K_{24}^0] \qquad (6.13)$$

$$\frac{dz}{d\eta} = -1 \quad \text{at} \quad \eta = 0, \quad \frac{dz}{d\eta} \to 0 \quad \text{as} \quad \eta \to \infty.$$

The boundary conditions at $\eta = 0$ and $\eta \to \infty$ were obtained from matching with the inner layer [2], and the post-flame zone, respectively. The parameter S^0 is evaluated at $T = T^0$, while the parameters K_{III}, and K_{24} will be evaluated at $T = T_b$, and is asymptotically justified because they differ from their values at $T = T^0$ by quantities of $O(\varepsilon)$. Evaluation of K_{III} and K_{24} at $T = T_b$ ensures that $d^2 z/d\eta^2 = 0$ at $\eta \to \infty$, and from the definition these parameters as shown in Eq. (6.11) it follows that at $T = T_b$

$$K_{\mathrm{III}} = a^{1/2} b^{3/2}, \quad K_{24} = ab^2.$$

To complete the description of the structure of the oxidation layer it is necessary to derive a differential equation for z_{CO}. The source terms ω_{II} appearing in Eq. (6.2)$_1$, can be written from Eqs. (3.1) and (4.4) in terms of the expansions shown in Eqs. (6.4) as

$$\omega_{\mathrm{II}} = (2q/\varepsilon)\, D_{\mathrm{II}}\, G_{\mathrm{II}} \,(z+a)^{1/2}\,(z+b-0.5z_{\mathrm{CO}})^{1/2} \left(z_{\mathrm{CO}} - \frac{2\alpha}{1+\alpha}\, z \right) \tag{6.14}$$

where

$$D_{\mathrm{II}} = \frac{q(1+\alpha)^0\, k^0_{9f}\, X^0_{\mathrm{H_2O}}\, L_{\mathrm{CO}}}{\varepsilon\, k^0_{5f}\, C^0_{\mathrm{M}}\, K^0_3\, L_{\mathrm{H_2}}\, L_{\mathrm{O_2}}}$$

$$G_{\mathrm{II}} = \frac{(1+\alpha)\, k_{9f}\, (K^0_1\, K^0_2)^{1/2}(\lambda/c_p)}{(1+\alpha)^0\, k^0_{9f}\, (K_1\, K_2)^{1/2}(\lambda/c_p)} \left(\frac{T^0}{T} \right)^2. \tag{6.15}$$

In the inner layer $G_{\mathrm{II}} = 1$. Introducing Eqs. (6.4) into Eq. (6.2)$_1$, followed by use of Eqs. (6.6), (6.7), (6.11), (6.12), (6.14) and (6.15) with $G_{\mathrm{II}} = 1$, results in the leading-order problem.

$$\frac{d^2 z_{\mathrm{CO}}}{d\eta^2} = D_{\mathrm{II}}(z+a)^{1/2}\,(z+b-0.5z_{\mathrm{CO}})^{1/2} \left(z_{\mathrm{CO}} - \frac{2\alpha}{1+\alpha}\, z \right)$$

$$\frac{d z_{\mathrm{CO}}}{d\eta} = -1 \quad \text{at} \quad \eta = 0, \quad \frac{d z_{\mathrm{CO}}}{d\eta} \to 0 \quad \text{as} \quad \eta \to \infty. \tag{6.16}$$

The boundary conditions at $\eta = 0$ and $\eta \to \infty$ were obtained from matching with the inner layer [2], and the post-flame zone, respectively. Equations (6.13) and (6.16) imply that

$$2 \int_0^{z^0} \left\{ (z+a)^{3/2}\,(z+b-0.5z_{\mathrm{CO}})^{3/2} - K_{\mathrm{III}}\,(z+a) \right.$$

$$\left. + S^0[(z+a)\,(z+b-0.5z_{\mathrm{CO}})^2 - K_{24}] \right\} dz = 1 \tag{6.17}$$

$$2D_{\mathrm{II}} \int_0^{z^0_{\mathrm{CO}}} (z+a)^{1/2}\,(z+b-0.5z_{\mathrm{CO}})^{1/2} \left(z_{\mathrm{CO}} - \frac{2\alpha}{1+\alpha}\, z \right) d z_{\mathrm{CO}} = 1.$$

The solution of Eq. (6.17) to determine z^0 and z^0_{CO} as a function of a, b, S^0, and D_{II} must be obtained numerically. For a given value of Φ, thermochemical calculations would yield the values of T_b, $X_{\mathrm{O_2 b}}$, $X_{\mathrm{H_2 b}}$, $X_{\mathrm{CO b}}$, $X_{\mathrm{CO_2 b}}$, and $X_{\mathrm{H_2O b}}$. The quantities a, b and D_{II} depend on ε which may be expressed as a function of T^0 according to

$$\varepsilon = \frac{T_b - T^0}{t^0(T_c - T_u)}. \tag{6.18}$$

The quantity t^0 is related to z^0, and z^0_{CO} as shown in Eq. (6.6). In addition S^0 depends on T^0; hence all parameters in Eq. (6.17) depend on results from the structure of the inner layer. The quantity L defined by Eq. (5.3) can be written in terms of the expansions shown in Eq. (6.5) as

$$L = \frac{k^{0^2}_{1f}\, k''^0_{13}\, (z^0+a)^{5/2}(z^0+b-0.5z^0_{\mathrm{CO}})^{3/2}\, L_{\mathrm{O_2}}}{k^0_{5f}\, C^0_{\mathrm{M}}\, k'^0_{11f}\, k'^0_{13f}\, L_F\, (1-\kappa/2-\theta/2)}, \tag{6.19}$$

where use has been made of Eqs. (5.1), (5.4), (6.4), (6.6), (6.7), (6.11) and (6.12). Eq. (6.19) expresses L in terms of T^0, z^0 and z_{CO}^0. Since Eq. (5.10) provides an independent expression for L as a function of T^0, the quantities T^0, z^0, and z_{CO}^0 can be calculated numerically when Eqs. (6.13) and (6.16) are integrated. In view of Eqs. (6.12) and (6.18), the burning velocity can be calculated by rewriting the first of Eqs. (5.4) and (6.11) as

$$
v_u^2 = \frac{Y_{Fu}}{W_F} \left[\frac{\lambda}{c_p}\right]^0 \left[\frac{T_u}{T^0}\right]^2
$$
$$
\frac{(k_{5f} \, C_M \, K_1^{1/2} \, K_2^{1/2} \, K_3)^0 \, (1 - \kappa/2 - \theta/2) \, (L_{H_2} \, L_{O_2})^{3/2} \, (T_b - T^0)^4}{2^{5/2} \, X_{H_2O}^0 \, (T_c - T_u)^4 \, (t^0)^4 \, q^4} . \tag{6.20}
$$

6.1 Analytical Solution of Eqs. (6.13) and (6.16) in the Limit $D_{II} \to \infty$

with $a = b = 0$

In the previous asymptotic analyses [1,2] the water-gas shift reaction II was assumed to be in chemical equilibrium everywhere in the oxidation layer except in a thin sublayer located between the inner layer and the oxidation layer. To facilitate comparison between the value of $z_{H_2}^0$ obtained from the present model and that outlined in Ref. 3, analytical solution of Eqs. (6.13) and (6.16) in the limit $D_{II} \to \infty$, with $a = b = 0$ will be obtained. In this limit Eq. (6.16) yields the leading order solution $(1 + \alpha)z_{CO} = 2\alpha \, z$. It follows from the definition of z shown in Eq. (6.7) that $(1 + \alpha)z_{H_2} = 2z$. Substituting these results into the first of Eq. (6.17) and integrating yields the leading order result

$$
z^0 = 2^{1/4} \, (1 + \alpha^0)^{3/8} \left[1 + \frac{S^0}{(1 + \alpha^0)^{1/2}}\right]^{-1/4} . \tag{6.21}
$$

As in the previous analyses the influence of reactions III and IV will be neglected in the non-equilibrium layer. Hence, $d^2 z / d\eta^2 = 0$ everywhere in this layer. To obtain corrections to the leading order solution of z_{H_2} and z_{CO} the following expansions are introduced

$$
z_{H_2} = \frac{2z^0}{(1 + \alpha^0)} + \nu \, z_{H_2}^1
$$
$$
z_{CO} = \frac{2\alpha \, z^0}{(1 + \alpha^0)} + \nu \, z_{CO}^1 \tag{6.22}
$$
$$
\eta = \nu \varsigma ,
$$

where

$$
\nu = D_{II}^{-(1/2)} \tag{6.23}
$$

represents the thickness of the layer where the water-gas shift reaction is not in equilibrium. Introducing the expansions Eq. (6.22) into Eq. (6.16) yields the equations

$$
-\frac{d^2 \, z_{H_2}^1}{d\varsigma^2} = \frac{d^2 \, z_{CO}^1}{d\varsigma^2} = \frac{z^0 [z_{CO}^1 - \alpha^0 \, z_{H_2}^1]}{(1 + \alpha)^{3/2}} . \tag{6.24}
$$

Boundary conditions for Eq. (6.24) obtained from matching with the inner layer at $\varsigma = 0$, and the oxidation layer at $\varsigma \to \infty$ are

$$\frac{d\,z_{\mathrm{H_2}}^1}{d\varsigma} = \frac{d\,z_{\mathrm{CO}}^1}{d\varsigma} = -1 \quad \text{at} \quad \varsigma = 0$$

$$\frac{d\,z_{\mathrm{H_2}}^1}{d\varsigma} = -\frac{2}{1+\alpha^0}, \quad \frac{d\,z_{\mathrm{CO}}^1}{d\varsigma} = -\frac{2\alpha^0}{1+\alpha^0} \quad \text{at} \quad \varsigma \to \infty. \tag{6.25}$$

The solutions of Eqs. (6.24) together with the boundary conditions of Eqs. (6.25) are

$$z_{\mathrm{H_2}}^1 = -\frac{1-\alpha^0}{(1+\alpha^0)^{3/4}\,(z^0)^{1/2}} \exp\left[-\frac{(z^0)^{1/2}}{(1+\alpha^0)^{1/4}}\varsigma\right] - \frac{2}{1+\alpha^0}\varsigma$$

$$z_{\mathrm{CO}}^1 = \frac{1-\alpha^0}{(1+\alpha^0)^{3/4}\,(z^0)^{1/2}} \exp\left[-\frac{(z^0)^{1/2}}{(1+\alpha^0)^{1/4}}\varsigma\right] - \frac{2\alpha^0}{1+\alpha^0}\varsigma. \tag{6.26}$$

Combining the first of Eq. (6.21) and (6.26) the value of $z_{\mathrm{H_2}}^0$ to the first order in ν is

$$z_{\mathrm{H_2}}^0 = \frac{2z^0}{(1+\alpha^0)} - \nu\,\frac{1-\alpha^0}{(1+\alpha^0)^{3/4}\,(z^0)^{1/2}} + \mathrm{O}(\nu^2), \tag{6.27}$$

where z^0 is given by Eq. (6.21). The result for $z_{\mathrm{H_2}}^0$ obtained from Eq. (6.27) for various values of ν will be compared with that obtained from numerical evaluation of Eqs. (6.13) and (6.16).

The solid lines in Fig. 4 represent results of numerical evaluations of Eqs. (6.13) and (6.16) for various values of D_{II} with $a = b = 0$, and $S^0 = 0.23$, and $\alpha^0 = 0.15$. This result is compared with that calculated from Eq. (6.27), which is shown by the dotted line in Fig. 4. For values of $D_{\mathrm{II}} \geq 1$ the deviation of the asymptotic value of $z_{\mathrm{H_2}}^0$ from the numerical value is less than 10 %. However, for near stoichiometric flames at $p = 1$ bar the value of D_{II} is around 0.35, and at this value of D_{II} the asymptotic results deviates from the corresponding numerical value by approximately 20 %. Hence, the values of the burning velocity calculated from the present model can be expected to be different from those predicted previously [1,2].

7. Results and Discussions

Equation (6.20) was used by the procedure described in section 6 to calculate the burning velocity as a function of the equivalence ratio, initial temperature, and pressure. In these calculations the value of λ/c_p appearing in Eq. (6.20) was expressed as $\lambda/c_p = 2.58 \times 10^{-4}(T/298)^{0.7}$ g/(cm/s). The Lewis numbers for the various species were presumed to be constant, with $L_{\mathrm{F}} = 0.97$, $L_{\mathrm{O_2}} = 1.1$, $L_{\mathrm{H_2O}} = 0.83$, $L_{\mathrm{CO_2}} = 1.39$, $L_{\mathrm{H_2}} = 0.3$, $L_{\mathrm{H}} = 0.18$ and $L_{\mathrm{CO}} = 1.11$. Calculations were performed for values of ϕ between 0.5 and 1.0 for values of p between 1 atm to 40 atm, and for $T_u = 300$ K.

Numerical calculations of the coupled system of Eqs. (6.13) and (6.16) together with Eqs. (5.10) and (6.19) show that the value of D_{II} appearing in Eq. (6.16) is small. Hence, it was necessary to consider a rather large integration domain. To improve the accuracy and speed of the numerical computations, the differential equations were rescaled by defining a new small parameter ε_m which is based on the characteristic Damköhler Number of reaction II. Thus

$$\varepsilon_m = \left[\frac{Ak_{9f}^0(K_1^0 K_2^0)^{1/2}(L_{\mathrm{O_2}}L_{\mathrm{H_2}})^{1/2}L_{\mathrm{CO}}(1+\alpha^0)(1-\kappa/2-\theta/2)}{2^{5/2}q^3}\right]^{-1/3},$$

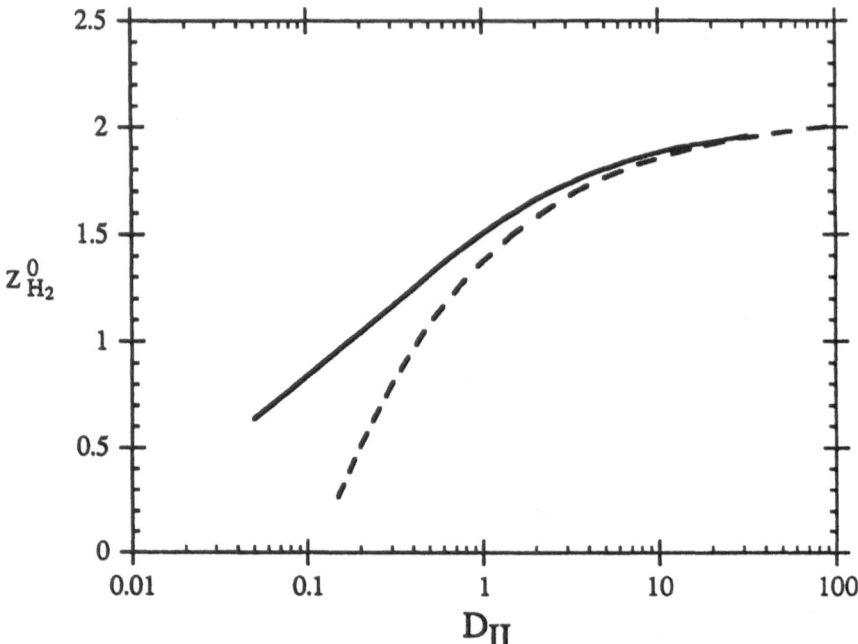

Figure 4. Results for $z_{H_2}^0$ obtained from numerical evaluation of Eq. (6.30) and (6.33) for $a = b = 0, S^0 = 0.23$, and $\alpha = 0.15$ shown as solid lines and that calculated from Eq. (6.44) shown as dotted lines.

where the subscript m will be used to identify the variables and parameters in the modified formulation. The expansions introduced for the dependent and the independent variables which are now identified with the same symbols with the addition of the subscript m, are similar to that shown in Eq. (6.4) with ε replaced by ε_m, and the coupling relations of Eq. (6.6) are recovered. If, as before the variations of the values of the chemical reaction rate coefficients, transport coefficients and X_{H_2O} are neglected in the oxidation layer it can be easily verified that the modified form of Eq. (6.16) does not contain any parameters, while the r.h.s. of the modified form of Eq. (6.13) must be multiplied by a parameter D_{IIIm} which is to be evaluated from the expression

$$D_{IIIm} = \frac{\varepsilon_m k_{5f}^0 C_M^0 K_3^0 L_{H_2} L_{O_2}}{q X_{H_2O}^0 k_{9f}^0 L_{CO}(1 + \alpha^0)}.$$

The quantity D_{IIIm} is the reciprocity of D_{II} defined in Eq. (6.15) with ε replaced by ε_m. It can be verified that $D_{II} = (D_{IIIm})^{-3/4}$, and $\varepsilon = \varepsilon_m(D_{IIIm})^{-1/4}$.

The parameters S_m^0, K_{IIIm}, and K_{24m} appearing in the modified form of the differential equation (6.13) are the same as those defined in Eq. (6.11) with ε replaced by ε_m. The r.h.s. of the modified form of Eq. (6.19) with z^0 and z_{CO}^0 replaced by the modified variables z_m^0 and z_{COm}^0 must be multiplied by D_{IIIm} to obtain L_m. The modified form of the expression for calculating the burning velocity is that shown in Eq. (6.20) with t^0 replaced by t_m^0 and divided by D_{IIIm}. The dependent (z_m, z_{COm}) and independent (η_m) variables in this modified formulation must be multiplied by $(D_{IIIm})^{1/4}$ to recover the dependent (z, z_{CO}) and independent (η) variables shown in Eq. (6.13) and (6.16).

Figure 5 shows profiles of z_{CO} and z_{H_2} obtained from numerical integration of the modified forms of Eqs. (6.13) and (6.16) in terms of the original variables shown in those

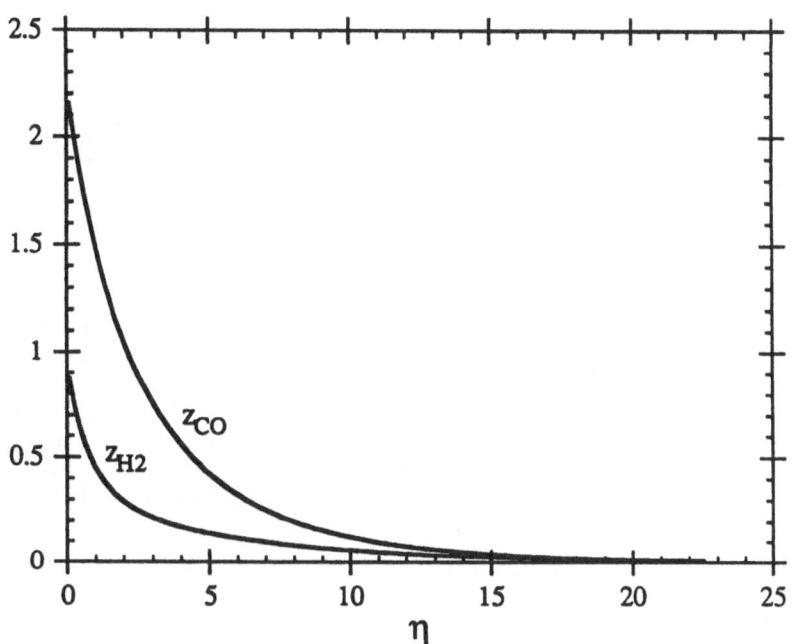

Figure 5. Results of the numerical integration of Eqs. (6.30) and (6.33) for $D_{II} = 0.35$, $S^0 = 0.23$, $\alpha^0 = 0.15$, $a = 0.5$, and $b = 0.5$.

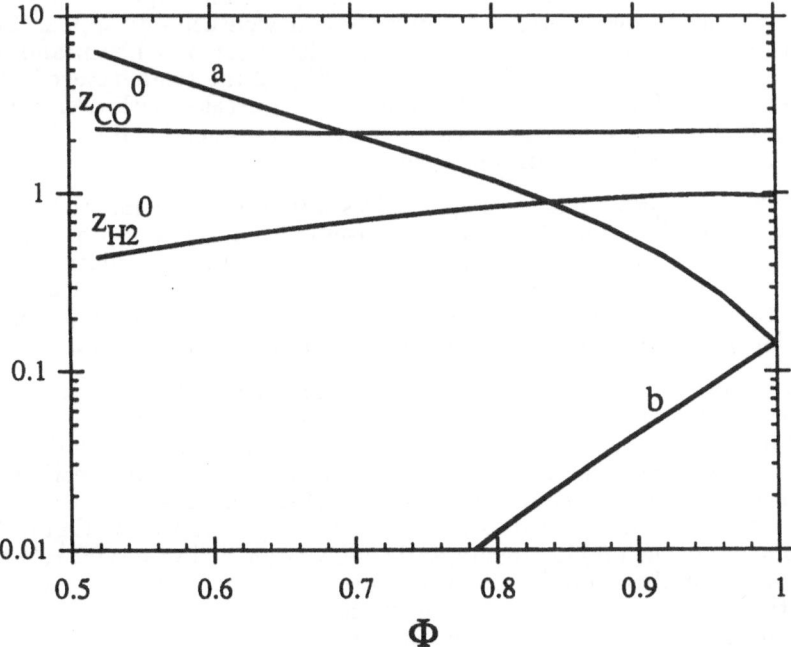

Figure 6. Results of numerical calculations showing z^0, z^0_{CO}, a, and b as functions of ϕ for $p = 1$ atm and $T_u = 300$ K.

equations. The calculations were performed for $D_{II} = 0.35, S^0 = 0.23, \alpha^0 = 0.15, a = b = 0.5$ which roughly correspond to the values of these quantities at $\phi = 1, p = 1$ atm, and $T_u = 300$ K. Further details of the procedure employed in performing the numerical integration are discussed in the Appendix. Figure 5 shows that the value of z_{CO} is larger than the value of z_{H_2} everywhere in the oxidation layer. Since $\alpha^0 < 1$, the values of z_{CO} approaches $\alpha^0 z_{H_2}$ only for large values of η. Hence, reaction II is not in chemical equilibrium in a major part of the oxidation layer.

Figure 6 shows results for $z_{CO}^0, z_{H_2}^0, a$ and b for various values of ϕ for $p = 1$ atm and $T_u = 300$ K. The increase in the values of a and the exponential decrease in the values of b with decreasing values of ϕ is due to the increase in the concentration of oxygen, and the decrease in the concentration of hydrogen and carbon monoxide in the post flame zone. However, Fig. 6 shows only a small decrease in the values of $z_{H_2}^0$ and z_{CO}^0 with decreasing values of ϕ. Fig. 7 shows the values of L and ω as functions of ϕ, demonstrating that ω increases with decreasing ϕ. It has been shown previously [2] that for stoichiometric flames the value of L decreases and the value of ω increases with increasing pressure. Hence, the expression for L given by Eq. (5.8) becomes more accurate for fuel lean flames at high pressure, and the structure of such flames would resemble that shown in Fig. 3b.

In Fig. 8 the variations of κ, θ and σ with ϕ are shown and in Fig. 9 the variation of δ and ε, which represent respectively the thickness of the inner layer and the oxidation layer shown is Figs. 2, 3a and 3b are plotted as functions of ϕ. The former shows that the current expansions, treating κ, θ, and σ as small, are reasonably accurate, and the latter shows the relative orderings to be reasonable. Since the ratio ε/q appear in the expansions shown in Eq. (6.4), this quantity effectively represents the thickness of the oxidation layer, and from Fig. 9 it can be verified that $\delta \ll \varepsilon/q$ for all values of ϕ. Fig. 10 shows that the values of Ψ and χ as a function of ϕ are not small, so there is some inaccuracy in the expansions in Eqs. (5.6) and (5.8), although the effects tend to be mitigated by other terms; at high pressures these inaccuracies disappear [2]. Fig. 11 shows the variations of the quantities S^0 and D_{II} with ϕ for $p = 1$ atm and $T_u = 300$ K, and the values of these quantities are small. In Fig. 12 the temperature in the inner layer T^0 is plotted as a function of ϕ. The value of T^0 decreases with decreasing values of ϕ. Since Eq. (6.20) shows that the burning velocity is proportional to $(T_b - T^0)^4$, the burning velocity must decrease with decreasing ϕ.

In Fig. 13 the burning velocity v_u is plotted as a function of ϕ for various values of p, and for $T_u = 300$ K. The solid lines represent results from the asymptotic analysis described here and the points represent results from full numerical calculations performed with the rate data of Table 1 for $p = 1$ atm and $T_u = 300$ K. For $p = 1$ atm the solid curves in Fig. 13 indicate that the burning velocity attains a value of 31.5 cm/s at $\phi = 1.0$, and is below the value obtained from detailed numerical calculations.

In Fig. 14 results for the burning velocity obtained from the previous asymptotic analysis [2] for stoichiometric flames are shown as dotted lines, and are compared with that obtained from the present model which are shown as solid lines, for various values of p. Results for the burning velocity shown in Fig. 4 of Ref. 2 were incorrectly calculated and the corrected results are shown in Fig. 14. This computational error does not influence the values of other quantities plotted in Figs. 3–10 of Ref. 2. Figure 14 shows that around $p = 1$ atm the previous asymptotic analysis yielded values for v_u which were generally higher than those predicted by the present model. However, both models predict that the values of v_u decrease with increase in the value of p in accord with experimental measurements. In particular, the current model predicts the experimental finding that $d\ln v_u/(d\ln p)$ is roughly equal to -0.5 for values of p between 5 atm and 40 atm.

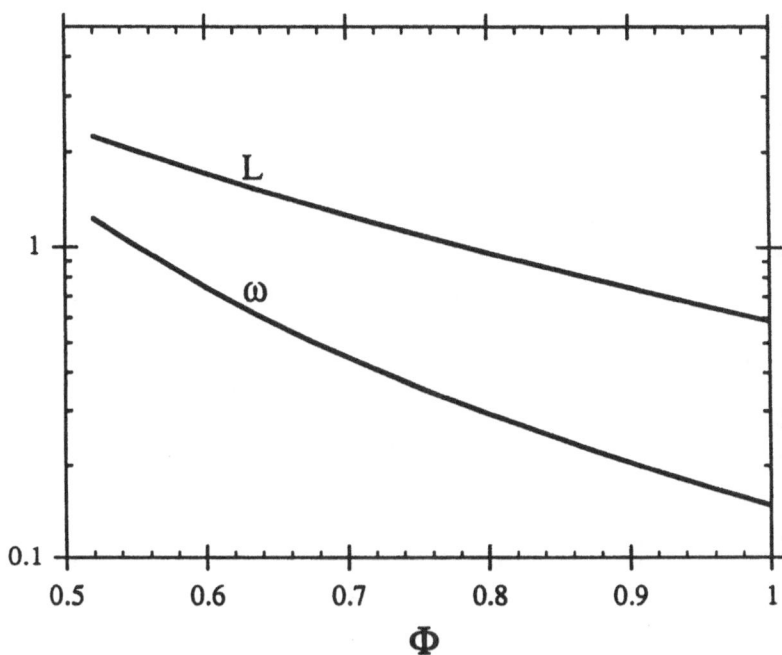

Figure 7. Variation of the parameters L and ω with the equivalence ratio ϕ for $p = 1$ atm and $T_u = 300$ K.

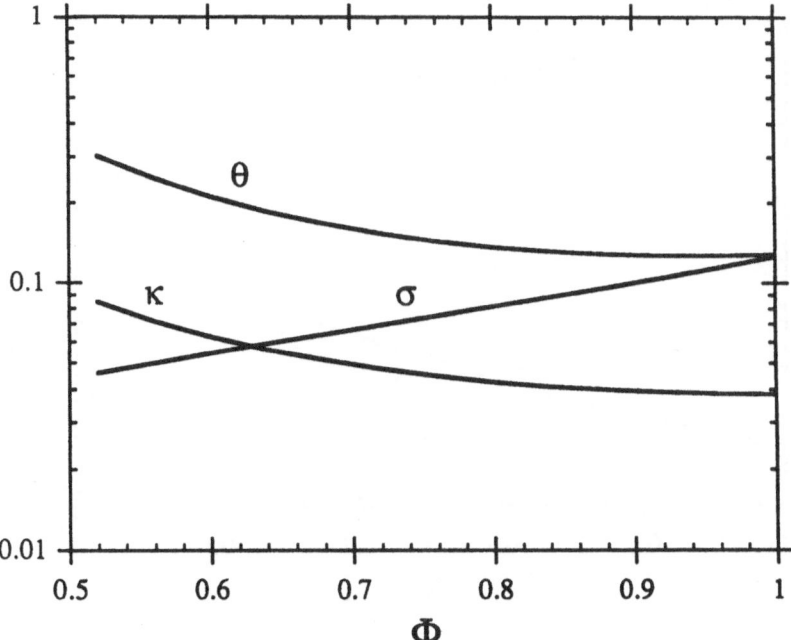

Figure 8. Variation of the parameters θ, σ and κ with ϕ for $p = 1$ atm and $T_u = 300$ K.

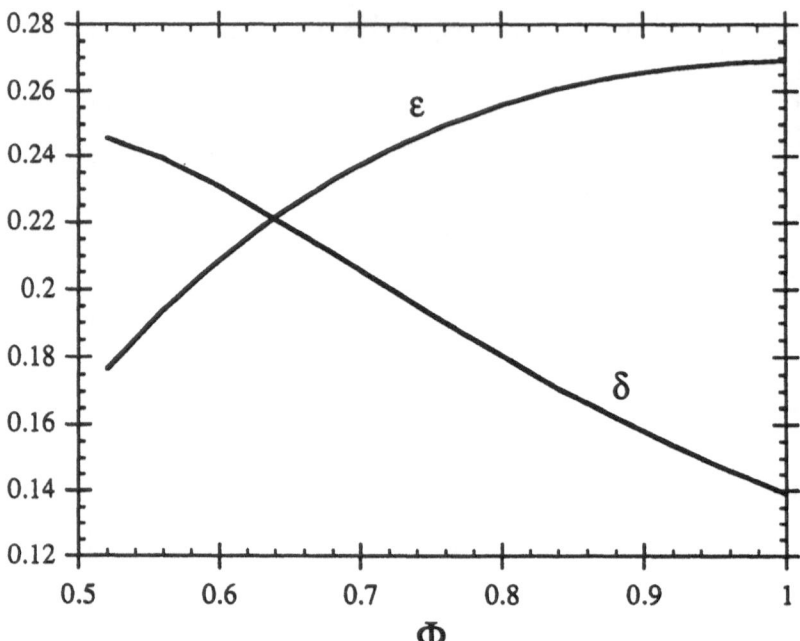

Figure 9. Variation of the quantities δ and ε with ϕ for $p = 1$ atm and $T_u = 300$ K.

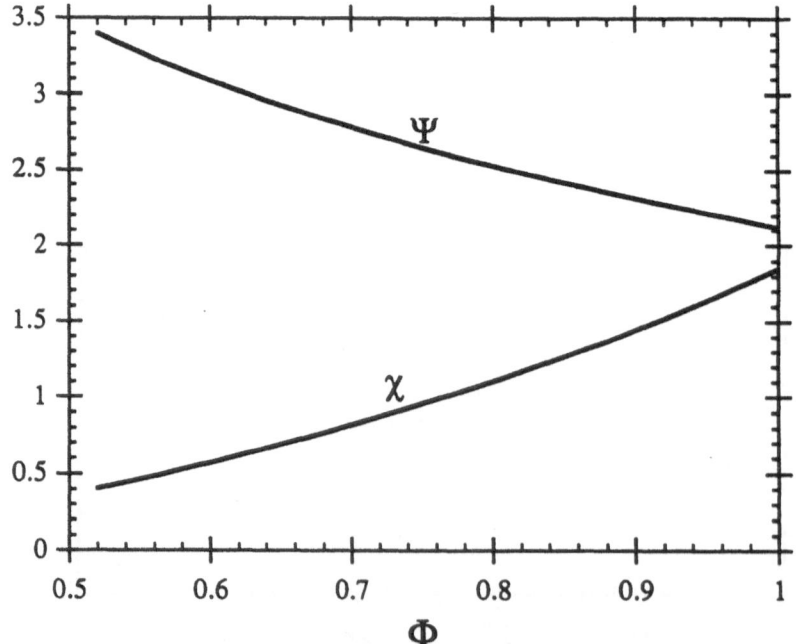

Figure 10. Variation of the quantities Ψ and χ with ϕ for $p = 1$ atm and $T_u = 300$ K.

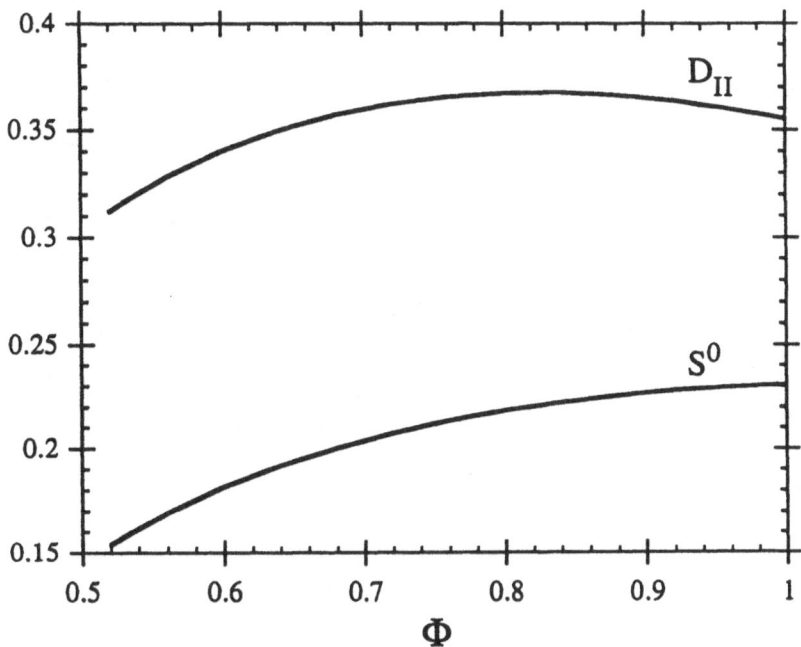

Figure 11. Variation of the quantities S^0 and D_{II} with ϕ for $p = 1$ atm and $T_u = 300$ K.

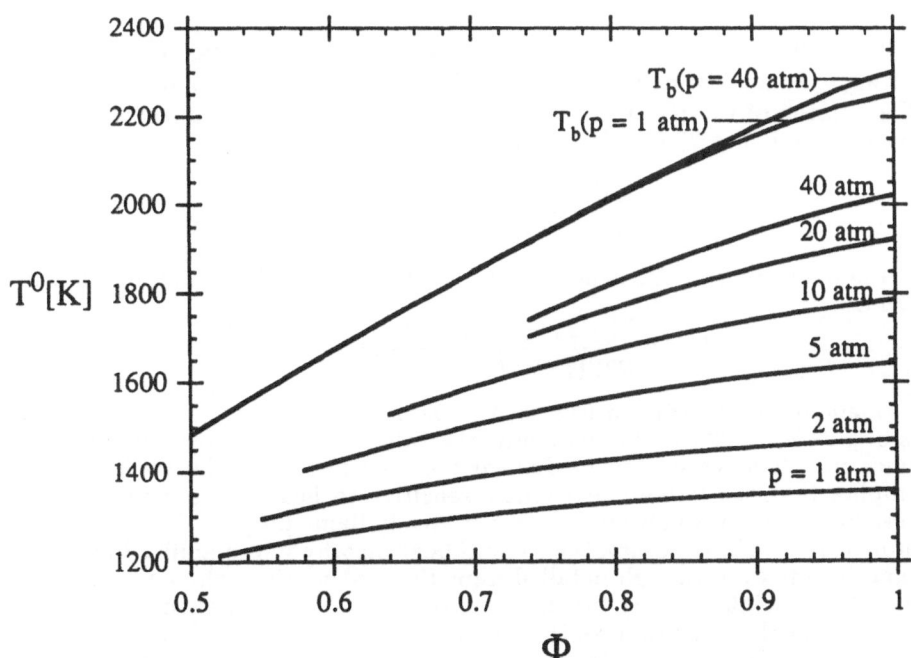

Figure 12. Variation of the temperature at the inner layer T^0 with ϕ for various values at the pressure p, and for $T_u = 300$ K.

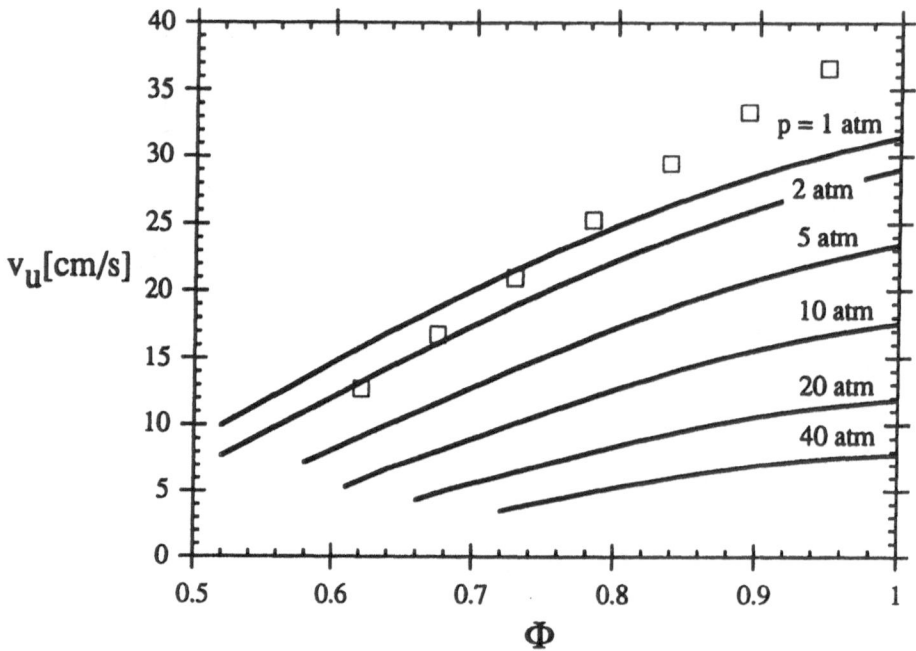

Figure 13. The burning velocity v_u as a function of ϕ for various values of the pressure p at $T_u = 300$ K from the present theory (lines) and from numerical integrations using the mechanism of Table 1 for $p = 1$ atm and $T_u = 300$ K (points).

It has been shown previously [1] that an effective activation energy E_{eff} may be obtained by fitting v_u to an Arrhenius form in T_b. Hence, if $E_{eff} \equiv 2\hat{R}T_b^2 d[ln(\rho_u v_u)]/dT_b$, then from differentiaton of Eq. (6.20) with respect to T_b under the assumptions that T^0 is independent of T_b and that T_b equals T_c, the relationship

$$E_{eff} = \frac{4\hat{R}T_b^2}{T_b - T_u}\left[\frac{T^0 - T_u}{T_b - T^0} + \frac{dT_u}{dT_b}\right]$$

is obtained. If the quantity dT_u/dT_b is neglected, an effective Zel'dovich number [5] can then be defined as

$$Ze = \frac{E_{eff}(T_b - T_u)^2}{\hat{R}T_b^2(T^0 - T_u)} = \frac{4}{\varepsilon\, t^0},$$

where use was made of Eq. (6.18). In Fig. 15, this Ze is plotted as a function of ϕ for various values of p at $T_u = 300$ K. Interestingly, the value of Ze increases rapidly with decreasing values of ϕ. The large value of Ze near the experimentally observed flammability limits implies that the flame is extremely sensitive to heat losses, and hence it would be increasingly difficult to obtain in practice steady flame propagation in increasingly fuel-lean mixtures. These observations could bear on why flammability limits are observed in experiments even though detailed numerical calculations show that steady flame propagation is possible for very fuel-lean mixtures [6]. It must be emphasized, as discussed in Ref. 1, that the large value of the Zel'dovich number obtained here is related to T^0, which is determined by the relative rates of important elementary reactions, and is unrelated to one-step activation-energy asymptotics.

8. Conclusions

This paper attempts to refine the asymptotic analysis of methane flames initiated in [1,2] by considering non-equilibrium of reaction II everywhere in the oxidation layer. It shows that such an analysis in principle can be done but that many parameters enter into the formulation. Nevertheless, the essence of the structure originally proposed in [1] remains valid in the entire range of equivalence ratios and pressures considered here. A particularly useful result is the possibility of defining and calculating an effective Zel'dovich number and thereby establishing a link to previous large-activation-energy analyses.

Acknowledgements

We thank Professor N. Peters at RWTH Aachen, and Professor F. A. Williams at UCSD for many valuable suggestions. We also thank Ms. Sabine Tillmann and Ms. Ute Bennerscheidt for preparing this manuscript and Ms. Alexandra Kees for preparing the figures. The international collaboration was partially supported by the U. S. National Science Foundation Grant number NSF-INT-86-09939.

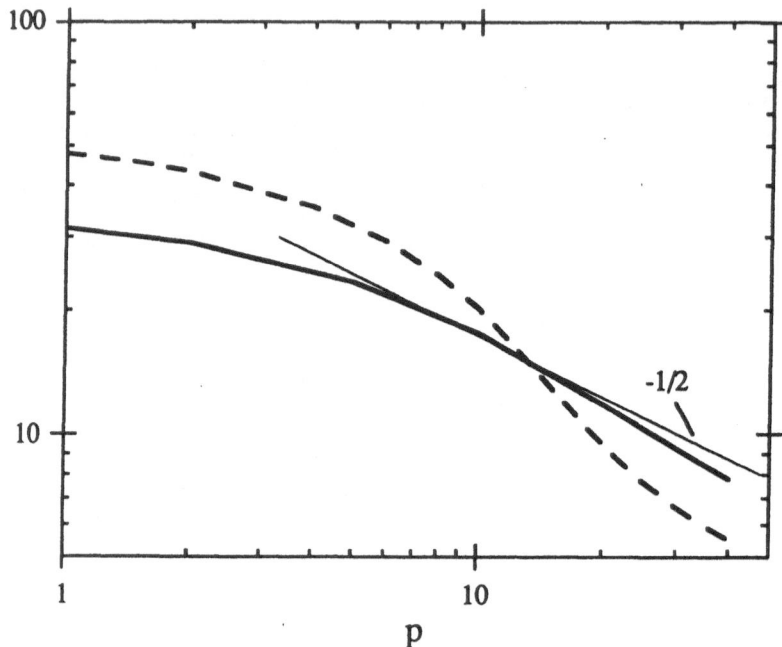

Figure 14. Comparison between the values of v_u obtained using the previous asymptotic model [2] shown as dotted lines, and the present model shown as solid lines for stoichiometric flames with $T_u = 300$ K.

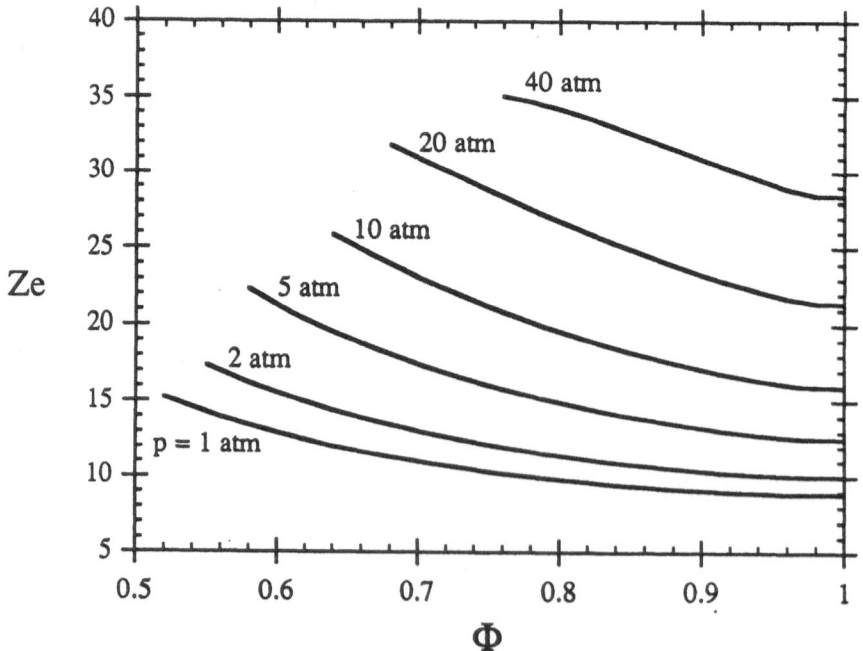

Figure 15. Variation in the value of the effective Zel'dovich number with ϕ for various values of p at $T_u = 300$ K.

References

1. Peters, N., and Williams, F. A., "Combustion Flame" **68** (1987), 185–207.

2. Seshadri, K., and Peters, N., "The Inner Structure of Methane-Air Flames", to appear in Combustion Flame (1990).

3. Peters, N. , "Numerical Simulation of Combustion Phenomena" (R. Glowinski, B. Larrouturo, and R. Teman Eds.), Lecture Notes in Physics **241**, Springer Verlag (1985), 90–109.

4. Kennel, C., Göttgens, J. and Peters, N., "The Basic Structure of Lean Propane Flames", Twenty-Third Symposium (International) on Combustion, The Combustion Institute, Pittsburgh, to appear (1990).

5. Williams, F. A., "Combustion Theory", 2nd Edition, Blaisdell Publishing Company (1985).

6. Lakshmisha, K. N., Paul, P. J., Rajan, N. K. S., Goyal, G. and Mukunda, H. S., "Behaviour of Methane-Oxygen-Nitrogen Mixtures near Flammability Limits", Twenty-Second Symposium (International) on Combustion, The Combustion Institute 1988, 1573–1578.

Appendix

The following is a brief description of the numerical algorithms employed in solving the present asymptotic model. Since the temperature T^0 is unknown and has to be computed as part of the solution, it is necessary to solve a set of non-linear equations to obtain the burning velocity. However, for a given value of T^0 most parameters can be evaluated readily. Therefore a feasible procedure is to assume a value of T^0 and test whether the values for L from the equations (5.10) and (6.19) agree, i.e. one needs to solve one non-linear equation as a function of a single independent variable. Additional complications arise because the quantities z_m^0, $z_{CO_m}^0$, D_{IIm}, α, a, b_m and S^0 are mutually dependent, which requires solving the ODEs (6.13) and (6.16) as part of the function evaluation. Here this is done using a pseudotransient method, where the second derivatives are approximated by the standard second order space-centered formula and evaluated implicitly. The source terms are both evaluated at the previous time step, which leads to two independent linear systems with tridiagonal matrices and (almost) constant coefficients. Asymptotic boundary conditions are used at the right boundary, which are non-linear for small values of b_m, as shown below. An advantage of the pseudotransient method is that the computation of the parameters $D_{(II)m}$ and S^0 can be easily embedded within the pseudotime iterations.

Instead of using homogeneous Neumann or Dirichlet boundary conditions at very large values of η, boundary conditions which are compatible with the physics of the problem are derived (which consequently significantly reduce the computational time). For nearly stoichiometric mixtures, a_m and b_m are small, but of the same order of magnitude. The dominant eigenvalue can then be estimated to be

$$s^* \approx \max \left[-(a_m b_m)^{1/4}, -\frac{1}{2} \sqrt{\frac{6(1-\alpha)}{(1+\alpha)D_{IIm}}} a_m^{3/4} b_m^{1/4} \right] \qquad (A1)$$

and the boundary condition used is

$$\frac{z'_m}{z_m} = \frac{z'_{COm}}{z_{COm}} = s^* .$$ (A2)

For lean mixtures b_m becomes very small such that the linearized equations are valid only at very large values of η. Also a_m is then no longer small, so that in the region of interest we can neglect z_m in the term $a_m + z_m$. In this case it is better to expand the equations about $b_m = 0$. To leading order this implies that z_{COm} is proportional to z_m, namely $z_{COm} = rz_m$, where r is a constant between $2\alpha/(1+\alpha)$ and 2. Substituting $z_{COm} = rz_m$ into one of the differential equations leads to

$$r = 1 - \frac{D_{IIm}}{a_m} + \sqrt{(1 - \frac{D_{IIm}}{a_m})^2 + 2\frac{2\alpha}{1+\alpha}\frac{D_{IIm}}{a_m}}$$ (A3)

and the boundary conditions now are

$$z'_m = -\frac{2z_m^{5/4}}{\sqrt{5\tilde{D}_{IIm}}}$$

$$z'_{COm} = -r^{-1/4}\frac{2z_{COm}^{5/4}}{\sqrt{5\tilde{D}_{IIm}}}$$ (A4)

$$\tilde{D}_{IIm} = \frac{D_{IIm}}{[a_m(1 - r/2)]^{3/2}}$$

which are linearized with respect to pseudotime (which requires to update only the last main diagonal element in the iteration matrix).

CHAPTER 7

ASYMPTOTIC ANALYSIS OF METHANE-AIR DIFFUSION FLAMES

Harsha K. Chelliah
Department of Mechanical and Aerospace Engineering
Princeton University
Princeton, NJ 08544

César Treviño
Facultad de Ingenieria
Universidad Nacional Autonoma de Mexico
04510 Mexico

Forman A. Williams
Department of Applied Mechanics and Engineering Sciences
University of California, San Diego
La Jolla, CA 92093

1. Introduction

Activation energy asymptotic analyses based on one-step, irreversible reaction models have made significant contributions to our understanding of diffusion-flame structures and extinction conditions [1-4]. Some of these studies have indicated that for methane-air diffusion flames, increasing strain produces increasing leakage of the fuel through the reaction zone, which results in lower flame temperatures and subsequent extinction of the flame. This leakage of fuel, however, is contrary to the oxygen leakage observed in experimental measurements [5-9] and also in numerical calculations employing full reaction mechanisms [9-13] and simplified reaction mechanisms [14,15]. In order to reconcile such differences between analytical results with a one-step reaction model and numerical calculations with detailed chemistry, asymptotic analyses involving reduced reaction mechanisms have been performed [16,17]. More recently, additional aspects of the approach adopted by Treviño and Williams [17] have been considered, and detailed comparisons have been made with experiments [18].

Although they begin with basically the same four-step reduced mechanism, the two asymptotic approaches [16,17] are quite different in that they emphasize two different limiting cases. Both approaches have identified two principal reaction zones, a thin fuel-consumption zone and a somewhat broader but still asymptotically thin oxygen-consumption zone. The fuel-consumption zone is located (at Z_f on the mixture fraction coordinate) on the fuel-rich side of the stoichiometric point (Z_s). All of the complex fuel chemistry occurs in this zone by reaction of fuel species with the radicals diffused from the oxygen-consumption zone to form CO and H_2, as well as H_2O and CO_2. The oxygen-consumption zone extends from the fuel-consumption zone to the fuel-lean side of stoichiometry, where O_2, H_2 and CO are consumed, while chain carriers are produced by the $H_2 - O_2$ branching reactions, with additional H_2O and CO_2 also formed. The chain carriers are removed through termination mechanisms beginning with three-body steps on the fuel-lean side of the fuel-consumption zone and also through reactions with fuel species in the fuel-consumption zone. The two analytical approaches differ in the way that the radical-nonequilibrium layer at the interface between the fuel- and oxygen-consumption zones is treated. In one approach [16] it is assumed that the radical-nonequilibrium layer is thin and is embedded in the fuel-consumption zone, with the ratio of the thickness of the fuel-consumption zone to that of radical-nonequilibrium zone (ω) treated as ∞. In the other approach [17], the opposite limiting case, $\omega \to 0$ is assumed, implying a narrow fuel-consumption zone which can be approximated by a reaction sheet. In a recent study, Seshadri and Peters [19] have shown that the actual value of ω is about 0.1 for low pressures, but becomes of order unity at elevated pressures. Except for these differences in the treatment of the radical-nonequilibrium layer, the other aspects of the analytical approximations [16-18] are similar and they have both predicted qualitatively correct diffusion-flame structures and extinction conditions that are consistent with the results of numerical integrations employing full kinetics [9-13]. The intention here is to present the detailed structure and extinction results employing the reaction mechanism given in Table II of Chapter 1, since the previous studies employed somewhat different rate data.

After a discussion of the formulation of the problem, the simplified reaction mechanism and the reduced conservation equations, attention is focused on the fuel-consumption zone. Because of the complexities of the chemistry in the fuel-consumption zone and since the limit $\omega \to 0$ seems to be reasonable for low pressures, the subsequent consideration of the structure adopts the reaction-sheet approximation for this zone. Next, a simplified treatment of the oxygen-consumption zone is presented and is employed to calculate extinction curves that are then compared with results obtained by the approach of Seshadri and Peters [16], by experiments [8] and by numerical integrations [20]. The inaccuracies associated with employing a constant-density approximation in relating the experimentally measured strain rates to the scalar dissipation rates are demonstrated by considering a variable-density mixing layer, and new comparisons are made with the corrected values. Agreements are improved and results help in interpretations to the meanings of observed extinction parameters.

2. Formulation

The present formulation and notation parallel those employed previously [17,18]. As shown in [21], for example, it is convenient to use the mixture fraction Z as an independent variable since this allows us to perform *most* of the analyses without any reference to specific flow configurations. With a well-known set of reasonable approximations [22], the conservation of species and energy in the N-component mixture can be written for locally planar flames as

$$-\frac{\lambda}{c_p}|\nabla Z|^2 \frac{d^2}{dZ^2}(Y_i/L_i) = w_i, \quad i = 1, \ldots, N, \tag{2.1}$$

$$-\frac{\lambda}{c_p}|\nabla Z|^2 \frac{d^2 T}{dZ^2} = \sum_{i=1}^{N} \frac{h_i^0 w_i}{c_p} \tag{2.2}$$

where λ is the thermal conductivity of the mixture, c_p is the specific heat at constant pressure, which is treated as a constant, Y_i is the mass fraction of species i, w_i is the chemical source term of species i, h_i^0 is the chemical enthalpy of species i, and L_i is an effective, constant, Lewis number for species i in the mixture (inversely proportional to a diffusion coefficient of species i). For example, for combustion in air, this can be considered as a Lewis number based on the binary diffusion coefficient of species i with N_2, as detailed in Chapter 1. As is the common practice in analytical studies of diffusion flames, in Eq. (2.1), we have chosen not to expand L_i about unity in the hope of achieving better accuracy, even though the formulation, involving only one Z, formally is restricted to value of all L_i unity.

With the subscripts 0 and ∞ identifying the conditions in the oxidizer and fuel streams respectively, the nondimensional temperature θ and normalized mole fractions

y_i are introduced, according to the formulas

$$\theta = (T - T_0)c_p W_{CH_4} L_{CH_4}/(Q_T Y_{CH_4\infty}), \tag{2.3}$$

$$y_i = Y_i W_{CH_4} L_{CH_4}/(L_i W_i Y_{CH_4\infty}), \quad i = 1,\ldots,N. \tag{2.4}$$

Here Q_T (192 kcal) is the heat released in the overall reaction

$$CH_4 + 2O_2 \rightarrow CO_2 + 2H_2O \tag{2.5}$$

and W_i is the molecular weight of species i. With these nondimensional variables, the nondimensional forms of Eqs. (2.1) and (2.2) are

$$\frac{d^2\theta}{dZ^2} = -\sum_{j=1}^{M} q_j r_j \equiv \frac{d^2}{dZ^2}(\sum_{i=1}^{N} h_i y_i), \tag{2.6}$$

$$\frac{d^2 y_i}{dZ^2} = -\sum_{j=1}^{M} \nu_{ij} r_j, \quad i = 1,\ldots,N \tag{2.7}$$

where there are M reactions, each having a nondimensional rate

$$r_j = \omega_j(W_{CH_4} L_{CH_4}/Y_{CH_4\infty})(c_p/\lambda)/|\nabla Z|^2, \quad j = 1,\ldots,M, \tag{2.8}$$

a nondimensional heat release $q_j = \sum_{i=1}^{N} h_i \nu_{ij} = Q_j/Q_T$, and a net increase $\nu_{ij} (\equiv \nu_{ij}'' - \nu_{ij}')$ in the number of moles of species i. Here ω_j is the molar rate per unit volume for step j, h_i ($= h_i^0 W_i/Q_T$) is the nondimensional heat of destruction of species i, and Q_j ($= \sum_{i=1}^{N} h_i^0 W_i \nu_{ij}$) is the energy released in step j. With fixed two-point boundary conditions for θ and all y_i, specified at $Z = 0$ and $Z = 1$, there is a one-parameter family of solutions dependent on a strain-rate parameter that increases monotonically with a representative value of $(\lambda/\rho c_p)|\nabla Z|^2$. The only flow-dependent parameter in this formulation is the common factor $|\nabla Z|^2$ and its description involves approximations somewhat at the discretion of the investigator. The implications of the simplified transport and thermodynamic descriptions employed in Eqs. (2.6)-(2.8) have been discussed in detail elsewhere [18].

For the case of steady, planar, counterflow flames established between two planar, parallel, porous plates, boundary-layer approximations are unnecessary if appropriate boundary conditions are applied at the feed exits. Introduction of the mixture fraction as the independent variable is helpful in automatically including convective effects in a formally purely diffusive-reactive formulation [1], but strictly speaking this can be done only for Lewis numbers of unity, since otherwise there is more than one mixture fraction. The procedure can be motivated by a formal expansion about Lewis numbers of unity, for example, followed by stretching about the stoichiometric mixture fraction,

$$Z_s = [1 + (2Y_{CH_4\infty} W_{O_2})/(W_{CH_4} Y_{O_2 0})]^{-1}, \tag{2.9}$$

which itself is a small number for these flames. Comparisons are then made in the *thermodynamic* coordinates of temperature and concentrations as functions of Z, and the problem of predicting profiles in the physical coordinate, which is a separate and largely unrelated problem, is not addressed. For comparison with results of experiment and of more detailed numerical integrations, Bilger's definition of the mixture fraction (which enforces a fixed Z_s) is adopted (see Chapter 2). As earlier [17], for formal consistency of the development, exact for a stagnant diffusion layer but only approximate for other configurations, Eq. (2.9) will be modified by multiplying $Y_{O_2 0}$ by L_{CH_4}/L_{O_2}, a 10% correction of Z_s, within profile uncertainties, giving $Z_s = 0.051$ for air.

3. Reaction Mechanism

The elementary reaction mechanism adopted here is given in Table II of Chapter 1. In addition, the elementary reaction step

(Step 26) $$HO_2 + H \rightarrow H_2O + O$$

has been included to investigate the influence of the additional branching effects of the HO_2 reaction path. Under the assumptions of partial equilibrium for step 3 and steady states for CH_3, H_2CO, HCO, HO_2, and O, motivating neglect of the concentrations of these species, the reduced four step-mechanism

$$CH_4 + \left[\frac{2}{1+\gamma'}\right] OH + \left[\frac{2\gamma'}{1+\gamma'}\right] H \rightarrow CO + \left[\frac{2+4\gamma'}{1+\gamma'}\right] H_2 + \left[\frac{1-\gamma'}{1+\gamma'}\right] H_2O, \qquad I$$

$$O_2 + \left[\frac{1+3\gamma'}{1+\gamma'}\right] H_2 \rightleftharpoons \left[\frac{2}{1+\gamma'}\right] OH + \left[\frac{2\gamma'}{1+\gamma'}\right] H + \left[\frac{2\gamma'}{1+\gamma'}\right] H_2O, \qquad II$$

$$\left[\frac{2}{1+\gamma'}\right] OH + \left[\frac{2\gamma'}{1+\gamma'}\right] H + \left[\frac{1-\gamma'}{1+\gamma'}\right] H_2 \rightarrow \left[\frac{2}{1+\gamma'}\right] H_2O, \qquad III$$

$$CO + H_2O \rightleftharpoons CO_2 + H_2 \qquad IV$$

can be derived [17], where

$$\gamma' = K_3 c_{H_2}/c_{H_2O}. \qquad (3.1)$$

In general c_i denotes the concentration of the species i and $K_j \equiv k_{ja}/k_{jb}$ the equilibrium constant for step j. The rates for each of the four steps are found by reductions now considered standard to be

$$\omega_I = \omega_{11a} + \omega_{12a} - \omega_{10b} - \omega_{11b}, \qquad (3.2)$$

$$\omega_{II} = \omega_{1a} + \omega_6 + \omega_{26} - \omega_{1b} + \omega_{HO_2}, \qquad (3.3)$$

$$\omega_{III} = \omega_5 + \omega_{10b} + \omega_{16} + \omega_{24} + \omega_{25} + \frac{1}{2}\omega_{HO_2} + \frac{1}{2}\omega_{HCO} + \omega_O - \frac{1}{2}\omega_{CH_3}, \qquad (3.4)$$

$$\omega_{IV} = \omega_{9a} - \omega_{9b}. \qquad (3.5)$$

The net molar rates of production ω_i of intermediates HO_2, HCO, CH_3 and O have been retained to identify the subtractions that were made in deriving the Eqs. (3.2)-(3.5).

Further reductions of the four-step mechanism can be achieved by introducing the partial-equilibrium approximation for step 9 [16-18], leading to a simpler three-step mechanism. Whenever the approximation leading to this three-step mechanism is valid, the water-gas shift reaction [step (IV)] is in equilibrium, along with the partial equilibrium for step 3. Although these two partial equilibria are helpful for facilitating analysis, they tend to be inaccurate in certain regions and therefore are in need of further study.

4. Reduced Conservation Equations

The steady-state and partial-equilibrium approximations yield relationships among various species so that the number of independent species are reduced. With steady states for CH_3, H_2CO, HCO, HO_2, and O, the mechanism of Table II in Chapter 1 gives

$$\omega_{11a} + \omega_{12a} - \omega_{10b} - \omega_{11b} - \omega_{13} = 0, \qquad (4.1)$$

$$\omega_{13} - \omega_{14} - \omega_{15} = 0, \qquad (4.2)$$

$$\omega_{14} + \omega_{15} - \omega_{16} - \omega_{17} = 0, \qquad (4.3)$$

$$\omega_5 - \omega_6 - \omega_7 - \omega_8 - \omega_{26} = 0, \tag{4.4}$$

$$\omega_{1a} + \omega_{2b} + \omega_{4a} + \omega_{26} - \omega_{1b} - \omega_{2a} - \omega_{4b} - \omega_{13} = 0. \tag{4.5}$$

Solution of these equations for the concentrations of c_i of the intermediates gives

$$c_{CH_3} = \frac{(k_{11a}c_H + k_{12}c_{OH})c_{CH_4}}{k_{10b}c_{M_{22}}c_H + k_{11b}c_{H_2} + k_{13}c_O}, \tag{4.6}$$

$$c_{H_2CO} = \frac{k_{13}c_Oc_{CH_3}}{k_{14}c_H + k_{15}c_{OH}}, \tag{4.7}$$

$$c_{HCO} = \frac{k_{13}c_Oc_{CH_3}}{k_{16}c_H + k_{17}c_{M_{17}}}, \tag{4.8}$$

$$c_{HO_2} = \frac{k_5 c_{M_5} c_H c_{O_2}}{k_8 c_{OH} + (k_6 + k_7 + k_{26})c_H}, \tag{4.9}$$

$$c_O = \frac{(k_{1a}c_{O_2} + k_{2b}c_{OH} + k_{26}c_{HO_2})c_H + k_{4a}c_{OH}^2}{k_{1b}c_{OH} + k_{2a}c_{H_2} + k_{4b}c_{H_2O} + k_{13}c_{CH_3}}, \tag{4.10}$$

where c_{M_j} denotes the sum of the concentrations of all species, weighted according to their chaperon efficiencies for step j, and the k's are specific reaction-rate constants parameterized in Table II of Chapter 1. The partial-equilibrium for step 3 results in

$$c_{OH} = c_H c_{H_2}/(K_3 c_{H_2}). \tag{4.11}$$

Since elimination of c_O through Eq. (4.10) is more complicated, simplifications have been introduced in the analysis of the oxygen-consumption layer. The revision of step (I) when the c_{CH_3} was not eliminated through its steady state has been considered in a previous study [18]; here however, for simplicity, the steady state for c_{CH_3} is employed everywhere.

The concentrations are related to the y_i of Eq. (2.4) by

$$c_i = y_i L_i (Y_{CH_4\infty}/L_{CH_4})(\bar{W}/W_{CH_4})p/(RT), \quad i = 1, \ldots, N, \tag{4.12}$$

where p is the pressure, $\bar{W}$ is the average molecular weight, and in terms of the chaperon efficiencies η_{ij} of each species i in reaction j

$$c_{M_j} = \sum_{i=1}^{N}(\eta_{ij}Y_i/W_i)\bar{W}p/(RT) \equiv \eta_j p/(RT). \tag{4.13}$$

Here η_j is defined as a molar-weighted chaperon efficiency for species j. Equations (4.6) through (4.11) allow us to eliminate six concentrations from the conservation equations. The remaining seven concentrations must be obtained by solving the differential equations. From the mechanism of Table II in Chapter 1 it may be shown from Eq. (2.7) that

$$d^2 y_{CH_4}/dZ^2 = r_I, \tag{4.14}$$

$$d^2 y_{O_2}/dZ^2 = r_{II}, \tag{4.15}$$

$$d^2 y_R/dZ^2 = 2r_I - 2r_{II} + 2r_{III}, \tag{4.16}$$

$$d^2 y_{CO_2}/dZ^2 = -r_{IV}, \tag{4.17}$$

$$d^2(y_{CH_4} + y_{CO} + y_{CO_2})/dZ^2 = 0, \tag{4.18}$$

$$d^2(y_H + y_{OH} + 2y_{H_2} + 2y_{H_2O} + 4y_{CH_4})/dZ^2 = 0, \tag{4.19}$$

$$d^2(y_O + y_{OH} + y_{H_2O} + y_{CO} + 2y_{CO_2} + 2y_{O_2})/dZ^2 = 0, \qquad (4.20)$$

where the r_I, r_{II}, r_{III} and r_{IV} are related to ω_I, ω_{II}, $(\omega_{III} - \omega_O)$ and ω_{IV} through Eq. (2.8), respectively, and the overall radical-pool concentration variable is

$$y_R \equiv y_H + y_{OH} + 2y_O. \qquad (4.21)$$

Here the concentrations of CH_3, H_2CO, HCO, and HO_2 have been assumed to be small enough to be neglected. Equations (4.18)-(4.20) express conditions of atom conservation in the chemistry.

5. General Flame Structure

As indicated in the introduction, the asymptotic approaches [16-17] have identified separate fuel-consumption and oxygen-consumption zones, where the fuel and oxidizer chemistry appear through the steps (I) and (II), respectively. The main difference between these two approaches arises in the way steps (III) and (IV) are distributed in the structure. In the approach of Seshadri and Peters [16] (where the approximations parallel those employed in the premixed-flame structure analysis [23]), steady states for H and OH are introduced, so that a three-step mechanism is obtained, and only the overall fuel-consumption step $(CH_4 + O_2 \rightarrow CO + H_2 + H_2O)$ occurs in the fuel-consumption zone, while the overall oxygen-consumption step $(O_2 + 2H_2 \rightarrow 2H_2O)$ occurs in a broader oxygen-consumption zone, in which the water-gas step (IV) maintains partial equilibrium, except near its fuel-consumption boundary, where nonequilibrium of step (IV) is included by a perturbation method. In the approach of Treviño and Williams [17], partial equilibrium of step (IV) was imposed everywhere for simplicity, and step (III) was confined to a narrow recombination zone at the oxygen-side boundary of the oxygen-consumption zone (at a location defined by equality of the rates of step 1a and 5, ie. a radical cut-off point), so that step (II) occurred in the oxygen-consumption zone and produced the intermediates H and OH, which reacted with CH_4 in the fuel-consumption zone as a diffusion flame within the diffusion flame. Qualitatively, these two approaches represent the two limiting cases where $\omega \rightarrow 0$ and $\omega \rightarrow \infty$. Extension of the latter approach to include the influences of the water-gas nonequilibrium and the distributed radical recombination has been considered recently [18], and the results indicate that the distributed recombination can influence the structure and extinction predictions significantly, while the water-gas nonequilibrium has a higher-order effect. On the other hand, within the context of the alternate approach [16] that corresponds to $\omega \rightarrow \infty$ it has been found that the water-gas nonequilibrium has a significant influence on the extinction predictions because of its influence on the fuel chemistry when the radical-nonequilibrium zone is embedded therein. Nevertheless, both asymptotic approaches have shown that the extinction is caused by the finite rates in the oxygen-consumption zone.

A recent asymptotic analysis of the premixed-flame structure with an arbitrary value for ω has shown that for atmospheric or low-pressure flames, the limit $\omega \rightarrow 0$ may be a reasonable approximation, while for higher pressures ω order of unity applies [19]. In order to test further the occurrence of these different limits and to have a better understanding of the distribution of the four steps across the flame, numerical calculations with the detailed reaction mechanism given in Table II of Chapter 1 have been employed. Figures 1 and 2 show a plot of the four rates ω_I, ω_{II} (neglecting ω_{HO_2}), $\omega_{III} - \omega_O$ (neglecting ω_{HO_2}, ω_{HCO} and ω_{CH_3}) and ω_{IV}, for low and high strain rates, respectively, for methane-air diffusion flames in potential counterflows at atmospheric pressure and stream temperatures of 300 K. There is seen to be no significant change in the structure for different strain rates, except for the increase in the overall rates because of the oxygen leakage and somewhat narrower flame thickness at high strain rate. Although, the fuel-consumption zone is broader in mixture fraction than previously assumed [17,18],

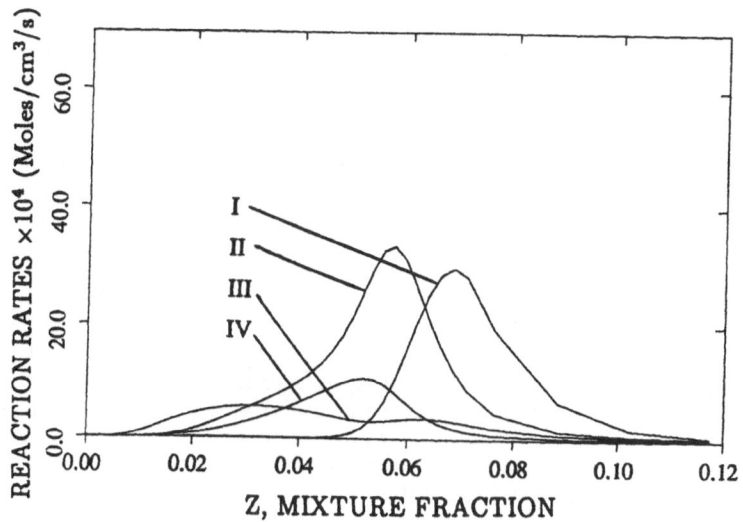

Figure 1: Distribution of the four overall steps for low-strain rate.

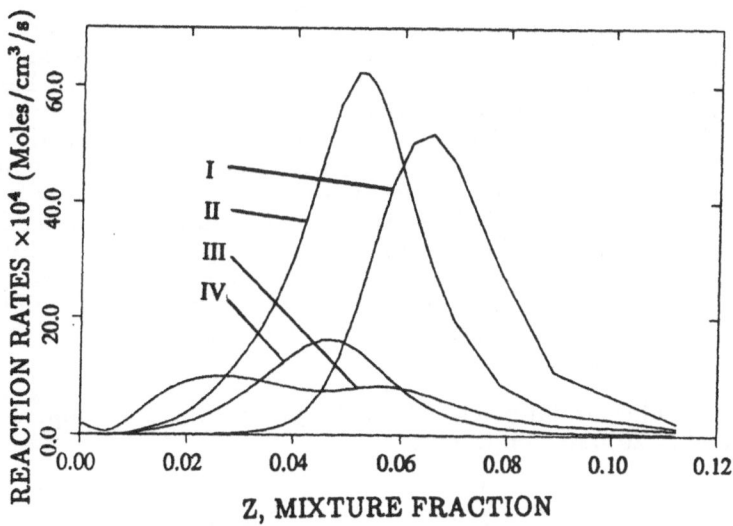

Figure 2: Distribution of the four overall steps for high-strain rate.

the distributed-recombination approximation [18] is seen to be reasonable for all strain rates. The water-gas equilibrium assumption is seen to break down on the wings of the oxygen-consumption zone, but the departure at the fuel-consumption boundary is the most important and has been analyzed in the earlier works [16,18].

Because of the significant simplifications that the limit $\omega \to 0$ offers and its apparent reasonable accuracy for low pressure flames, and because extinction apparently is caused by finite rates of the oxygen-consumption zone, subsequent considerations in the present chapter retain the reaction-sheet limit for the fuel-consumption zone.

6. The Oxygen-Consumption Zone

For flames at low-strain rates, flame structures have been previously [17,18] calculated based on the assumption that partial equilibrium of step II applies throughout the oxygen-consumption zone. Furthermore, for simplicity, it is assumed that the water-gas reaction is in equilibrium ($\omega_{9a} = \omega_{9b}$) everywhere in this zone; influences of water-gas nonequilibrium have been investigated elsewhere [18], and the results have indicated that in the present approach to the leading order it has no effect on the extinction condition.

In the limit of large Damköhler number for step (I), $D_I \to \infty$, the flame structure consists of a thin diffusion flame at Z_f where the radicals are consumed by fast fuel chemistry, and a somewhat broader oxygen consumption zone, which extends to the lean side of Z_f where partial equilibrium for step II is maintained. In this latter zone, oxygen is consumed while radicals are produced through the H_2-O_2 branching reactions. Under these approximations the lean end of this zone is determined by the crossover of the branching (ω_{II}) and recombination (ω_{III}) steps. The temperature and the location on the mixture-fraction coordinate at which this occurs is identified here as T_c and Z_c, respectively. Because of the large activation energy of step (II) (primarily of step 1a) and low activation of step (III), below T_c the branching is essentially negligible. Under the present assumptions, at Z_f and Z_c the radical concentration vanishes, and these two locations define the boundaries of the oxygen-consumption zone. At Z_f, H_2O and CO_2 are formed by fuel chemistry, and H_2 and CO are formed as well due to incomplete combustion associated with freezing of the water-gas shift on the fuel side. This H_2 and CO are consumed across the oxygen-consumption zone so that their concentrations vanish at Z_c, while O_2 is depleted in this zone. Therefore, the nonvanishing concentrations at Z_c are those of O_2, CO_2, and H_2O, while those at Z_f are those of CO_2, H_2O, CO, and H_2.

Under these conditions, integrals of Eqs. (2.6) (with $h_{O_2} \equiv 0$ adopted for brevity), (4.18), (4.19) and (4.20) are seen to give

$$\theta = \sum_{i=1}^{N} h_i y_i + Z(\theta_\infty - h_{CH_4}), \tag{6.1}$$

$$y_{CO} + y_{CO_2} = Z, \tag{6.2}$$

$$2y_{H_2O} + 2y_{H_2} + y_{OH} + y_H = 4Z, \tag{6.3}$$

$$2y_{O_2} + 2y_{CO_2} + y_{CO} + y_{H_2O} + y_{OH} + y_O = 4Z_s(1 - Z)/(1 - Z_s), \tag{6.4}$$

with y_{CH_4} set equal to zero for this zone. Equation (6.4) has employed Eq. (2.9) with the near-unity factor L_{CH_4}/L_{O_2} inserted, as indicated in the last paragraph of Section 2. At Z_f, with $y_{Hf} = y_{OHf} = y_{Of} = 0$, these equations give

$$\theta_f = Z_f(1 + \theta_\infty) - (h_{H_2O} - h_{H_2})y_{H_2f} - (h_{CO_2} - h_{CO})y_{COf}, \tag{6.5}$$

$$y_{COf} + y_{CO_2f} = Z_f, \tag{6.6}$$

$$y_{H_2f} + y_{H_2Of} = 2Z_f, \tag{6.7}$$

$$2y_{CO_2f} + y_{COf} + y_{H_2Of} = 4Z_s(1 - Z_f)/(1 - Z_s) - 2y_{O_2f}. \tag{6.8}$$

Thus, the concentrations and temperature at Z_f can be determined in terms of Z_f from Eqs. (6.5) through (6.8), with the use of $y_{O_2f}=0$ for partial equilibrium of step II and with water-gas equilibrium, given by

$$y_{CO_2}y_{H_2} = K_{CO_2}y_{H_2O}y_{CO} \tag{6.9}$$

where

$$K_{CO_2} \equiv (K_9/K_3)L_{CO}L_{H_2O}/(L_{H_2}L_{CO_2}), \tag{6.10}$$

while Eqs. (6.1) through (6.4) provide the following relationships at Z_c

$$y_{CO_2c} = Z_c, \tag{6.11}$$

$$y_{H_2Oc} = 2Z_c, \tag{6.12}$$

$$y_{O_2c} = 2(Z_s - Z_c)/(1 - Z_s), \tag{6.13}$$

$$Z_c = \theta_c/(1 + \theta_\infty). \tag{6.14}$$

As mentioned before, the crossover temperature T_c (or θ_c) is determined by setting the ratio

$$\Phi \equiv \frac{\omega_{II}}{\omega_{III} - \omega_O} = \frac{\omega_{1a} - \omega_{1b} + \omega_5\kappa/(1 + \kappa)}{\omega_5 + \omega_{24} + \omega_{25}} \tag{6.15}$$

equal to unity. This temperature has been calculated before [18] exhibiting the influence of the competition of the recombination and branching effects on T_c. In Eq. (6.15), the parameter κ is a measure of additional chain branching and recombination reactions through HO_2 and is given by

$$\kappa \equiv (\omega_6 + \omega_{26})/(\omega_7 + \omega_8) = (k_6 + k_{26})/[k_7 + k_8(c_{OH}/c_H)]. \tag{6.16}$$

The calculations for T_c as a function of pressure are repeated here for the mechanism given in Table II of Chapter 1 and the results are shown in Fig. 3. At atmospheric pressure, for the case $c_{OH}/c_H = 1$ the values $T_c = 940K$ and $Z_c = 0.2$ are obtained. These results agree well with earlier [18] results and are consistent with numerical integrations employing detailed kinetics, which gives $T_c = 1305$ K and $Z_c = 0.27$ at the crossover of ω_{II} and $(\omega_{III} - \omega_O)$ (see Figs. 1 and 2).

To describe the structure in the mixture-fraction region between Z_c and Z_f, we need to solve for the eight independent species and temperature. Equations (6.1) through (6.4) (with fuel chemistry neglected), steady state for O-atom and the partial-equilibrium approximation for steps 3, II, and IV provide eight relationships for the nine unknowns. The other additional relationship can be readily obtained by a linear combination of Eqs. (4.14), (4.15), (4.16), and (4.21) as

$$\frac{d^2}{dZ^2}(y_H + y_{OH} + 2y_O + 2y_{O_2}) = 2r_{III}. \tag{6.17}$$

Because of the low-activation energies of the reactions contributing to ω_{III} and the particular variation of H and O_2 concentrations in this zone, r_{III} can be approximated by a constant, r. For these flames, this argument is supported by numerical results shown in Figs. 1 and 2. With the boundary conditions at Z_c and Z_f [17,18], integration

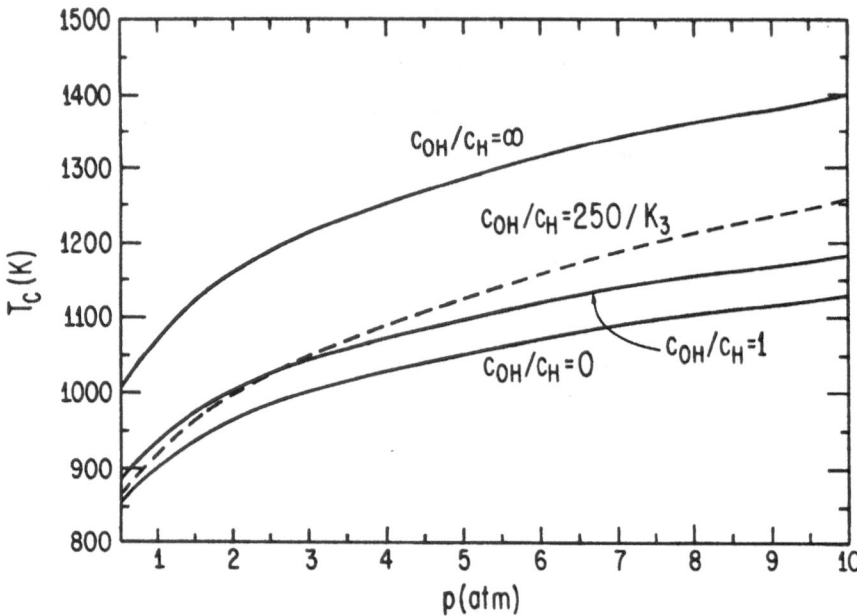

Figure 3: The crossover temperature T_c as a function of pressure for some limiting values of c_{OH}/c_H.

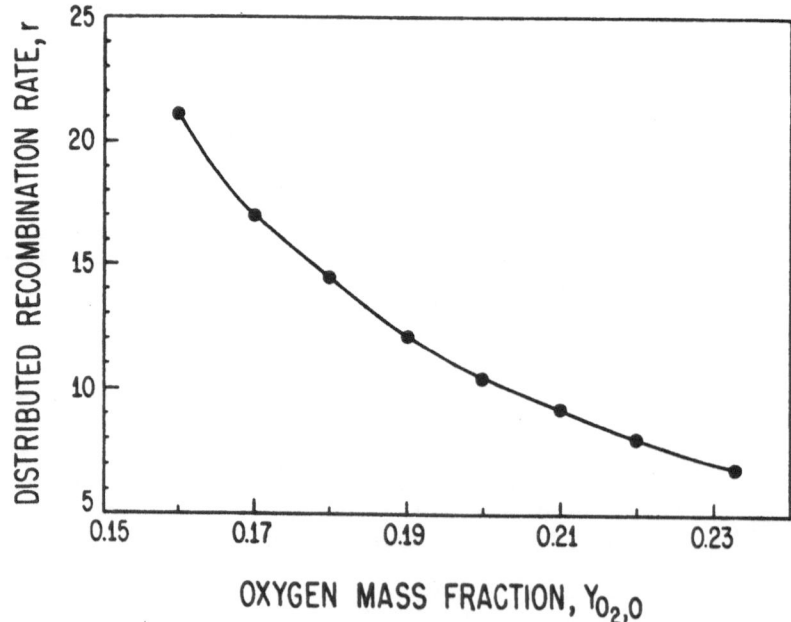

Figure 4: The effect of dilution on the average distributed recombination rate.

of Eq. (6.17) yields

$$y_H + y_{OH} + 2y_O + 2y_{O_2} = r(Z_f - Z)^2 - r(Z_f - Z_c)^2(1 - Z)/(1 - Z_c)$$
$$+ 2[y_{O_2c} + Z_c - (1 + y_{O_2c})Z]/(1 - Z_c) \qquad (6.18)$$

for $Z_c < Z < Z_f$. Use of $y_H + y_{OH} + 2y_O = 0$ at $Z = Z_f$ in Eq. (6.18) gives a cubic equation for Z_f, an approximate solution to which, for small $r(Z_f - Z_c)^2$, is

$$Z_f = \frac{y_{O_2c} + Z_c - y_{O_2f}(1 - Z_c)}{1 + y_{O_2c}} - \frac{r(1 - Z_c)^3(1 + y_{O_2f})(y_{O_2c} - y_{O_2f})^2}{2(1 + y_{O_2c})^4}. \qquad (6.19)$$

For the case where partial equilibrium for step (II) is valid, ie. for low strain rates, there is no oxygen leakage through the flame, thus $y_{O_2f} = 0$ applies. In applying Eqs. (6.18) and (6.19), for simplicity, the contributions from steps 24, 25 and 26 to r_{III} have been neglected and the value of r has been estimated by use of y_H and y_{O_2} profiles calculated for $r = 0$; from these profiles, the integral of the right-hand side of Eq. (6.17) from $Z = Z_c$ to $Z = Z_f$ can be calculated and is set equal to $2r(Z_f - Z_c)$ to estimate r. The resulting estimate of r for different dilutions of the oxidizer and fuel streams with nitrogen in stoichiometric proportions is shown in Fig. 4. Comparisons of Fig. 4 with Fig. 6 of Chelliah and Williams [18] shows somewhat higher values of r at the lower oxygen mass fractions with the present, improved rate data.

From Eq. (3.3) with ω_{HO_2} neglected it is seen that the partial equilibrium requires $\omega_{1b} = \omega_{1a} + \omega_6 + \omega_{26}$, which, from the relationship following Eq. (6.16), can be written as

$$c_{OH}c_O = c_H c_{O_2} \left[k_{1a} + \left(\frac{\kappa}{1 + \kappa}\right) k_5 c_{M_s}\right]/k_{1b}. \qquad (6.20)$$

In Eq. (6.20) c_{OH} is related to c_H through the partial equilibrium of step 3, given by Eq. (4.1), and when this applied with Eq. (4.12), Eq. (6.20) results in

$$y_{H_2O}y_O = K_{II}y_{H_2}y_{O_2} \qquad (6.21)$$

where

$$K_{II} \equiv \left[k_{1a} + \left(\frac{\kappa}{1 + \kappa}\right) k_5 c_{M_s}\right] \frac{K_3}{k_{1b}} \frac{L_{H_2}L_{O_2}}{L_O L_{H_2O}}. \qquad (6.22)$$

In applying Eq. (4.10), the step 26 has been neglected, and the steady state for O reduces to

$$c_O = \frac{(k_{1a}c_{O_2} + k_{2b}c_{OH})c_H + k_{4a}c_{OH}^2}{k_{1b}c_{OH} + k_{2a}c_{H_2} + k_{4b}c_{H_2O}}. \qquad (6.23)$$

These results enable us to calculate species and temperature profiles in mixture-fraction space.

Figure 5 shows representative results for the flame structure with partial equilibrium of step II. The solid lines correspond to $\kappa = 0$ and were calculated as described earlier [17,18]. Shown by dashed lines for some of the curves in Fig. (7.2) are the modifications produced if $\kappa \neq 0$, with the selection $c_{OH}/c_H = 1$. The present results with the mechanism in Table II of Chapter 1 are seen to differ negligibly from those obtained earlier [18]. If the fuel-consumption layer is identified by the maximum value of the rate of step (I) in Fig. 1, then the corresponding value of Z_f from numerical integrations is smaller than that predicted here, but the width of the zone seen in Fig. 1 causes appreciable chemistry to extend to values of Z greater than the Z_f obtained here.

With increasing strain rate, the partial-equilibrium approximation for step (II) breaks down so that the differential equation in Eq. (4.5) must be employed instead of

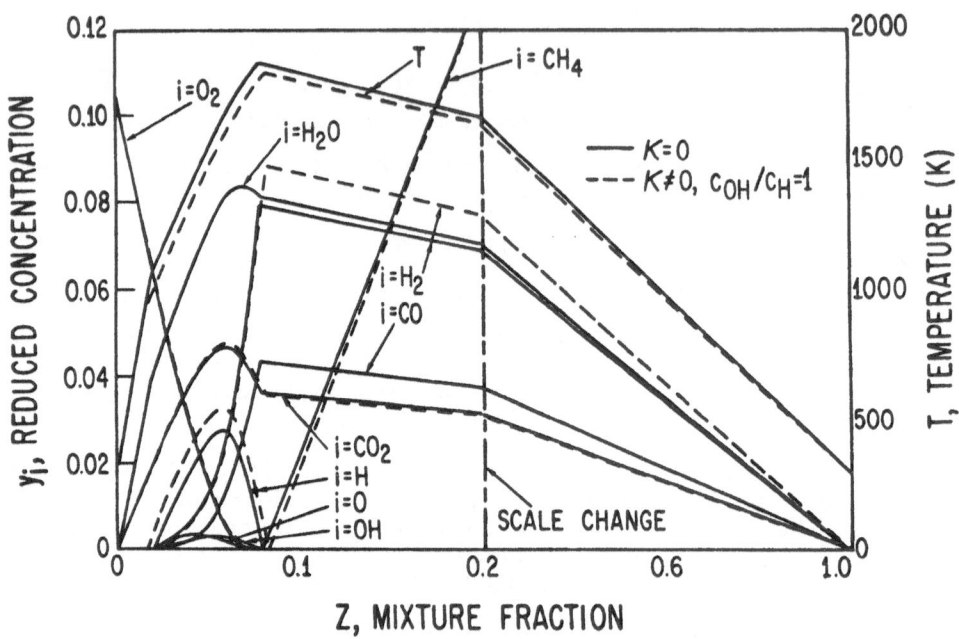

Figure 5: The low-strain structure for the undiluted methane-air flame with and without the influence of the additional HO_2 reactions.

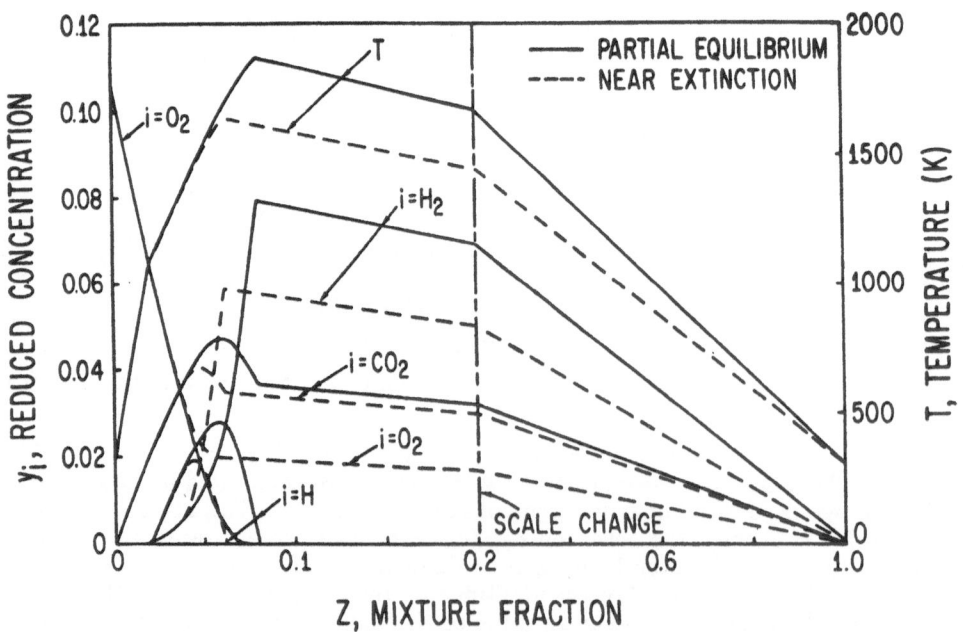

Figure 6: A comparison of the low-strain and high-strain structure for $\kappa = 0$ and $r = 0$.

Eq. (6.21). Following the earlier work [17], Eq. (4.15) is written as

$$\frac{d^2 y_{O_2}}{dZ^2} = R_{II} y_H [K_{II} y_{O_2} - (y_{H_2O}/y_{H_2}) y_O],$$ (6.24)

where

$$R_{II} = \frac{c_p/\lambda}{|\nabla Z|^2} \left[\frac{Y_{CH_4 \infty}}{W_{CH_4}}\right] \left[\frac{p\bar{W}}{RT}\right]^2 \frac{k_{1b}}{K_3} \frac{L_H L_O L_{H_2O}}{L_{H_2} L_{CH_4}}.$$ (6.25)

The breakdown of partial equilibrium of ω_{II} will modify the conditions obtained earlier at $Z = Z_e$ and $Z = Z_f$. Because of the finite rates of ω_{II}, it is necessary to allow for $y_{O_2 f} \neq 0$ due to the diffusion of O_2 into the inert zone on the fuel side, even with infinite rates of fuel consumption at $Z = Z_f$.

Since the effective activation energy in R_{II} is large and negative, it is found that the partial-equilibrium approximation for step II, resulting in the relationship given by Eq. (6.22), is best at the lowest temperatures and will become inaccurate as the temperature increases, especially near Z_f [17]. Therefore to attain simplifications in analyzing the oxygen-consumption zone with breakdown of the partial-equilibrium approximation, a sudden-freezing approximation is introduced [17], in which Eq. (6.22) is recovered from Eq. (6.24) in the region $Z_e < Z < Z_i$ where Z_i is the mixture fraction at the freezing point, while the term involving y_O of Eq. (6.24) (reverse of reaction step II) is neglected for $Z_i < Z < Z_f$. The accuracy of this sudden-freezing approximation is found to improve with increasing effective activation energy for R_{II} [17].

The calculations for conditions for near extinction were repeated here by a method similar to the methods described in the previous studies [17,18]. A comparison between partial-equilibrium and near-extinction profiles is shown in Fig. 6, and the differences between the present predictions and those in ref. [18] again remain negligible. Oxygen leakage, as well as reduced temperatures, radical concentrations, and extent of the oxygen-consumption zone, are evident in approaching extinction and are consistent with detailed numerical calculations. For this limiting strain rate, the fuel-consumption layer is predicted to lie at a mixture fraction quite close to that of the peaks of ω_I found from numerics and shown in Fig. 2. Although the numerical calculations show little variation in the location of this peak with increasing strain rate, the results in Fig. 6 show an appreciable decrease in Z_f, roughly the same as the decrease in the center of gravity of the ω_I curves in Figs. 1 and 2 from numerics.

Figure 7 and 8 shows the influence of r on the flame structure with partial equilibrium of step II and near extinction, respectively, for the undiluted methane-air flame. It is seen that the decrease in radical concentrations (mainly H-atoms) with increasing r results in a decreased Z_f. In the case of the partial-equilibrium structure, this shift of Z_f toward the stoichiometric position Z_s is associated with an increase in flame temperature by about 100K. For near extinction these effects are much smaller. The tendency for the influences of κ and r to cancel is evident in Figs. 7 and 8, although the effect of r generally is the larger. All of these results are in agreement with our earlier [18] findings.

7. Influence of the Variable-Density Mixing Layer on χ

The scalar dissipation rate χ, is defined here and earlier as

$$\chi = (2\lambda/\rho c_p)|\nabla Z|^2,$$ (7.1)

where we have used a Lewis number of unity. For counterflow flames, in terms of the nondimensional coordinate η and nondimensional stream function f, it can be shown

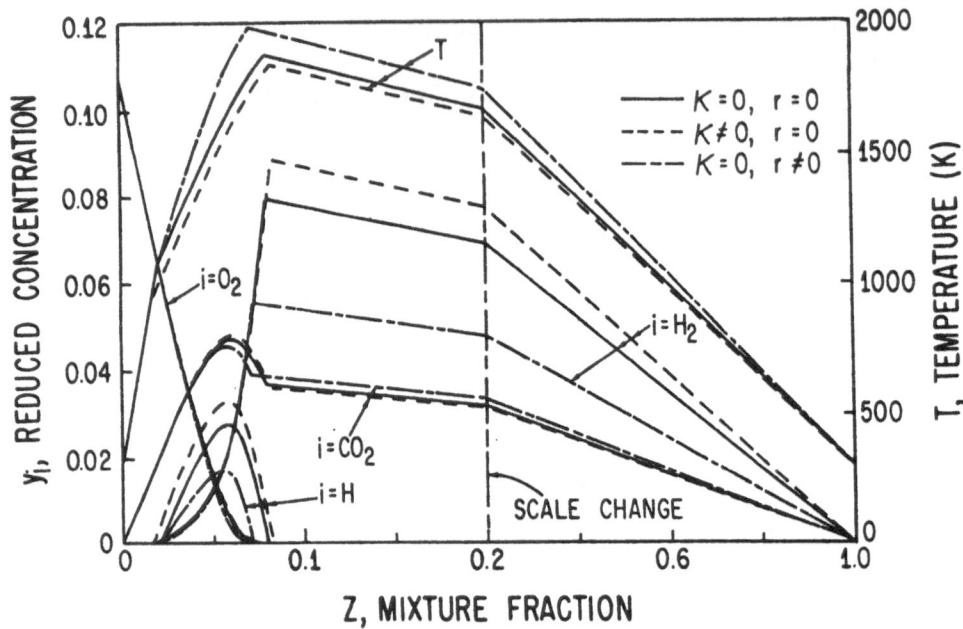

Figure 7: The influence of the distributed recombination (r) and the additional HO_2 reactions (κ) on the low-strain structure of the undiluted methane-air flame.

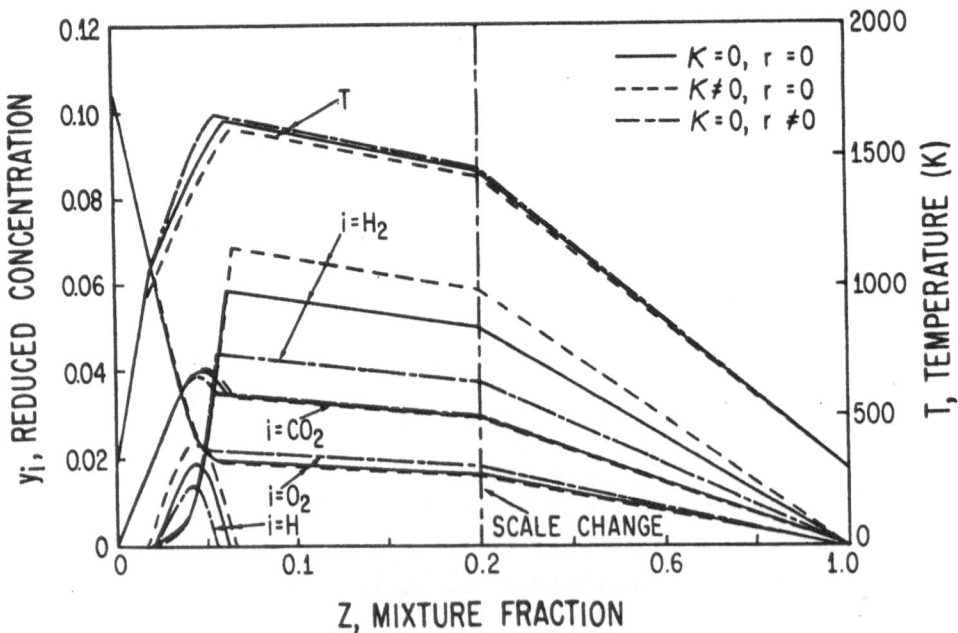

Figure 8: The influence of the distributed recombination (r) and the additional HO_2 reactions (κ) on the high-strain structure of the undiluted methane-air flame.

that near $Z = Z_f$, with the Chapman-Rubesin parameter and the Schmidt number set equal to unity,

$$\chi = 2(dv/dx)_0/[I^2 \exp(2 \int_{\eta_f}^{\eta} f d\eta)], \tag{7.2}$$

where

$$I \equiv \int_{-\infty}^{\infty} \exp(- \int_{\eta_f}^{\eta} f d\eta) d\eta, \tag{7.3}$$

and $(dv/dx)_0$ is the normal gradient of the normal component of velocity in the external oxidizer stream, the imposed strain rate. For flames of interest here, small values of Z_f afford simplification in calculating variations with Z. Near $Z = Z_f$ it is found that Z varies smoothly with η, the first discontinuity appearing in its fourth derivative at $\eta = \eta_f$. It is then found for small Z_f that near $Z = Z_f$

$$\chi = \chi_f \exp[2(Z - Z_f)/Z_f], \tag{7.4}$$

$$\chi_f \equiv 2(dv/dx)_0/I(\infty)^2. \tag{7.5}$$

In order to obtain χ_f from measured strain rates through Eq. (7.5), the value of I is needed. A parametric study of the factor I has been performed recently [24] by considering the mixing of a variable-density, nonreacting counterflow of hot products (at temperature T_f) against cold fuel (at temperature T_∞). For an axisymmetric, counterflow configuration, with the Prandtl number, the Schmidt number and the Chapman-Rubesin parameter set equal to unity, the conservation equations for momentum and energy (or mixture fraction) can be written along the axis of symmetry in the self-similar form

$$f''' + ff'' + \{[\alpha - (\alpha - 1)Z] - f'^2\}/2 = 0, \tag{7.6}$$

$$Z'' + fZ' = 0, \tag{7.7}$$

where the prime denotes differentiation with respect to η. The corresponding boundary conditions for potential flow on both sides of the mixing layer [20] are

$$\begin{aligned} f' &= 1, \quad Z = 1 \text{ as } \eta \to +\infty \\ f' &= \alpha^{1/2} \quad Z = 0 \text{ as } \eta \to -\infty, \end{aligned} \tag{7.8}$$

where $\alpha = T_f/T_\infty$. The numerical solution of this system of ordinary differential equations provides the necessary solution for the nondimensional stream function f as a function of η to evaluate the parametric dependence of I.

Integration of Eq. (7.7) once, with use of Eq. (7.8), yields

$$I = 1/Z_f', \tag{7.9}$$

and a further integration gives a relationship between Z and η as

$$Z = [\int_{-\infty}^{\eta} \exp(- \int_{-\eta_f}^{\eta} f d\eta) d\eta]/I. \tag{7.10}$$

As seen from Eqs. (7.3) and (7.9), I depends on α and η_f (or Z_f). Figure 9 shows a plot of I as a function of Z_f for a range of values of α of interest. With decreasing temperature of the hot products, the estimate of I approaches the constant-density value, given by

$$I_{const} = \sqrt{2\pi} \exp(\eta_f^2/2) \equiv \sqrt{2\pi} \exp([\text{erfc}^{-1}(2Z_f)]^2). \tag{7.11}$$

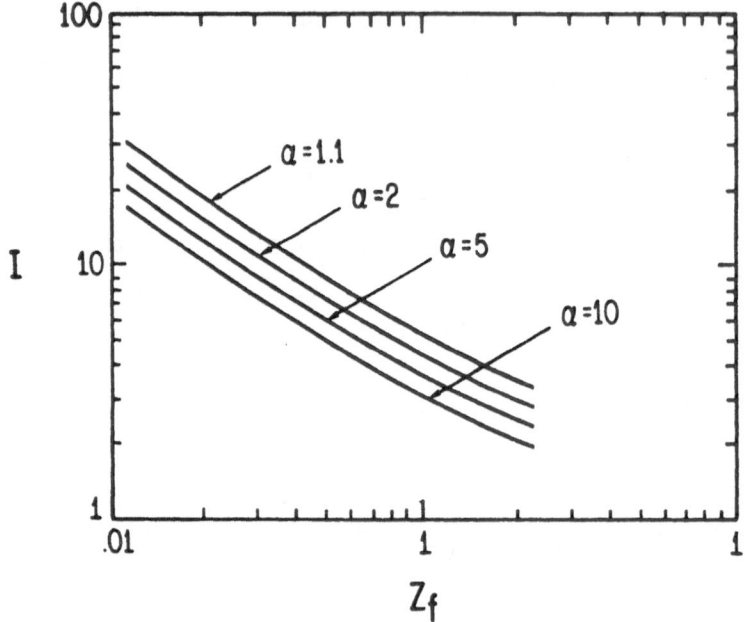

Figure 9: The integral I as a function of Z_f and α.

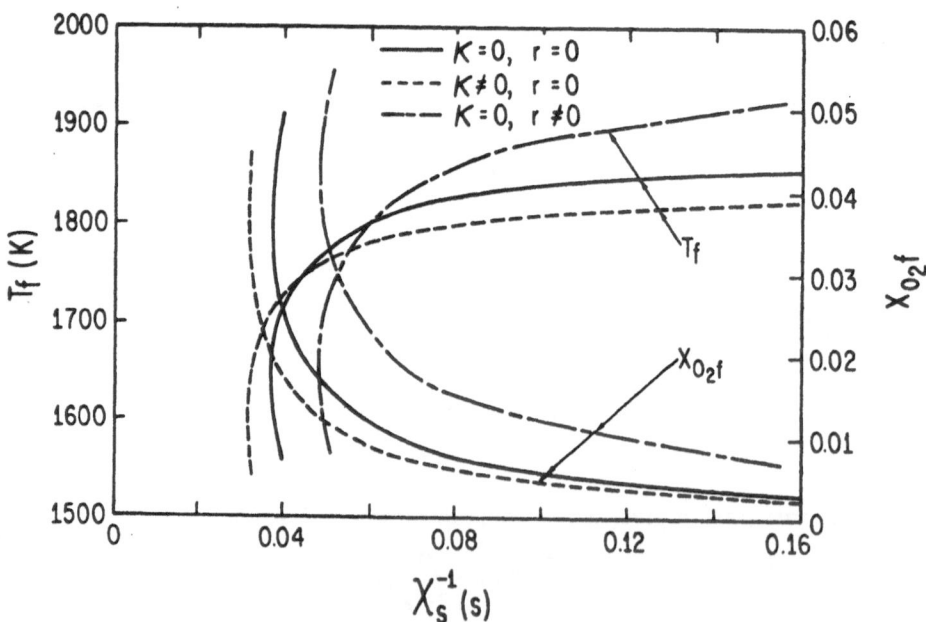

Figure 10: The variation of the flame temperature and oxygen mole fraction at the flame with the scalar dissipation, including the additional influences κ and r.

Since I appears as a square in Eq. (7.5), for methane-air diffusion flames the ratio of I_{const}^2/I^2 can introduce a factor of two error in relating the experimentally measured strain rates to the scalar dissipation rates calculated here. The effect of these results in interpreting the experimental measurements is addressed in the next section. Additional discussion from a different perspective is given elsewhere [24].

8. Extinction Results and Discussion

Methods for predicting the dependences of flame structure on the scalar dissipation rate and for determining the extinction limits have been described in detail elsewhere [18]. After presenting dependences of the flame temperature and the oxygen leakage on the stoichiometric scalar dissipation rate χ_s, calculated with the mechanism in Table II of Chapter 1, detailed comparisons of the extinction predictions with the two asymptotic approaches, with experiments [8] and with numerical integrations [20] are presented. As far as the structure in the Z variable is concerned, there is no ambiguity of the stretch effects because for given χ_s the Damköhler number associated with step (II) is well defined. However, relating χ_s to the imposed strain rate, as seen from Eq. (7.5), can have significant influence depending on the procedure adopted to evaluate the integral I.

Figure 10 shows the influences of κ and r on the variations of T_f and $X_{O_2,f}$ with χ_s. Recombination in the oxygen-consumption zone is seen to lower the extinction strain rate substantially, while the additional HO_2 chain channels increase it somewhat. The two effects cause a shift in Z_f in such a way that increasing r leads to a stronger variation of flame temperature with strain rate, while increasing κ makes it weaker. Since the effect of r exceeds than that of κ, the combined effect is to decrease the extinction strain rate. It has been shown [18] that as extinction is approached, the selection of reaction-rate data affects these predictions appreciably, but the differences between the present predictions with Table II of Chapter 1 and those with the best of the previous data [18] are small.

All preceding results (except Figs. 3 and 4) concern the methane-air diffusion flame for the conditions of the test problem. Further instructive comparisons can be achieved by investigating the dependence of extinction on dilution. Two independent sets of experiments have now been performed [8,20] on dilution effects in the counterflow mixing layer at atmospheric pressure with both ambient streams at about 300 K. In these experiments, methane and air were both diluted with nitrogen in proportions needed to keep the stoichiometric mixture fraction Z_s fixed. Although the experimental extinction results are in rather good agreement with each other [20], somewhat premature extinction at low dilutions (that is for fuel-air systems) appears to have occurred from disturbances in one experiment [8] and slightly delayed extinction at high dilutions, possibly from burner heating, in the other [20]. Both experiments corresponded more closely to the plug-flow boundary conditions of flow between parallel plates than to the potential-flow boundary conditions of Eq. (7.8) [20]. This difference introduces uncertainty in application of the preceding variable-density correction to I for the experiments.

Figure 11 compares various sets of results on the influence of dilution on the scalar dissipation rate χ_s at extinction. The results of the detailed numerical integrations with plug-flow boundary conditions are the most accurate ones for the given experimental conditions. The numerical calculations employed the data in Table II of Chapter 1, without the revision of the rate constant for step 10 described in Chapter 1. In the experiments considered here [8] only the injection velocities were measured, and rotational channel-flow theory [25] was employed to calculate the velocity gradient $(dv/dx)_0$ needed in Eq. (7.5) for obtaining the scalar dissipation rate. Moreover, the χ_f of Eq. (7.5) was assumed to be representative of χ_s, which introduces an error in evaluating I

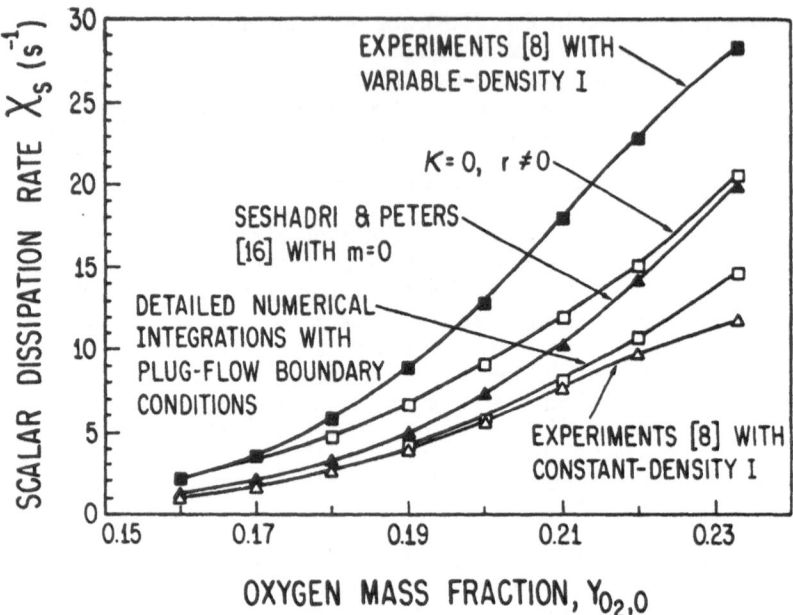

Figure 11: Influences of dilution on the scalar dissipation rate at extinction.

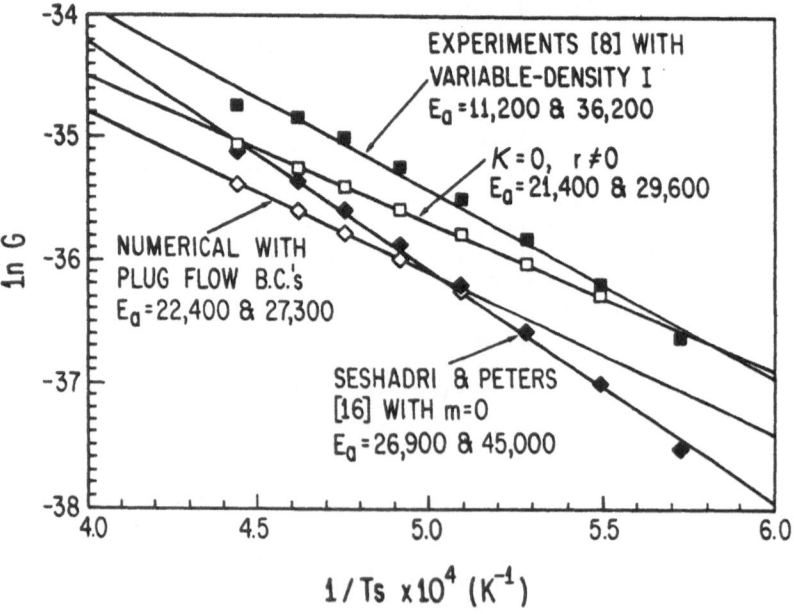

Figure 12: Arrhenius plot for the effective overall activation energy at extinction.

and is discussed in the next paragraph. With this procedure, application of constant-density I to the experimental extinction results is seen from Fig. 11 to produce good agreement with the numerical results, especially in view of the deficiency of this experiment at the higher values of $Y_{O_2 0}$, as mentioned above. This agreement, however, is a fortuitous compensation of canceling errors. From LDV measurements [20], for example, application of the rotational-flow formula is known to produce values of the effective oxidizer-side potential-flow velocity gradient $(dv/dx)_0$ too large by an amount on the order of a factor of two [20]. Therefore an improved rotational-flow description accounting for the displacement effect of the hot-gas boundary layer at the flame, currently under development by Kim and Libby, may be expected to lower the experimental curve for constant-density I in Fig. 11 by about a factor of two and to correspondingly lower the experimental curve for variable-density I, so that the latter achieves closer agreement with the results of the numerical integrations, instead of being the poorest of all, as Fig. 11 erroneously suggests. With this in mind, we see from Fig. 11 that the present theory with $\kappa=0$ and $r \neq 0$ in general is in poorer agreement with the correct extinction results than is the theory of Seshadri and Peters, although the predictions of the two asymptotic theories are quite close to each other at low dilutions.

There are other approaches to testing predictions of scalar dissipation at extinction. One is to employ $(dv/dx)_0$ determined experimentally by LDV [20], thereby removing inaccuracy associated with the use of the rotational-flow formula. It remains necessary to determine I to obtain χ_e from $(dv/dx)_0$, and when this is done with the I from the variable-density description there is rather good agreement with the χ_e from the numerical integration, except at low dilution, where the result from numerics falls approximately half way between the experimental results interpreted by variable-density and constant-density I's. To see whether this difference is a consequence of inaccuracy in application of the variable-density correction, Eq. (7.5) with both constant-density and variable-density I's was applied directly to the results of the numerical integrations, to use the value of $(dv/dx)_0$ obtained from the numerics for calculating χ_e through Eq. (7.5) in the same manner indicated above, which then can be compared with actual value of χ_e obtained from the numerics. When this is done for undiluted methane-air flames, with the result $(dv/dx)_0=391$ s^{-1} from plug-flow boundary conditions [20], it is found that under the assumption $\chi_e = \chi_f$ the variable-density I gives $\chi_e = 20.5$ s^{-1} and constant-density I gives $\chi_e=8.6$ s^{-1}, while the numerical value, $\chi_e = 14.7$ s^{-1}, lies between these values. For the potential-flow boundary conditions where $(dv/dx)_0=509$ s^{-1} [20], corresponding calculations with numerics still gave $\chi_e = 14.7$ s^{-1} as the correct value but $\chi_e=26.7$ and 12.1 s^{-1} with the variable-density and constant-density I, respectively. This last comparison clearly shows that an inherent uncertainty of about a factor of two must be anticipated in deducing χ_e from $(dv/dx)_0$ for real flames. This uncertainty is associated with the strong variation of χ in the vicinity of $Z = Z_s$, shown by numerical integrations [20]. This strong dependence also is seen in Eq. (7.4). Putting numbers into this expression for the conditions of Fig. 8, for example, shows that χ_e is less than χ_f by about a factor of two. The best way to use Eq. (7.5) for real flames therefore is in conjunction with results like those in Fig. 8 to evaluate χ_f from Eq. (7.5) and then relate χ_f to χ_e through Eq. (7.4). The uncertainties in precisely what values should be selected for Z_f for flames with real chemistry are main sources of inaccuracies in determining χ_e (which is inaccessible experimentally without elaborate laser-Raman procedures) from experimental values of $(dv/dx)_0$ by use of the flame-sheet results of Section 7.

Also shown in Fig. 11 is a curve obtained employing the alternative asymptotic approach of Seshadri and Peters [16]. In calculating this curve a short reaction mechanism is employed and the scalar dissipation rate is assumed to be constant across the reaction zone. The reason to select this simple case was that the extinction predictions (for the undiluted methane/air flame) are seen to agree well with the more general model where the variation of scalar dissipation rate is included and an extended reaction mechanism is employed. Comparisons of the two asymptotic approaches show close agreement of the predicted scalar dissipation rate at extinction for the undiluted case as indicated above, but with increasing dilution the differences between the predictions of the two approaches increase. The difference in slope of the two asymptotic approaches is the primary cause for the observed differences of the overall activation energy, seen in the next figure. The water-gas nonequilibrium had no effect in the present approach; however, in the alternate approach it has a significant influence on the extinction predictions.

The Arrhenius graph obtained by varying the dilution is shown in Figure 12. The method for construction of such a graph has been outlined previously [3,18]. It gives a effective overall activation energy on the basis of a one-step approximation as described in Chapter 4. In this figure the straight lines are least-square fits; the activation energies obtained from the two points at each end are shown in the figure, demonstrating that the theoretical and experimental curves exhibit curvature in the same direction. The results of detailed numerical integrations are available only for dilutions up to $X_{O_2,0}=0.19$. Therefore the value of 27,300 indicated in Fig. 12 is from the slopes corresponding to $X_{O_2,0}=0.20$ and 0.19. The present prediction of extinction activation energy is seen to agree well with the numerics for the range considered. The experimental curve is less accurate, as indicated above, while the alternate approach [16] predicts values of Ea much higher than the present approach at high dilutions. It is surprising that the present results agree well with the numerics concerning Ea, considering the fact that fuel chemistry was approximated by a flame sheet even though the Figs. 1 and 2 have shown that the fuel-consumption zone is fairly broad and extends much into the oxygen consumption zone. This agreement is fortuitous since Fig. 11 clearly shows that the asymptotic approximations of Seshadri and Peters [16] are the better of the two at high dilutions.

9. Conclusion

The purpose of this study has been to present an asymptotic approach to describe the structure and extinction of methane-air diffusion flames employing the mechanism in Table II of Chapter 1, to explore the effect of variable density in the interpretation of experimental results, and to make detailed comparisons with experiments and numerics. The present theory is seen to predict well the trend of the extinction scalar dissipation rates with dilution, while the alternative approach [16], in which the fuel-consumption zone is coupled to the rest of the structure, is seen to give better extinction predictions at higher dilutions.

The interpretation of experimentally measured strain rates is shown to depend on the method by which the integral I of Eq. (7.3) is evaluated. When the results for a variable-density mixing layer are used, the extinction scalar dissipation rates are increased by about factor of two over the values obtained with constant density. In addition, it appears that the detailed numerical predictions of scalar dissipation rates agree better with the experimental results when variable density is considered, if inaccuracies in currently available rotational-flow descriptions are taken into account.

It is very likely that rates of the fuel chemistry do influence extinction conditions but do not change the trend predicted here, which is seen to agree well with the numerical integrations. Based on present comparisons, the results suggest that the correct model lies between the two limiting cases considered here, and to confirm this asymptotic analyses of diffusion flames must be carried out for conditions under which the parameter ω of Seshadri and Peters [19] is order of unity.

References

[1] Liñán, A., "The asymptotic structure of counterflow diffusion flames for large activation energies", *Acta Astronautica*, **1**, (1974), p. 1007.

[2] Krishnamurthy, L., Williams, F.A., and Seshadri, K., "Asymptotic theory of diffusion flame extinction in the stagnation-point boundary layers", *Comb. and Flame*, **26**, (1976), p. 363.

[3] Williams, F.A., "A review of flame extinction", *Fire Safety Journal*, **3**, (1981), p. 163.

[4] Peters, N., "Local quenching due to flame stretch and non-premixed turbulent combustion", *Comb. Sci. and Tech.*, **30**, (1983), p. 1.

[5] Tsuji, H., and Yamaoka, I., "The structure of counterflow diffusion flames in the forward stagnation region of a porous cylinder", *Twelfth Symposium (International) on Combustion*, The Combustion Institute, (1969), p. 997.

[6] Tsuji, H., and Yamaoka, I., "The structure analysis of counterflow diffusion flames in the forward stagnation region of a porous cylinder", *Thirteenth Symposium (International) on Combustion*, The Combustion Institute, (1971), p. 723.

[7] Ishizuka, S., and Tsuji, H., "An experimental study of effect of inert gases on extinction of laminar diffusion flames", *Eighteenth Symposium (International) on Combustion*, The Combustion Institute, (1981), p. 695.

[8] Puri, I.K., and Seshadri, K., "Extinction of diffusion flames burning diluted methane and diluted propane in diluted air", *Comb. and Flame*, **65**, (1986), p. 137.

[9] Smooke, M.D., Puri, I.K., and Seshadri, K., "A comparison between numerical calculations and experimental measurments of the structure of a counterflow diffusion flame burning diluted methane in diluted air", *Twenty first Symposium (International) on Combustion*, The Combustion Institute, (1988), p. 1783.

[10] Dixon-Lewis, G., Fukutani, S., Miller, J.A., Peters, N., Warnatz, J., "Calculation of the structure and extinction limit of a methane-air counterflow diffusion flame in the forward region of a porous cylinder", *Twentieth Symposium (International) on Combustion*, The Combustion Institute, (1985), p. 1893.

[11] Miller, J.A., Kee, R.J., Smooke, M.D., and Grcar, J.F., "The computation of the structure and extinction limit of a methane-air stagnation point diffusion flame", Paper WSS/CI 84-10, *Western States Section of the Combustion Institute*, April, (1984).

[12] Dixon-Lewis, G., David, T., and Gaskell, P.H., "Structure and properties of methane -air and hydrogen-air counterflow diffusion flames", *Archivum Combustionis*, vol-6 (1986), No. 1.

[13] Puri, I.K., Seshadri, K., Smooke, M.D., and Keyes, D.E., "A comparison between numerical calculations and experimental measurments of the structure of a counterflow methane-air diffusion flame", *Comb. Sci. and Tech.*, **56**, (1987), p. 1.

[14] Bilger, R.W., and Kee, R.J., "Simplified kinetics for diffusion flames of methane in air", *Joint Conference Western States and Japanese Sections of the Combustion Institute*, Honolulu, Hawaii, (1987), p. 277.

[15] Peters, N., and Kee, R.J., "The computation of stretched laminar methane-air diffusion flames using reduced four-step mechanism", *Comb. and Flame*, **68**, (1987), p. 17.

[16] Seshadri, K., and Peters, N., "Asymptotic structure and extinction of methane-air diffusion flames", *Comb. and Flame*, **73**, (1988), p. 23.

[17] Treviño, C. and Williams, F.A., "An asymptotic analysis of the structure and extinction of methane-air diffusion flame", *Dynamics of Reactive Systems, Part I: Flames*, (A.L. Kuhl, J.R. Bowen, J.-C. Leyer, and A. Borisov, Eds.), Progress in Astronautics and Aeronautics, **113**, AIAA, Washington DC, (1988), p. 129.

[18] Chelliah, H.K., and Williams, F.A., "Aspects of the Structure and Extinction of Diffusion Flames in Methane-Oxygen-Nitrogen Systems", *Comb. and Flame*, **80**, (1990), p. 17.

[19] Seshadri, K., and Peters, N., "The Inner Structure of Methane-Air flames", *Comb. and Flame*, **81**, (1990), p. 96.

[20] Chelliah, H.K., Law, C.K., Ueda, T., Smooke, M.D., and Williams, F.A., "An experimental and theoretical investigation of of the dilution, pressure and flow-field effects on the extinction of methane-air-nitrogen diffusion flames", to appear in the proceedings of *Twenty-Third Symposium (Int.) on Combustion*.

[21] Williams, F.A., "Crocco variables for diffusion flames", in *Recent Advances in the Aerospace Science* (C. Casci, editor), Plenum Press, New York, p. 415.

[22] Williams, F.A., *Combustion Theory*, 2nd Ed., Addison-Wesley Publishing Co., Menlo Park, CA, (1985).

[23] Peters, N., and Williams, F.A., "The asymptotic structure of stoichiometric methane-air flames", *Comb. and Flame*, **68**, (1987), p. 185.

[24] Kim, J.S, and Williams, F.A., "Theory of counterflow mixing of fuel with hot products", *Comb. Sci. and Tech.*, **73**, (1990), p. 575.

[25] Seshadri, K., and Williams, F.A., "Laminar flow between parallel plates with injection of a reactant at high Reynolds number", *J. of Heat Mass Transfer*, **21**, (1978), p. 251.

CHAPTER 8

SENSITIVITY ANALYSIS OF LAMINAR PREMIXED CH$_4$-AIR FLAMES USING FULL AND REDUCED KINETIC MECHANISMS

Bernd Rogg
University of Cambridge, Department of Engineering,
Trumpington Street, Cambridge CB2 1PZ, UK
Cambridge, England

1. Introduction

The systematical reduction of detailed kinetic mechanism of elementary reactions to only a few global steps initiated by Peters [1] and Peters and Williams [2] has opened an avenue for chemically realistic modelling of a variety of reactive-flow problems which, from the combustion-chemistry point of view, previously could be dealt with only in an unsatisfactory manner. For instance, before the advent of systematically reduced kinetic mechanisms, simple global one-step and two-step models were employed to analyse laminar reacting flows in all but the simplest flow geometries, and in turbulent combustion outside the laminar-flamelet regime simple models of the eddy-breakup type were used to formulate turbulent reaction rates. Whilst, unjustifiedly, the situation is still unchanged with respect to modelling chemistry-flow interactions in turbulent combustion, much research has been devoted to derive reduced kinetic mechanisms for a variety of fuels, see e.g. [2-7], and to use them to analyse laminar combustion problems. In most of the reduced-mechanism-based laminar-flame studies theoretical methods were used as a tool for analysis, see e.g. [2,3,5,8-10] where also further references to theoretical work are given, whereas up to date only few numerical studies, e.g. [6,7,11,12], have been carried out even for the simplest geometries. With few exceptions [7,11], the various numerical studies using systematically reduced kinetic mechanisms have been performed with a computer program developed by Rogg.

The reasons why systematically reduced kinetic mechanism are not yet widely used in numerical simulations of combustion problems lie in various numerical difficulties associated with these mechanisms. It is the purpose of this chapter to *initiate* a systematical compilation and analysis of these difficulties and to give *first* hints to their resolution. Specifically, in this exploratory study we use a so-called "first-order sensitivity analysis" as tool to identify and analyze *some* of the numerical difficulties encountered when using systematically reduced mechanisms in numerical investigations of flame structures; further studies are in progress in which also other tools are employed.

Most applications of sensitivity analysis have centred about the sensitivity of laminar reacting systems with respect to the rates of the elementary steps appearing in the reaction mechanism, see, e.g., [13,14]. Notable exceptions are Smooke et al. [15], who investigated the sensitivity of laminar premixed flames with respect to transport coefficients, and Bockhorn [16] and Rogg [17], who investigated the sensitivity of turbulent combustion systems with respect to various input parameters. Herein we investigate the sensitivity of the structure of a freely propagating premixed laminar flame with respect to the kinetic data of a detailed mechanism of elementary steps and, alternatively, with respect to the kinetic data of a systematically reduced mechanism consisting of only four global reactions. From the results of the sensitivity analyses we draw preliminary conclusions on the suitability of Newton's method as the numerical-solution method in conjunction with a systematically reduced kinetic mechanism.

2. Conservation Equations and Boundary Conditions

The steady, planar, adiabatic deflagration is considered. Mass conservation is $\rho v = \rho_u v_u$, where ρ denotes the density and v the velocity, and where the subscript u is used to identify conditions in the unburnt gas. Momentum conservation is $p = \text{constant}$, where p denotes the pressure. The conservation equations for energy and species mass are

$$\rho v \frac{dT}{dx} = \frac{1}{c_p} \frac{d}{dx}\left(\lambda \frac{dT}{dx}\right) - \frac{dT}{dx} \sum_{i=1}^{N}\left(\frac{c_{pi}}{c_p}\rho Y_i V_i\right) - \frac{1}{c_p} \sum_{i=1}^{N} h_i w_i, \tag{2.1}$$

$$\rho v \frac{dY_i}{dx} = -\frac{d}{dx}\left(\rho Y_i V_i\right) + w_i, \quad i = 1, ..., N-1, \tag{2.2}$$

where Y_i is the mass fraction of species i, V_i its diffusion velocity, w_i its mass rate of production and h_i its enthalpy of formation at temperature T; c_p denotes the mixture's specific heat at constant pressure and λ its thermal conductivity. The mixture consists of N chemical species, and Eq. (2.2) is applied to the first $N-1$ of them, the mass fraction Y_N being obtained by difference. The boundary conditions for Eqs. (2.1) and (2.2) are

$$\begin{aligned} T = T_u, \quad Y_i = Y_{iu}, \quad i = 1, ..., N-1 \quad \text{at} \quad x = -\infty, \\ dT/dx = dY_i/dx = 0, \quad i = 1, ..., N-1 \quad \text{at} \quad x = \infty. \end{aligned} \tag{2.3}$$

The system of conservation equations (2.1) and (2.2) is closed by the ideal-gas equation of state,

$$\frac{p}{\rho} = R^0 T \sum_{i=1}^{N} \frac{Y_i}{W_i}, \tag{2.4}$$

where R^0 denotes the universal gas constant and W_i the molecular weight of species i, $i = 1, \ldots, N$.

3. Models for Thermochemistry and Molecular Transport

Although in numerical simulations it is straightforward (and therefore often done, see e.g. [6,12,15]) to employ detailed models for thermochemistry and molecular transport, the results presented in this chapter have been obtained using the simpler models recommended by Smooke in chapter 1 of this book. Specifically, the diffusion fluxes have been approximated by Fick's law,

$$\rho Y_i V_i = -\rho D_i \frac{dY_i}{dx} = -\frac{\lambda/c_p}{Le_i} \frac{dY_i}{dx}, \tag{3.1}$$

where to derive the second equality we have defined the Lewis number of species i as $Le_i = \lambda/(\rho c_p D_i)$. The Lewis numbers are taken as constants with numerical values of 0.97, 1.10, 0.83, 1.39, 0.30, 0.18, 0.70, 0.73, 1.10 1.12, 1.11, 1.00, 1.28, 1.27, and 1.30 for CH_4, O_2, H_2O, CO_2, H_2, H, O, OH, HO_2, H_2O_2, CO, CH_3, CH_2O, CHO, and CH_3O, respectively. In the Workshop announcement no numerical value was given for Le_{N_2}. It was found that the computed flame structures are nearly insensitive with respect to the value of Le_{N_2} for which, finally, the value 1.0 was adopted.

For the ratio λ/c_p appearing in Eq. (3.1) the relationship

$$\frac{\lambda}{c_p} = 2.58 \times 10^{-5} \left(\frac{T}{298}\right)^{0.7} \qquad (3.2)$$

was adopted, with T in Kelvin and λ/c_p in units of kg/m.

The specific heat c_p, which appears in Eq. (3.2) as well as in the energy equation (2.1) has been calculated from the definition

$$c_p = \sum_{i=1}^{N} Y_i c_{pi} . \qquad (3.3)$$

The specific heats c_{pi} appearing in Eq. (3.3) and the specific enthalpies h_i appearing in the energy equation (2.1) have been calculated from the NASA thermochemical polynomials

$$c_{pi} = \sum_{k=1}^{5} a_{k,i} T^{k-1} \text{ and } h_i = a_{k,0} + \sum_{k=1}^{5} a_{k,i} T^k /k . \qquad (3.4)$$

The coefficients $a_{k,i}$ appearing in Eqs. (3.4) may be found in various sources, e.g. in Appendix C of Gardiner [18].

4. Chemistry of Methane-Air Flames

Detailed Mechanism

In this chapter both a mechanism of elementary reactions and a mechanism consisting of four global reaction steps are employed in the numerical simulations of freely propagating, premixed laminar methane-air flames. The mechanism of elementary reactions is given in Table 1. It is the so-called "skeletal" mechanism proposed for the Workshop on "Reduced Kinetic Mechanisms and Asymptotic Approximations for Methane-Air Flames" held at the University of California at San Diego, La Jolla, 14th and 15th March 1989; it is reproduced here to make this chapter self-contained. The skeletal

No.	Reaction		A	α	E
1	$H + O_2$	$\rightarrow OH + O$	2.00E14	0.00	70.30
2	$OH + O$	$\rightarrow H + O_2$	1.57E13	0.00	2.89
3	$O + H_2$	$\rightarrow OH + H$	1.80E10	1.00	36.93
4	$OH + H$	$\rightarrow O + H_2$	8.00E09	1.00	28.29
5	$OH + H_2$	$\rightarrow H_2O + H$	1.17E09	1.30	15.17
6	$H_2O + H$	$\rightarrow OH + H_2$	5.09E09	1.30	77.78
7	$OH + OH$	$\rightarrow H_2O + O$	6.00E08	1.30	0.00
8	$H_2O + O$	$\rightarrow OH + OH$	5.90E09	1.30	71.26
9	$H + O_2 + M$	$\rightarrow HO_2 + M$	2.30E18	-0.80	0.00
10	$HO_2 + H$	$\rightarrow OH + OH$	1.50E14	0.00	4.20
11	$HO_2 + H$	$\rightarrow H_2 + O_2$	2.50E13	0.00	2.93
12	$HO_2 + OH$	$\rightarrow H_2O + O_2$	2.00E13	0.00	4.18
13	$CO + OH$	$\rightarrow CO_2 + H$	1.51E07	1.30	-3.17
14	$CO_2 + H$	$\rightarrow CO + OH$	1.57E09	1.30	93.74
15	$CH_4 + M$	$\rightarrow CH_3 + H + M$	6.30E14	0.00	435.19
16	$CH_3 + H + M$	$\rightarrow CH_4 + M$	5.20E12	0.00	-5.48
17	$CH_4 + H$	$\rightarrow CH_3 + H_2$	2.20E04	3.00	36.61

162

18	$CH_3 + H_2$	$\rightarrow$	$CH_4 + H$	9.57E02	3.00	36.61
19	$CH_4 + OH$	$\rightarrow$	$CH_3 + H_2O$	1.60E06	2.10	10.29
20	$CH_3 + H_2O$	$\rightarrow$	$CH_4 + OH$	3.02E05	2.10	72.90
21	$CH_3 + O$	$\rightarrow$	$CH_2O + H$	6.80E13	0.00	0.00
22	$CH_2O + H$	$\rightarrow$	$CHO + H_2$	2.50E13	0.00	16.70
23	$CH_2O + OH$	$\rightarrow$	$CHO + H_2O$	3.00E13	0.00	5.00
24	$CHO + H$	$\rightarrow$	$CO + H_2$	4.00E13	0.00	0.00
25	$CHO + M$	$\rightarrow$	$CO + H + M$	1.60E14	0.00	61.51
26	$CH_3 + O_2$	$\rightarrow$	$CH_3O + O$	7.00E12	0.00	107.34
27	$CH_3O + H$	$\rightarrow$	$CH_2O + H_2$	2.00E13	0.00	0.00
28	$CH_3O + M$	$\rightarrow$	$CH_2O + H + M$	2.40E13	0.00	120.56
29	$HO_2 + HO_2$	$\rightarrow$	$H_2O_2 + O_2$	2.00E12	0.00	0.00
30	$H_2O_2 + M$	$\rightarrow$	$OH + OH + M$	1.30E17	0.00	190.39
31	$OH + OH + M$	$\rightarrow$	$H_2O_2 + M$	9.86E14	0.00	-21.22
32	$H_2O_2 + OH$	$\rightarrow$	$H_2O + HO_2$	1.00E13	0.00	7.53
33	$H_2O + HO_2$	$\rightarrow$	$H_2O_2 + OH$	2.86E13	0.00	137.21
34	$H + OH + M$	$\rightarrow$	$H_2O + M$	2.20E22	-2.00	0.00
35	$H + H + M$	$\rightarrow$	$H_2 + M$	1.80E18	-1.00	0.00

Table I: Skeletal mechanism of elementary reactions for laminar C_4-air flames. The rate constants are written as $k = AT^\alpha \exp(-E/R^0T)$, with the individual quantites expressed in cm, mol, s, kJ and K units.

The third-body efficiencies are 6.50 for CH_4, 0.40 for O_2, 1.50 for CO_2, 6.50 for H_2O, 0.75 for CO, 0.40 for N_2, and 1.00 for all other species.

The rate data given for steps 15 and 16 are for the high-pressure value k_∞. The rate constants k for these steps are given by the Lindemann form $k = k_\infty/[1 + (cR^0T/p)]$, where $c = 0.517\exp(-9000/T)$, with p in bar and T in Kelvin.

mechanism consists of 35 elementary steps amongst the 15 reacting species CH_4, O_2, CO_2, H_2O, CO, H_2, H, OH, O, HO_2, CH_3, CHO, CH_2O, CH_3O, H_2O_2, and nitrogen, which is taken as inert. For this mechanism, the rates of production of the chemical species involved may be written as

$$w_i = W_i \sum_{k=1}^{M} (\nu''_{i,k} - \nu'_{i,k})(AT^\alpha)_k \, exp(-\frac{E_k}{R^0T}) \prod_{j=1}^{N} \left(\frac{\rho Y_j}{W_j}\right)^{\nu'_{j,k}}, \qquad (4.1)$$

$i = 1, ..., N$, where it is understood that forward and backward chemical reactions are treated separately. Besides quantities already defined above, in Eq. (4.1) M denotes the number of elementary reactions contained in the mechanism, and $\nu'_{i,k}$ and $\nu''_{i,k}$ are the stoichiometric coefficients of species i in reaction k, $k = 1, ..., M$, representing there reactant and product, respectively; $(AT^\alpha)_k$ and E_k are the pre-exponential factor in the specific reaction-rate constant and the activation energy of reaction k, respectively.

Systematically Reduced Mechanism

The systematically reduced mechanism employed in the numerical simulations is the 4-step scheme derived by Peters [1], consisting of the global reactions

$$CH_4 + 2H + H_2O = CO + 4H_2, \tag{I}$$
$$CO + H_2O = CO_2 + H_2, \tag{II}$$
$$2H + M = H_2 + M, \tag{III}$$
$$O_2 + 3H_2 = 2H + 2H_2O. \tag{IV}$$

Apart from N_2, which is taken as inert, this mechanism comprises the seven reacting species CH_4, O_2, CO_2, H_2O, CO, H_2 and H. The mass rates of production of these species may be written as

$$
\begin{aligned}
w_{CH_4} &= -W_{CH_4}\omega_I, \\
w_{O_2} &= -W_{O_2}\omega_{IV}, \\
w_{CO_2} &= W_{CO_2}\omega_{II}, \\
w_{H_2O} &= -W_{H_2O}(\omega_I + \omega_{II} - 2\omega_{IV}), \\
w_{CO} &= W_{CO}(\omega_I - \omega_{II}), \\
w_{H_2} &= W_{H_2}(4\omega_I + \omega_{II} + \omega_{III} - 3\omega_{IV}), \\
w_H &= -W_H(2\omega_I + 2\omega_{III} - 2\omega_{IV}), \\
w_{N_2} &= 0.
\end{aligned}
\tag{4.2}
$$

Here ω_I, ω_{II}, ω_{III} and ω_{IV} denote the global-reaction rates, which are given by

$$\omega_I = k_{17}[CH_4][H]\left(1 + \frac{k_{19}}{k_{17}K_I}\frac{[H_2O]}{[H_2]}\right),$$

$$\omega_{II} = \frac{k_{13}}{K_I}\frac{[H]}{[H_2]}\left([CO][H_2O] - \frac{[H_2][CO_2]}{K_{II}}\right),$$

$$\omega_{III} = \frac{k_{17} + k_{19}[H_2O]/(K_I[H_2])}{k_{24}[H] + k_{25}[M]}k_{24}[CH_4][H]^2 \tag{4.3}$$
$$+ k_9[H][O_2][M] + \frac{k_{34}}{K_I}\frac{[H]^2[H_2O][M]}{[H_2]},$$

$$\omega_{IV} = k_1[H]\left([O_2] - \frac{[H]^2[H_2O]^2}{[H_2]^3 K_{IV}}\right),$$

where the equilibrium constants K_I, K_{II} and K_{III} are defined as

$$
\begin{aligned}
K_I &\equiv k_5/k_6, \\
K_{II} &\equiv k_6 k_{13}/(k_5 k_{14}), \\
K_{III} &\equiv k_1 k_3 k_5^2/(k_2 k_4 k_6^2).
\end{aligned}
\tag{4.4}
$$

Note that in the skeletal and, therefore, in the reduced mechanism the step

$$CHO + O_2 \rightarrow CO + HO_2$$

is not included, wheres in Peters [1] it is. Also note that, following [1], in the reduced mechanism we have set the rate of elementary step 10 to zero.

At this point it is instructive to examine the functional form of the global rates ω_I to ω_{IV}. It is seen that in ω_I there appears the term

$$\frac{k_{19}}{K_I}\frac{[CH_4][H][H_2O]}{[H_2]}, \tag{4.5}$$

which may cause numerical difficulties if during the course of the numerical solution of the governing equations, for instance in a particular iteration step, $[H_2]$ attains values sufficiently close to zero to make this term (numerically) singular. The same argument holds for the term

$$\frac{k_{13}}{K_I} \frac{[H][CO][H_2O]}{[H_2]} \tag{4.6}$$

appearing in the expression for ω_{II}, the term

$$\frac{k_{34}}{K_I} \frac{[H]^2[H_2O][M]}{[H_2]} \tag{4.7}$$

appearing in the expression for ω_{III}, and the term

$$\frac{k_1}{K_{IV}} \frac{[H]^3[H_2O]^2}{[H_2]^3} \tag{4.8}$$

appearing in the expression for ω_{IV}. Note that, as $[H_2]$ tends to zero, from a mathematical point of view the terms in Eqs. (4.5) to (4.8) should be of the type [0/0] and tend towards a well defined finite value. Clearly, in portions of a chemically reacting flow where zero molecular-hydrogen concentrations prevail, for instance in a stream of pure methane, the present formulation of the above reduced mechanism with $[H_2]$ appearing in the denominator of various terms is of no use and, therefore, alternative formulations must be sought. In the computations done up to date only such geometries have been considered which, eventually after applying a similarity transformation, have allowed a formulation of the governing equation with only a single space-like coordinate. As a consequence, as boundary conditions for H_2 zero mass-flux fractions rather than zero mass fractions could be imposed, thereby avoiding zero H_2 concentrations within the domain of integration. However, the introduction of the mass-flux fraction is a "trick" which, by definition of this quantity, works for one-dimensional problems only, and hence cannot be used in truly two-dimensional laminar reactive-flow simulations based on the above reduced kinetic mechanism. This problem, and other numerical issues, will be discussed further below.

5. Numerical-Solution Method

Equations (2.1) and (2.2) with the conditions of Eqs. (2.3) were solved numerically by the method of Rogg, see e.g. [19-21]. Specifically, Newton's method is applied to the system of nonlinear equations,

$$\mathbf{F}(\mathbf{U}) = \mathbf{0}, \tag{5.1}$$

which results from the discretization of the governing equations on a non-uniform grid. Thus, the linear system

$$J(\mathbf{U}^k)(\mathbf{U}^{k+1} - \mathbf{U}^k) = -\omega_k \mathbf{F}(\mathbf{U}^k), \quad k = 0, 1, \ldots , \tag{5.2}$$

is solved where $\mathbf{U}^k$ denotes the solution after k Newton iterations, and ω_k and $J(\mathbf{U}^k)$ are the damping parameter [22] and the Jacobian matrix, respectively, based on $\mathbf{U}^k$. The Jacobian is generated numerically, and is re-evaluated only periodically [23]. A static adaptive-gridding procedure [24] is implemented to bound the local discretization error.

6. Results for Flame Structure and Discussion

Shown in Figs. 1 to 9 is the structure of an atmospheric-pressure, stoichiometric methane-air flame with an initial temperature of 300 Kelvin. In these figures, filled

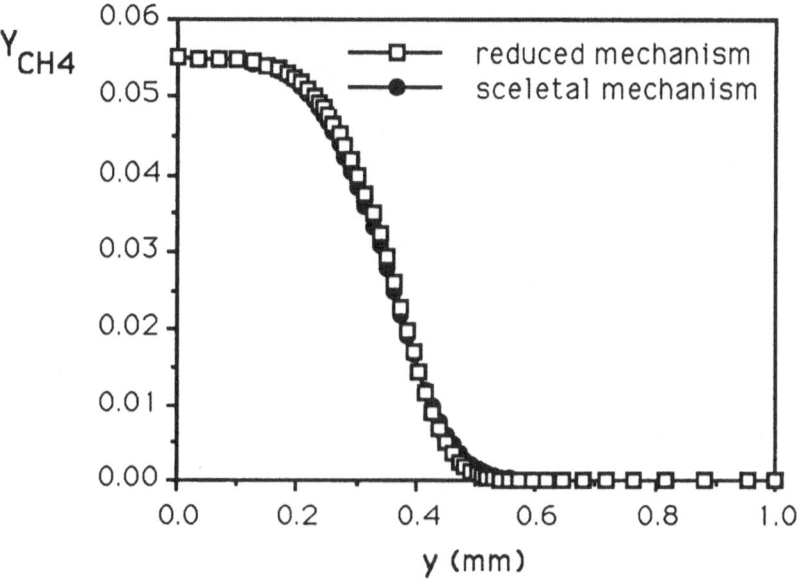

Figure 1: *Computed CH$_4$ mass fraction profiles for an atmospheric-pressure, stoichiometic methane-air flame with an initial temperature of 300 K. Filled circles: sceletal mechanism; void squares: reduced mechanism.*

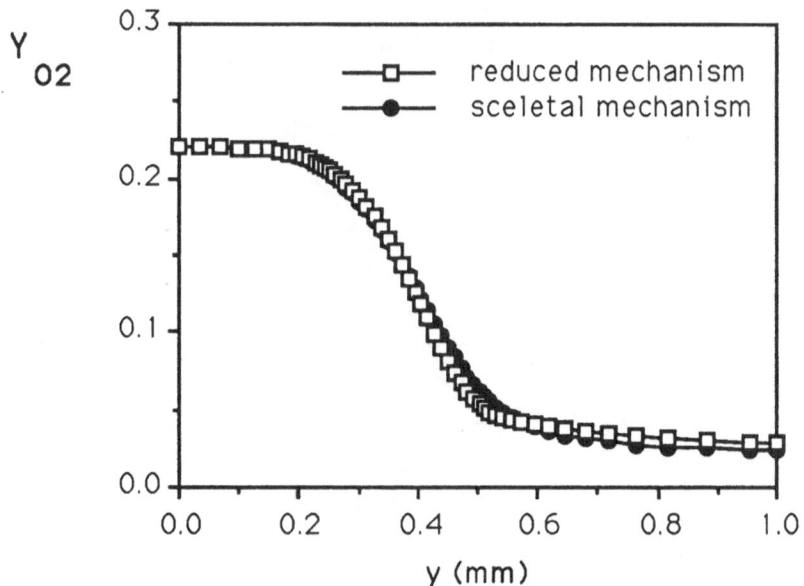

Figure 2: *As Fig. 1 but O$_2$ mass fraction profiles.*

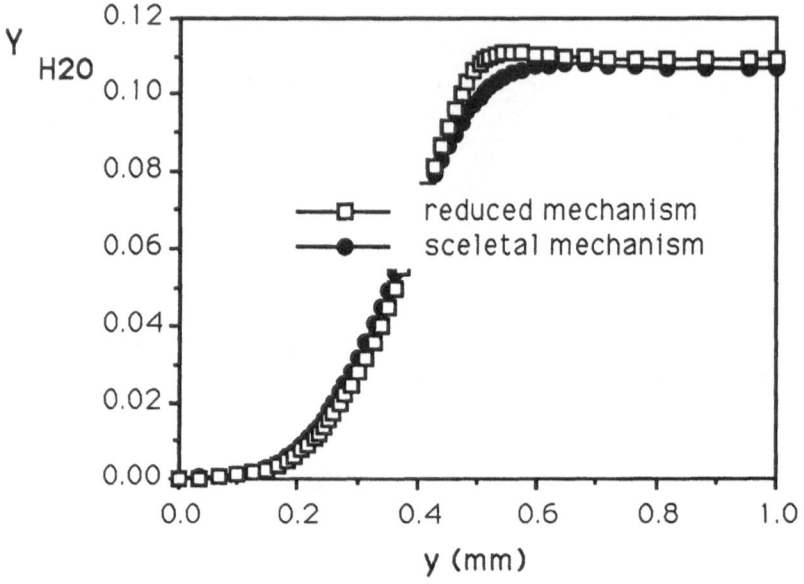

Figure 3: As Fig. 1 but H_2O mass fraction profiles.

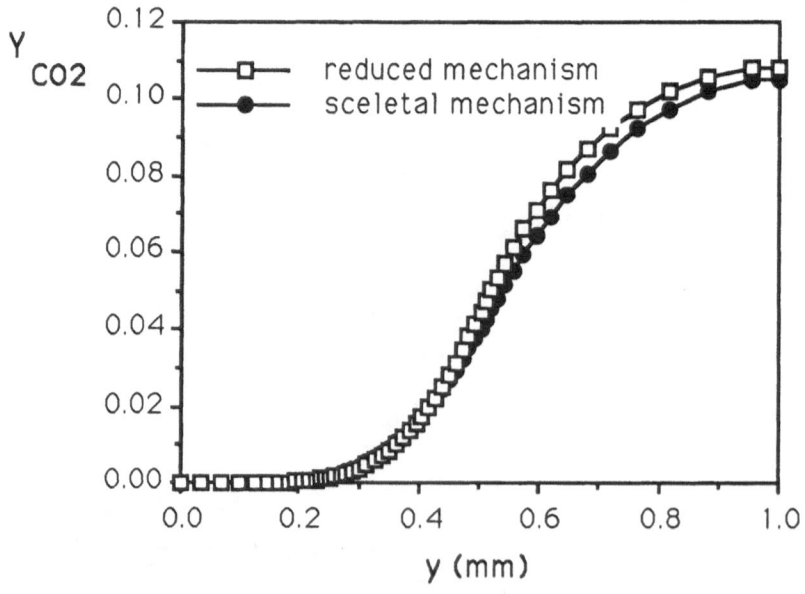

Figure 4: As Fig. 1 but CO_2 mass fraction profiles.

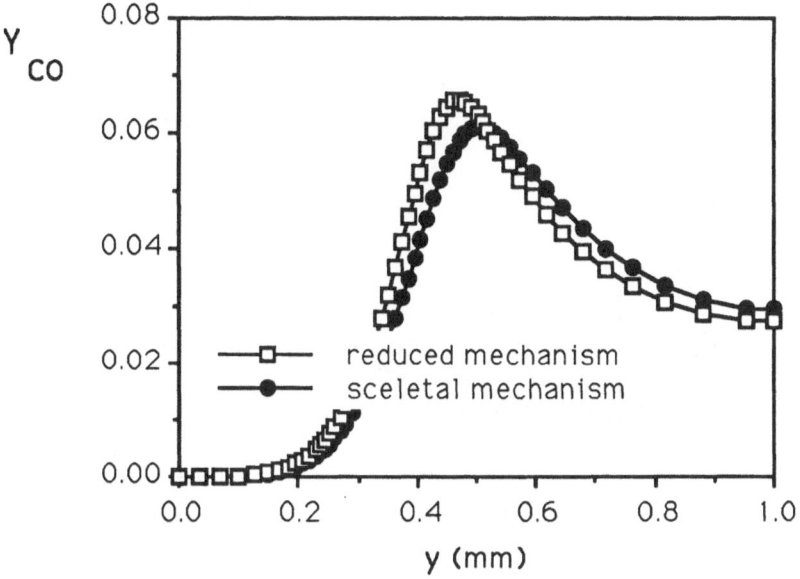

Figure 5: As Fig. 1 but CO mass fraction profiles.

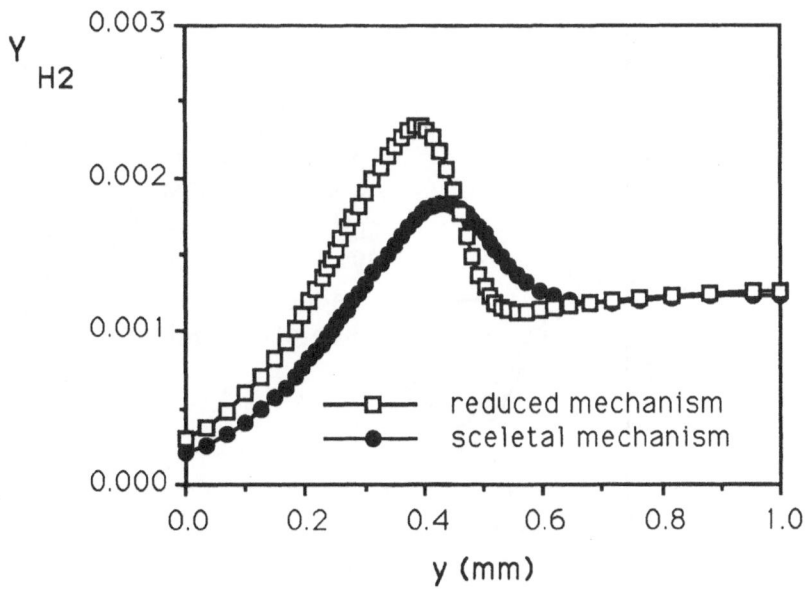

Figure 6: As Fig. 1 but H_2 mass fraction profiles.

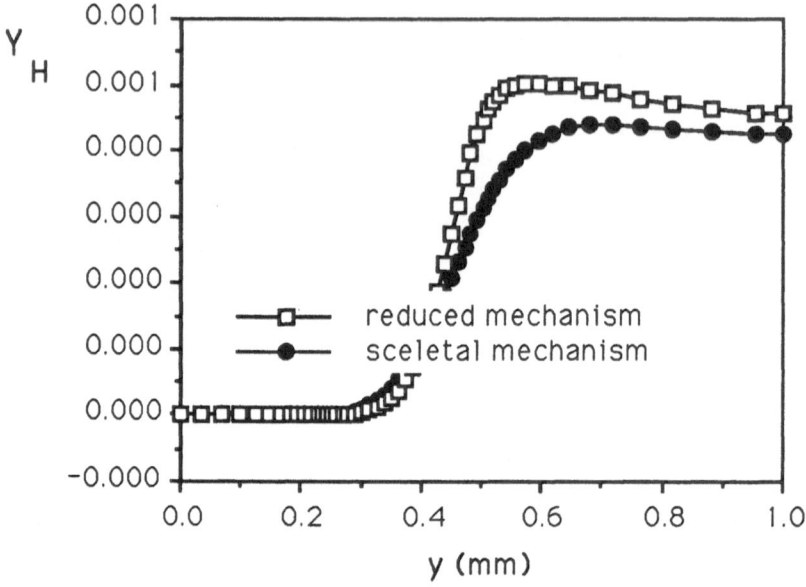

Figure 7: As Fig. 1 but H mass fraction profiles.

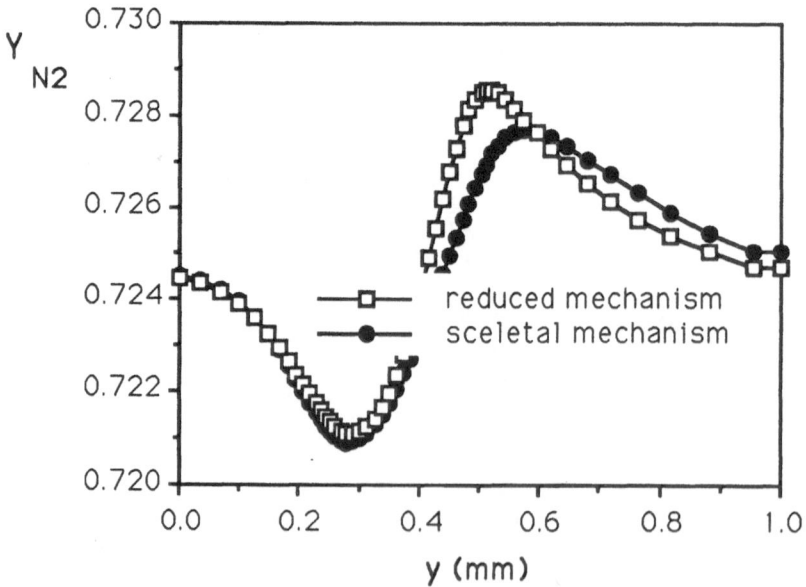

Figure 8: As Fig. 1 but N_2 mass fraction profiles.

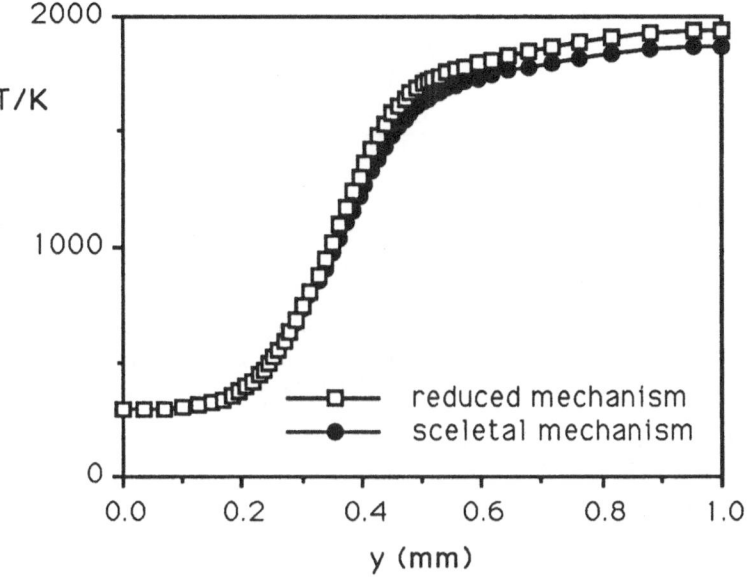

Figure 9: As Fig. 1 but temperature profiles.

circles identify results obtained with the skeletal mechanism, void squares denote results obtained with the systematically reduced mechanism. With both mechanisms for the burning velocity the (too) high value of 57 cm/s is predicted. Shown in Figs. 1 to 4 are the profiles through the flame of the reactants, methane and molecular oxygen, and of the two major products, carbon dioxide and water. For these chemical species, as well as for the temperature shown in Fig. 9, the agreement between the results obtained using the two different kinetic mechanisms is seen to be good to excellent. Figures 5 to 7 show some discrepancies between the predicted concentrations of CO, H_2 and H, particularly in the temperature range between roughly 1500 and 2000 K. However, discrepancies of this kind and order of magnitude are consequences of the approximations underlying the reduced mechanism, and were observed and discussed earlier [1,12]. Note that the cause for the prediction of the high burning velocity cannot be attributed to these discrepancies, because practically identical burning velocities are obtained from both mechanisms.

7. First-Order Sensitivity Analysis

To predict the effect of the variation of rate constants $k_1, k_2, \cdots, k_M$, which conveniently are written in vector form

$$\mathbf{C} \equiv (k_1, k_2, \cdots, k_M)^T, \tag{7.1}$$

on the vector of dependent variables $\mathbf{U}$, the vector of first-order sensitivity coefficients $\partial \mathbf{U}/\partial k_\alpha$, $\alpha=1,...,M$, must be calculated. The appropriate equation for the latter quantities is obtained by differentiating Eq. (5.1) with respect to $\mathbf{C}$, viz.,

$$\frac{d}{dk_\alpha}\left(\mathbf{F}(\mathbf{U};\mathbf{C})\right) = \frac{\partial \mathbf{F}}{\partial \mathbf{U}}\frac{\partial \mathbf{U}}{\partial k_\alpha} + \frac{\partial \mathbf{F}}{\partial k_\alpha} = \mathbf{0}\,, \tag{7.2}$$

$\alpha = 1\cdots,M$. Equation (7.2) can be rewritten as

$$J(\mathbf{U})\frac{\partial \mathbf{U}}{\partial k_\alpha} = -\frac{\partial \mathbf{F}}{\partial k_\alpha}. \tag{7.3}$$

Thus the first-order sensitivity coefficients $\partial \mathbf{U}/\partial k_\alpha$ can readily be calculated from Eq. (7.3) by solving a linear system of equations. In the remainder of this section we derive the sensitivities of various quantities with respect to the rate constants.

Mass-Fraction and Temperature Sensitivities

The sensitivities of the mass fractions, $\partial Y_i/\partial k_\alpha$, and the temperature, $\partial T/\partial k_\alpha$, are obtained by solving Eq. (7.3) without further manipulations. The relative sensitivities

$$S_i^\alpha \equiv \frac{k_\alpha}{Y_i}\frac{\partial Y_i}{\partial k_\alpha}\,, \tag{7.4}$$

$\alpha = 1,...,M$, and

$$S_T^\alpha \equiv \frac{k_\alpha}{T}\frac{\partial T}{\partial k_\alpha} \tag{7.5}$$

are obtained straightforwardly from the "raw" sensitivities by normalisation.

Mixture-Molecular-Weight Sensitivities

The mixture molecular weight $\overline{W}$ is given by

$$\overline{W} = \left(\sum_{j=1}^{N}\frac{Y_j}{W_j}\right)^{-1} = \sum_{j=1}^{N}X_jW_j\,, \tag{7.6}$$

where to derive the second equality in Eq. (7.6) we have used the relationship

$$Y_i = \frac{W_i}{\overline{W}}X_i\,, \tag{7.7}$$

Differentiation yields the raw sensitivity coefficient

$$\frac{\partial \overline{W}}{\partial k_\alpha} = -\overline{W}\sum_{j=1}^{N}\frac{\overline{W}}{W_j}\frac{\partial Y_j}{\partial k_\alpha}. \tag{7.8}$$

which upon normalisation can be written as

$$S_{\overline{W}}^\alpha \equiv \frac{k_\alpha}{\overline{W}}\frac{\partial \overline{W}}{\partial k_\alpha} = -\sum_{j=1}^{N}X_j\left(\frac{k_\alpha}{Y_j}\frac{\partial Y_j}{\partial k_\alpha}\right). \tag{7.9}$$

Density Sensitivities

The density ρ is related to the pressure p and the species mass fractions Y_j through the ideal-gas equation of state, Eq. (2.4). Differentiation of the latter equation with respect to k_α gives, after rearrangement,

$$\frac{\partial \rho}{\partial k_\alpha} = -\rho \Big[\sum_{j=1}^{N} \Big(\frac{\overline{W}}{W_j} \frac{\partial Y_j}{\partial k_\alpha} \Big) + \frac{1}{T} \frac{\partial T}{\partial k_\alpha} \Big], \tag{7.10}$$

where we have used Eq. (7.9). The relative sensitivity is given by

$$S_\rho^\alpha \equiv \frac{k_\alpha}{\rho} \frac{\partial \rho}{\partial k_\alpha} = -\Big[\sum_{j=1}^{N} \Big(X_j \frac{k_\alpha}{Y_j} \frac{\partial Y_j}{\partial k_\alpha} \Big) + \frac{k_\alpha}{T} \frac{\partial T}{\partial k_\alpha} \Big]. \tag{7.11}$$

Mole-Fraction Sensitivities

The species conservation equations given in section 2 are formulated in terms of the mass fractions of the chemical species. Upon using Eq. (7.7), the sensitivity of Y_i with respect to k_α can be written as

$$\frac{\partial Y_i}{\partial k_\alpha} = \frac{\partial X_i}{\partial k_\alpha} \frac{W_i}{\overline{W}} - \frac{X_i W_i}{\overline{W}^2} \frac{\partial \overline{W}}{\partial k_\alpha}, \tag{7.12}$$

which by virtue of Eq. (7.8), and after rearrangement, yields

$$\frac{\partial X_i}{\partial k_\alpha} = \frac{\overline{W}}{W_i} \frac{\partial Y_i}{\partial k_\alpha} - Y_i \sum_{j=1}^{N} \frac{\overline{W}^2}{W_i W_j} \frac{\partial Y_j}{\partial k_\alpha}. \tag{7.13}$$

By virtue of Eq. (7.7), from Eq. (7.13) the normalized sensitivity coefficient of species i with respect to parameter k_α is obtained as

$$\frac{k_\alpha}{X_i} \frac{\partial X_i}{\partial k_\alpha} = \frac{k_\alpha}{Y_i} \frac{\partial Y_i}{\partial k_\alpha} - \sum_{j=1}^{N} X_j \Big(\frac{k_\alpha}{Y_j} \frac{\partial Y_j}{\partial k_\alpha} \Big). \tag{7.14}$$

Velocity Sensitivities

For steady, unstrained, premixed laminar flames in low-Mach-number flows conservation of overall mass implies

$$m \equiv \rho v = \text{constant.} \tag{7.15}$$

Differentiation of Eq. (7.15) gives

$$\frac{\partial m}{\partial k_\alpha} = \rho \frac{\partial v}{\partial k_\alpha} + v \frac{\partial \rho}{\partial k_\alpha},$$

which by virtue of Eq. (7.10) can be written as

$$\frac{\partial v}{\partial k_\alpha} = \frac{m}{\rho} \Big[\sum_{j=1}^{N} \Big(\frac{\overline{W}}{W_j} \frac{\partial Y_j}{\partial k_\alpha} \Big) + \frac{1}{T} \frac{\partial T}{\partial k_\alpha} \Big] + \frac{1}{\rho} \frac{\partial m}{\partial k_\alpha}. \tag{7.16}$$

Thus, the relative velocity sensitivity is given by

$$S_v^\alpha \equiv \frac{k_\alpha}{v}\frac{\partial v}{\partial k_\alpha} = \sum_{j=1}^{N}\left(X_j\frac{k_\alpha}{Y_j}\frac{\partial Y_j}{\partial k_\alpha}\right) + \frac{k_\alpha}{T}\frac{\partial T}{\partial k_\alpha} + \frac{k_\alpha}{m}\frac{\partial m}{\partial k_\alpha}. \tag{7.17}$$

Note that the relative sensitivity of the burning velocity with respect to rate constant k_α,

$$\frac{k_\alpha}{v_u}\frac{\partial v_u}{\partial k_\alpha}, \tag{7.18}$$

is simply the value of S_v^α at the left boundary of the computational domain, i.e., the value of S_v^α as $x \to -\infty$.

8. Results of First-Order Sensitivity Analysis and Discussion

Shown in Fig. 10 are the relative sensitivities of the burning velocity v_u with respect to the rate constants for the flame discussed in section 6. The full bars represent the

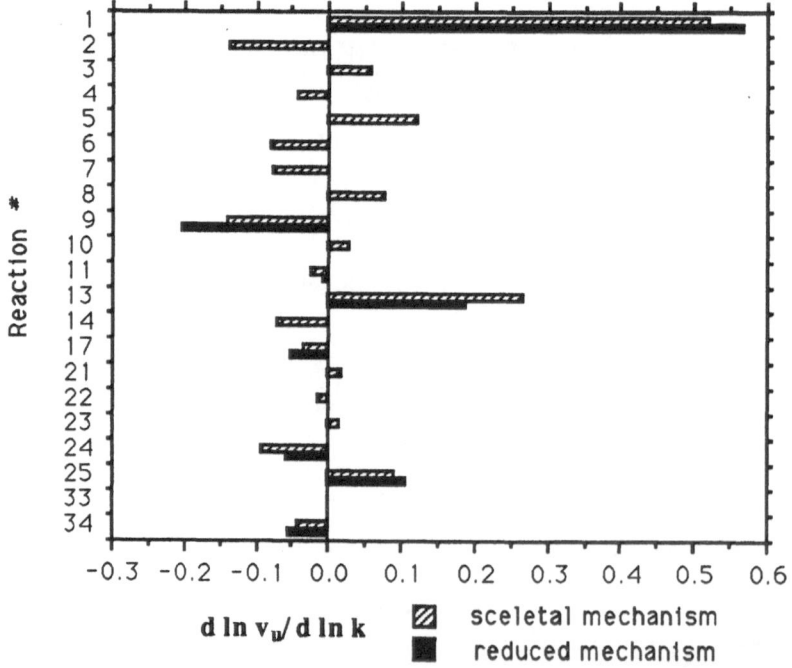

Figure 10: Computed relative first-order sensitivity coefficients of the burning velocity with respect to the rate constants for an atmospheric-pressure, stoichiometric methane-air flame with an initial temperature of 300 K. Only sensitivities greater than one percent are shown. Full bars: sceletal mechanism; hatched bars: reduced mechanism.

sensitivities based on the reduced mechanism, the hatched bars those based on the skeletal mechanism. Note that a sensitivity of x percent with respect to a particular rate constant k_α indicates that, as a response to a 100 percent change of k_α, v_u is estimated to change by x percent. Only burning-velocity sensitivities greater than 1 percent are included in the graph.

Figure 10 shows two interesting features. The first is that the sensitivities with respect to those elementary steps which are contained in both the detailed and the reduced mechanism are quite similar not only with respect to the sign but also with respect to the magnitude. In particular the sensitivities with respect to the key branching step 1,

$$H + O_2 \to OH + O \,,$$

and its low-temperature competitor, step 9,

$$H + O_2 + M \to HO_2 + M \,,$$

agree well, as do the sensitivities with respect to the watergas-shift reaction, step 13,

$$CO + OH \to CO_2 + H \,.$$

Note that for the reduced mechanism Fig. 10 shows zero sensitivities with respect to steps 2, 3, 4, 5, 6 and 14, because in this mechanism the rates of these steps are lumped into the equilibrium constants K_I, K_{II} and K_{III}. Herein sensitivities with respect to the equilibrium constants are not considered; this will be done, however, in future in studies.

The second interesting, although not too surprising, feature to be observed from Figure 10 is that steps are neglected in the reduced mechanism which have a *non-negligible* influence on the burning velocity. For instance, neglected steps are step 7,

$$OH + OH \to H_2O + O \,,$$

and its reverse, step 8,

$$H_2O + O \to OH + OH \,,$$

and steps 21, 22 and 23,

$$CH_3 + O \to CH_2O + H \,,$$
$$CH_2O + H \to CH_3 + O \,,$$
$$CH_2O + OH \to CHO + H_2O \,.$$

Thus, the good agreement in the predicted burning velocities using the detailed and the reduced mechanism is fortuitous in that the burning-velocity sensitivities with respect to steps 7 and 8 approximately cancel, as do those with respect to steps 21 to 23.

Shown in Figs. 11 to 40 are the profiles of the relative mass-fraction sensitivities of those species which are contained in both the skeletal and the reduced mechanism, and of the temperature and velocity. Figures 11 to 20 show the sensitivity profiles with respect to step 1, $H + O_2 \to OH + O$, Figs. 21 to 30 those with respect to its competitor 9, $H + O_2 + M \to HO_2 + M$, and Figs. 31 to 40 those with respect to the watergas-shift reaction 13, $CO + OH \to CO_2 + H$. Prior to examining Figs. 11 to 40 in detail it is worthwhile to recall that in general, as a consequence of the normalization, the absolute value of a relative sensitivity with respect to a particular quantity becomes large (typically greater than unity) when the quantity itself attains a small value (typically close to zero).

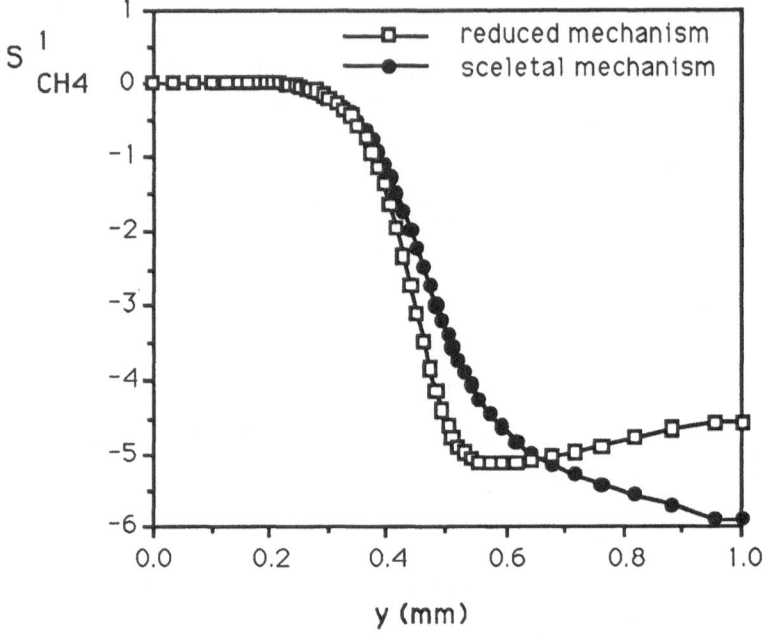

Figure 11: Computed profile of relative first-order sensitivity of the CH_4 mass fraction with respect to elementary step 1 for an atmospheric-pressure, stoichiometic methane-air flame with an initial temperature of 300 K. Filled circles: sceletal mechanism; void squares: reduced mechanism.

Step 1 involves the chemical species H, O_2, O and OH of which only H and O_2 appear explicitly in the reduced mechanism. Thus, as it is to be expected on physical grounds, Figs. 11 to 18 show that the mass fractions of H and O_2 (the sensitivities of Y_O and Y_{OH} are not shown as these species do not appear in the reduced mechanism) are most sensitive with respect to changes in the rate constant of this step, k_1, whereas the mass fractions of the other species depend only indirectly on k_1 through their interaction with H, O_2, O and OH in other reaction steps. It is seen from Figs. 12 and 17 that although both O_2 and H are consumed in step 1, an increase of k_1 leads to a local decrease of Y_{O_2} but to a local increase of Y_H. This increase of Y_H occurs because step 1 is branching, which implies that an increased k_1 entails increased concentrations of all the radicals contained in the hydrogen-oxygen radical pool. The change in sign of S_H^1 at approximately $y = 0.23$ mm, which for the reduced mechanism is seen in Fig. 17, is the consequence of an undershoot of the H-atom concentration to small negative values close to the cold boundary.

Since step 1 is a branching step with a high activation energy, an increase of its rate leads to a temperature increase in the high-temperature portion of the flame (Fig. 19) and, hence, to an increase of the velocity (Fig. 20).

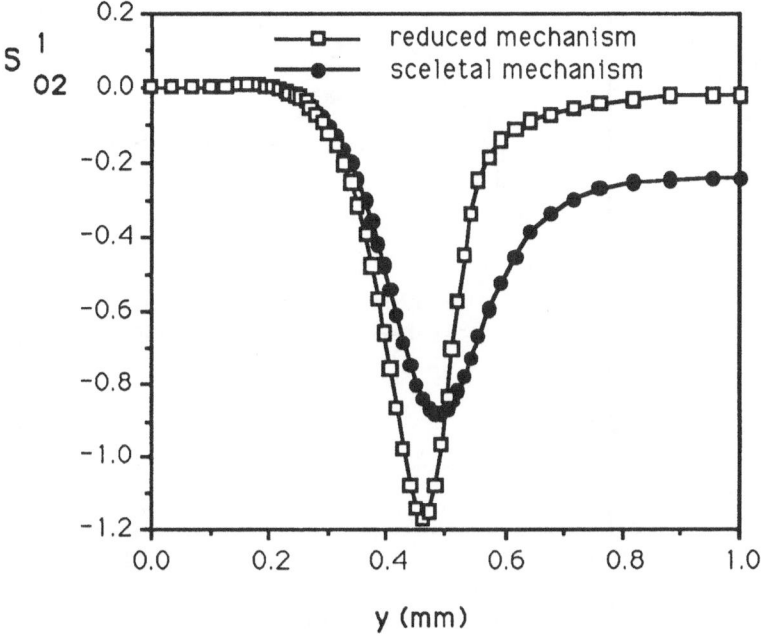

Figure 12: As Fig. 11, but for O_2.

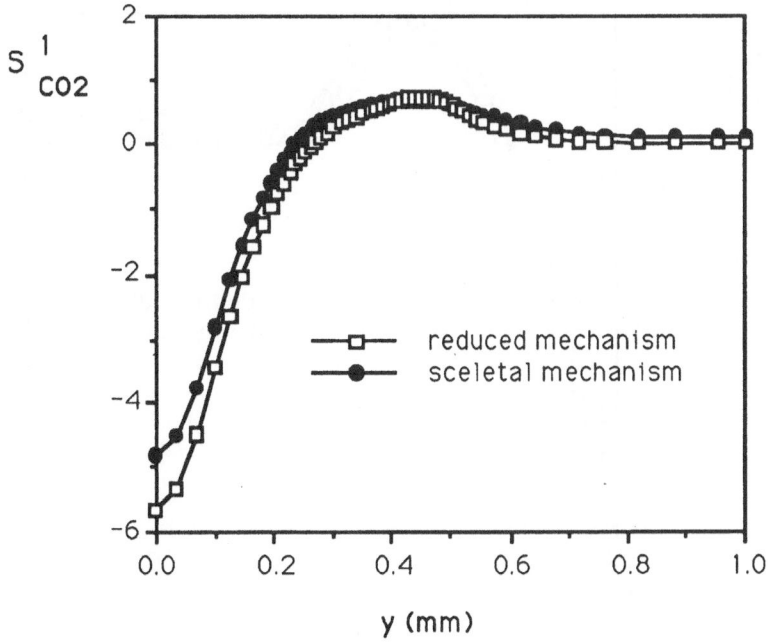

Figure 13: As Fig. 11, but for CO_2.

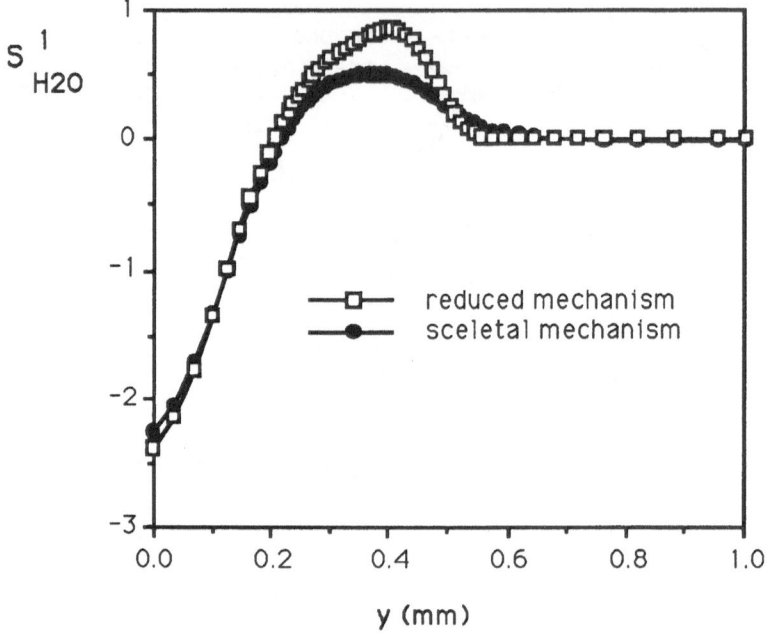

Figure 14: As Fig. 11, but for H_2O.

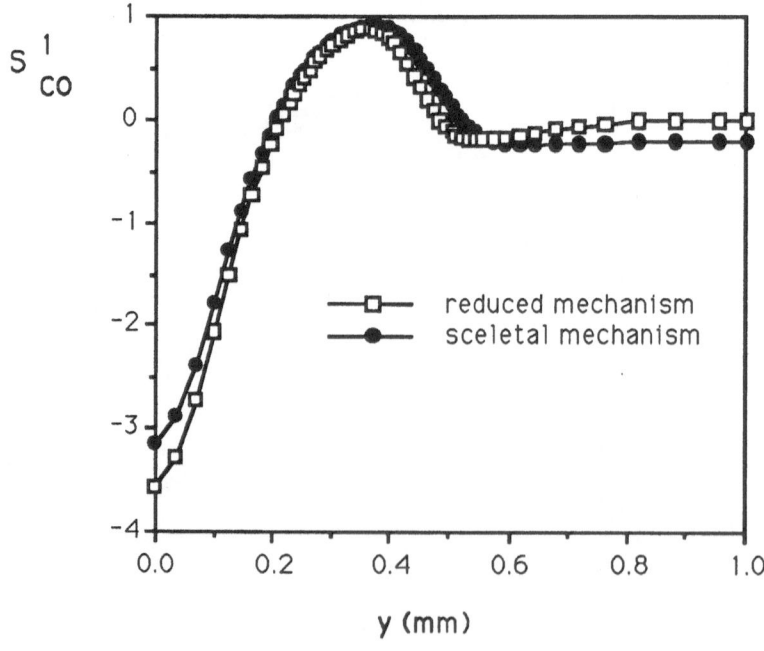

Figure 15: As Fig. 11, but for CO.

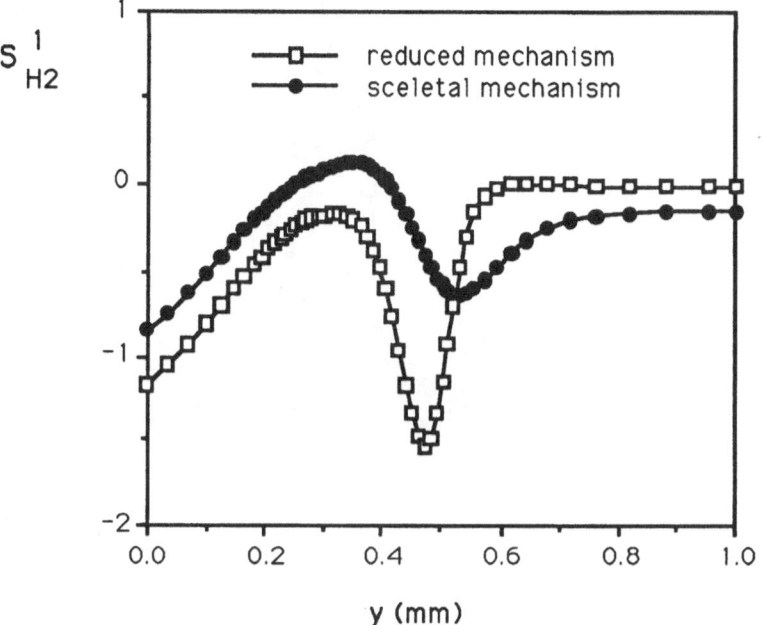

Figure 16: As Fig. 11, but for H_2.

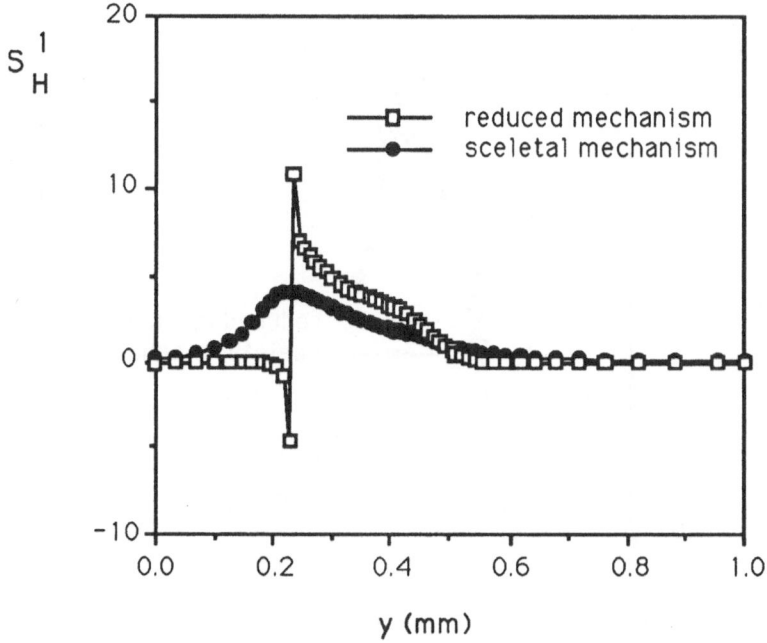

Figure 17: As Fig. 11, but for H.

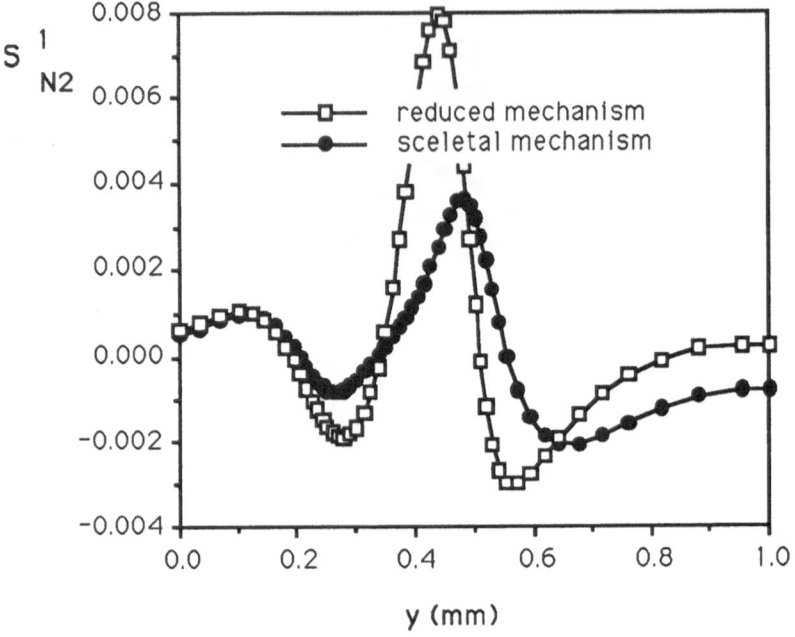

Figure 18: As Fig. 11, but for N_2.

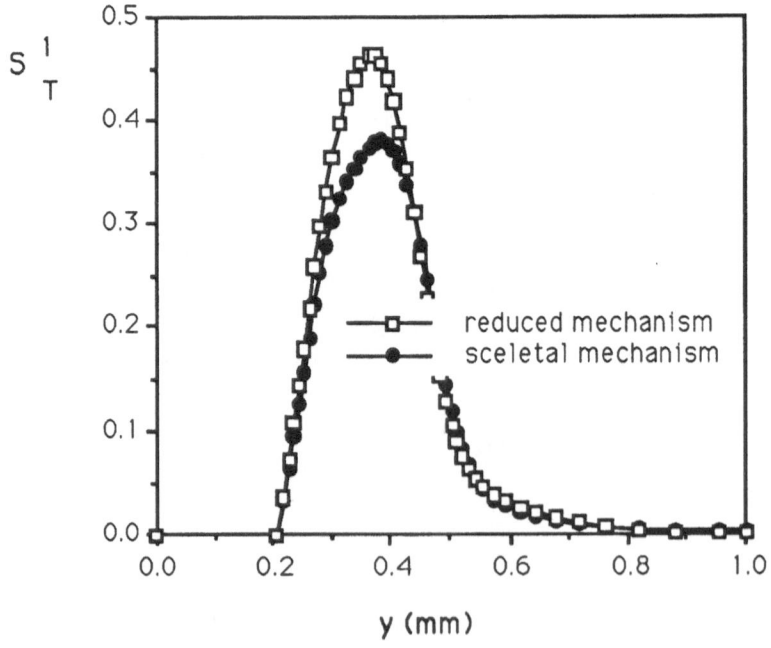

Figure 19: As Fig. 11, but for temperature.

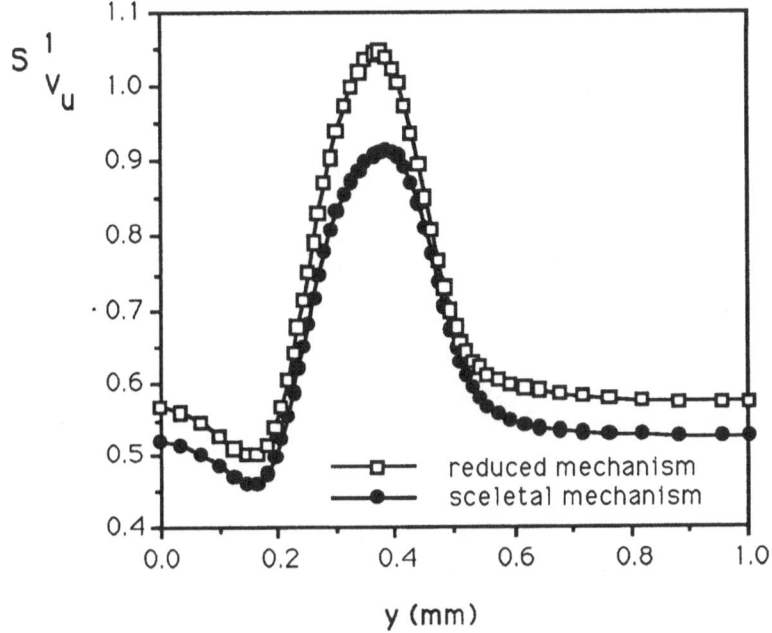

Figure 20: As Fig. 11, but for velocity.

Shown in Figs. 21 to 30 are the sensitivity profiles of mass fractions, temperature and velocity with respect to step 9, $H + O_2 + M \rightarrow HO_2 + M$, which has a zero activation energy and, therefore, dominates step 1 in the low temperature-portion of the flame. Step 9 involves the chemical species H, O_2 and HO_2 of which only H and O_2 appear explicitly in the reduced mechanism. Therefore, Figs. 21 to 28 show that the mass fractions of H and O_2 are most sensitive with respect to changes in the rate constant of this step, k_9, whereas the mass fractions of the other species depend only indirectly on k_9 through their interaction with H, O_2 and and HO_2 in other reaction steps. Similarly to step 1, since O_2 is consumed in step 9, an increase of k_9 leads to a local decrease of Y_{O_2}. However, since step 9 is terminating, an increase of k_9 leads to a local *decrease* of the concentrations of *all* the radicals contained in the hydrogen-oxygen radical pool as well as to a local decrease of the temperature (Fig. 29) and, hence, of the velocity (Fig. 30).

Shown in Figs. 31 to 40 are the sensitivity profiles of mass fractions, temperature and velocity with respect to the watergas-shift reaction 13, $CO + OH \rightarrow CO_2 + H$. Discussion of these profiles uses arguments similar to those used to interpret Figs. 11 to 30 and, therefore, is omitted here.

We are now in the position to reflect upon the suitability of Newton's method as the numerical-solution procedure in conjunction with systematically reduced kinetic mechanisms such as the one employed in the present study. Newton's method is one of several existing approaches to search for a vector **U** that makes a vector function **F** to zero; see Eq. (5.1). **U** and **F** have the same number of components; for the flames considered in this chapter, this number is $K \equiv (N - 1 + 1 + 1) \times J$, where N is the number of chemical species in the system and J the number of gridpoints used to cover the computational domain. Thus, Newton's method may be viewed as a procedure to

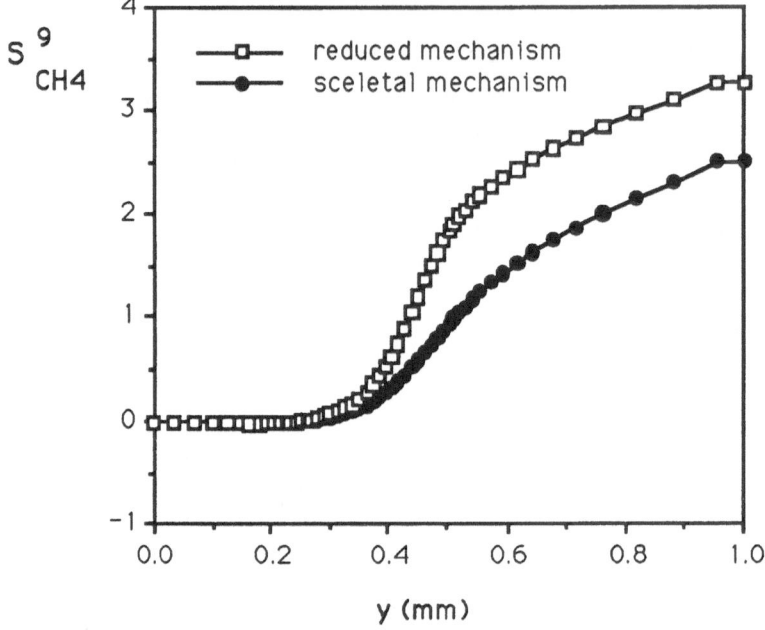

Figure 21: Computed profile of relative first-order sensitivity of the CH_4 mass fraction with respect to elementary step 9 for an atmospheric-pressure, stoichiometic methane-air flame with an initial temperature of 300 K. Filled circles: sceletal mechanism; void squares: reduced mechanism.

search in a K dimensional space for a vector $\mathbf{U}$ satisfying $\mathbf{F(U)} = \mathbf{0}$, the search being started with an initial guess $\mathbf{U} \equiv \mathbf{U^0}$. Furthermore, each iteration step k of Newton's method may be viewed as a move from $\mathbf{U^k}$ to $\mathbf{U^{k+1}}$ in this K dimensional space. The reduction of a detailed kinetic mechanism to a mechanism comprising only a few global steps effectively eliminates (a number of) species conservation equations, either through steady-state assumptions for selected species and/or partial-equilibrium assumptions for selected elementary reactions, and, therefore, diminishes the dimension of the space in which a solution vector $\mathbf{U}$ is to be sought. As a consequence, with the reduced mechanism, upon its iteration-step by iteration-step move towards the solution vector, Newton's method has fewer degrees of freedom, which manifests itself in convergence difficulties of the method. In addition, a reduced mechanism offers Newton's method less freedom on its move through space towards a solution through the algebraic constraints that result from steady-state and/or partial-equilibrium assumptions incorporated into the rate-expressions of the global rates. Finally, the appreciably enhanced and/or reduced concentration, temperature and velocity sensitivities with respect to the elementary rates which appear in the global-rate expressions of a systematically reduced kinetic mechanism, at each iteration step "bounce $\mathbf{U}$ forth and back through the solution space" rather then guiding it smoothly towards its final solution value as it usually does when a detailed mechanism of elementary reactions is employed. Attempts to overcome the described difficulties by solving the time-dependent version of the governing equations in order to bring $\mathbf{U}$ into the domain of convergence of Newton's method operating on the time-independent residuals are often not successful because of division by zero which occur, for instance, when the concentration of molecular hydrogen becomes numerically small; see section 4 of this chapter.

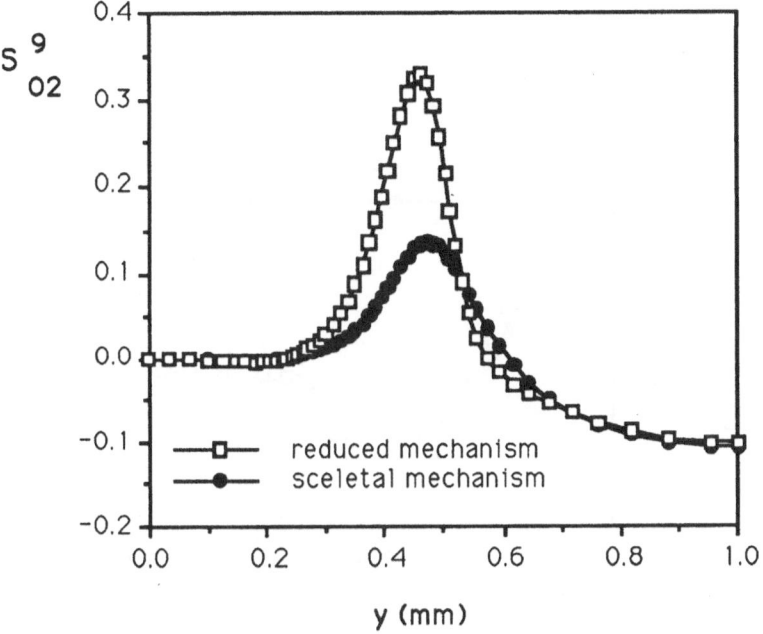

Figure 22: As Fig. 21, but for O_2.

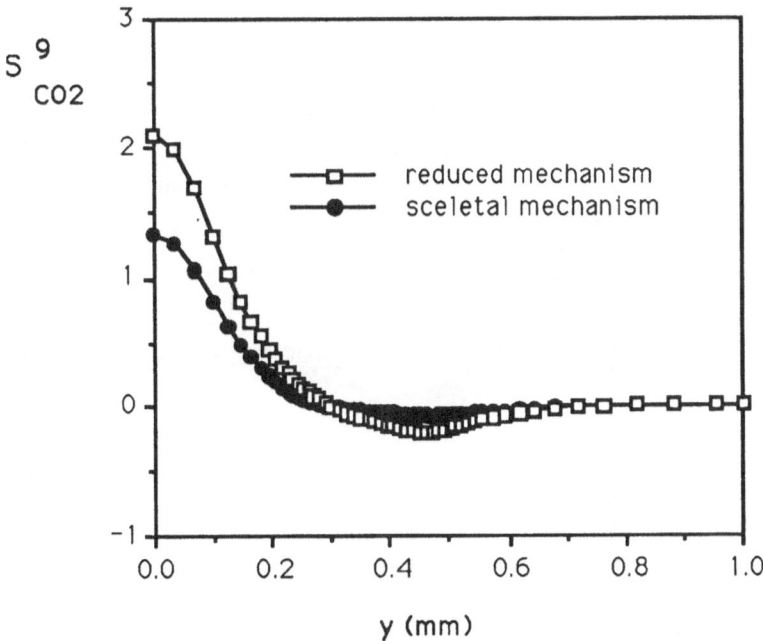

Figure 23: As Fig. 21, but for CO_2.

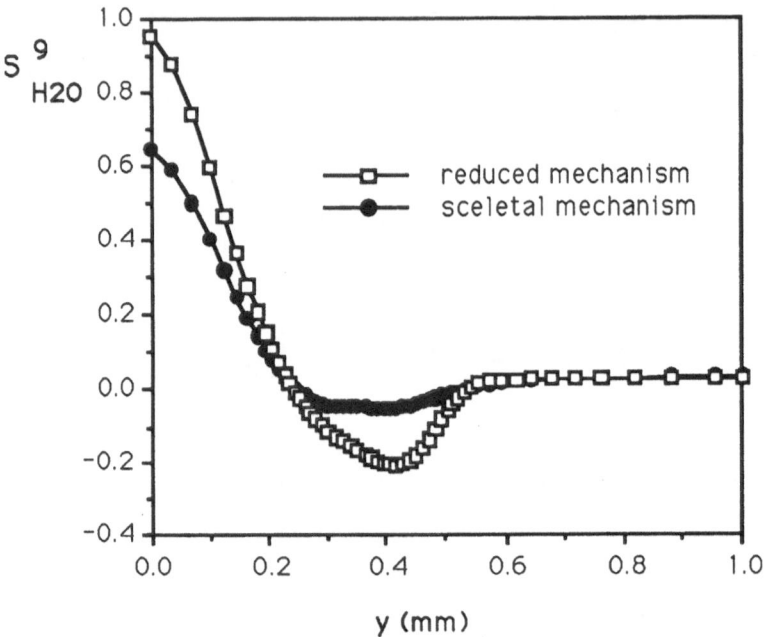

Figure 24: As Fig. 21, but for H_2O.

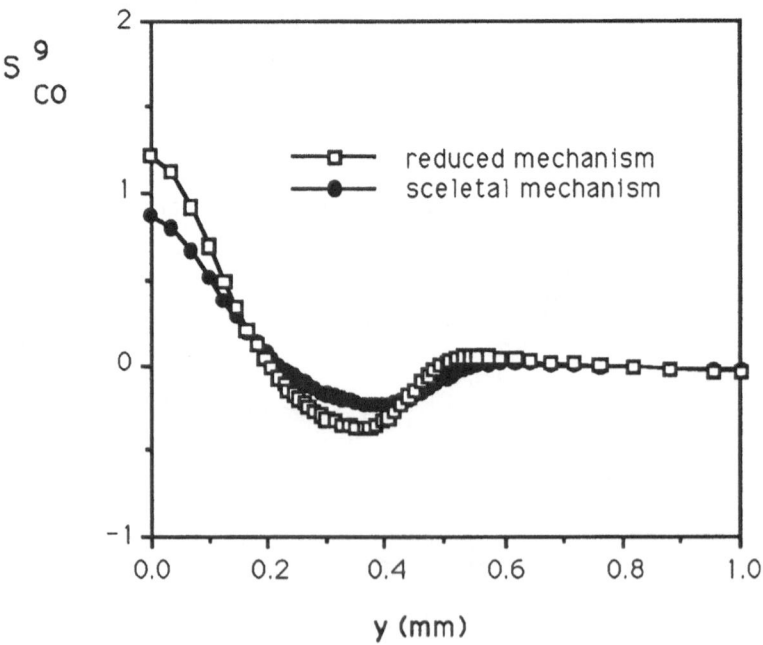

Figure 25: As Fig. 21, but for CO.

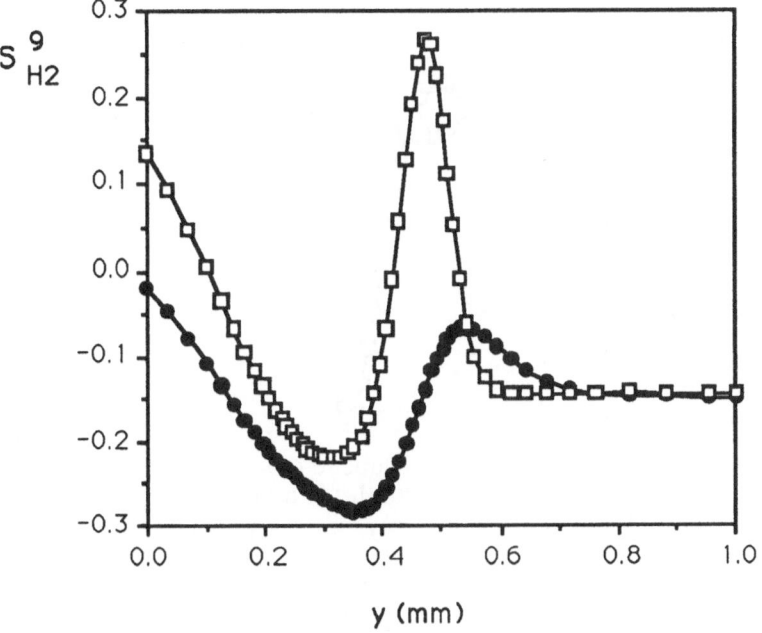

Figure 26: As Fig. 21, but for H_2.

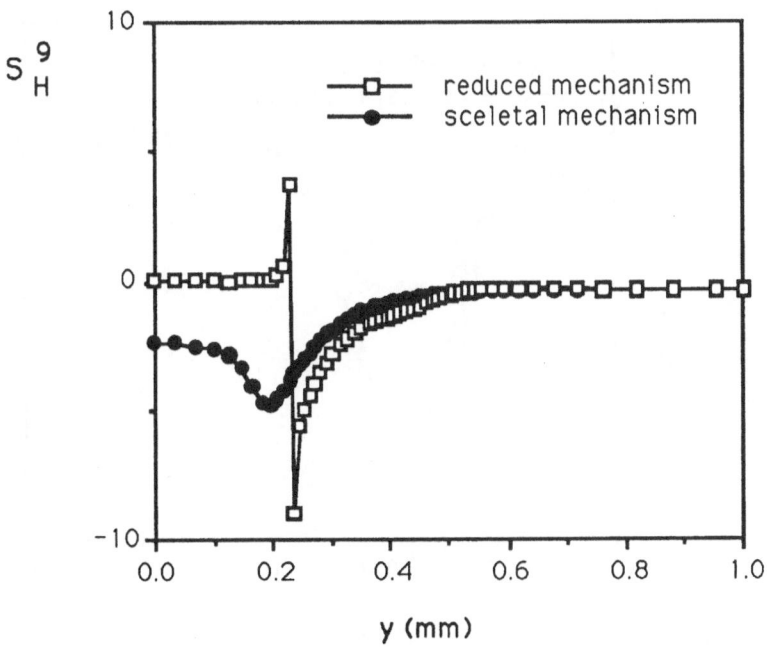

Figure 27: As Fig. 21, but for H.

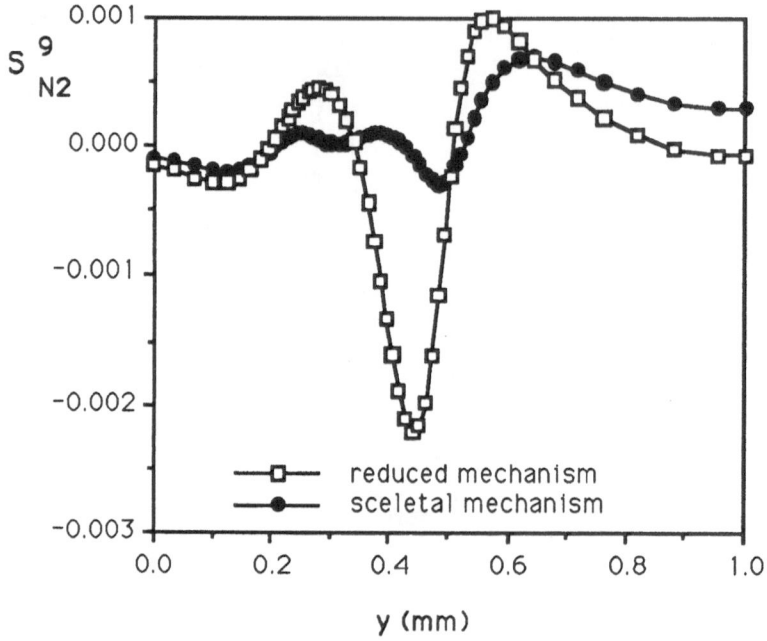

Figure 28: As Fig. 21, but for N_2.

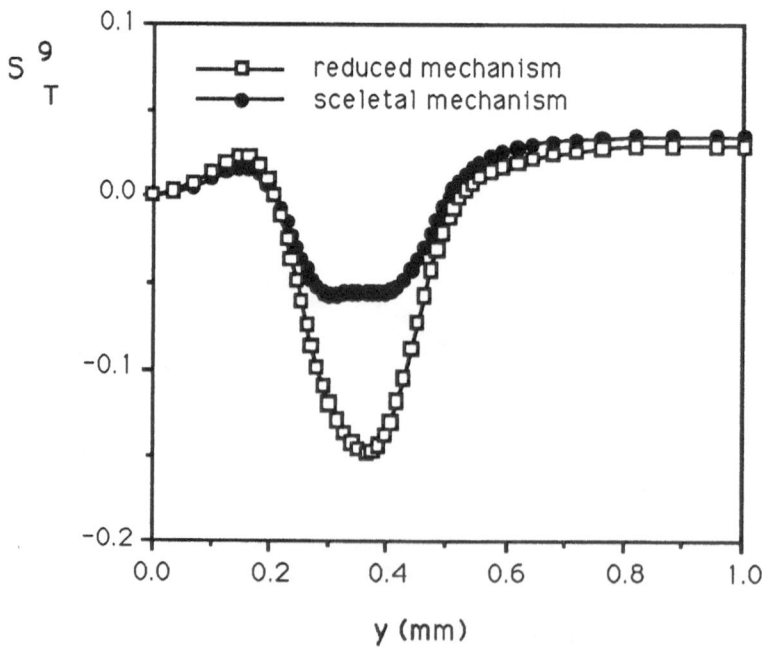

Figure 29: As Fig. 21, but for temperature.

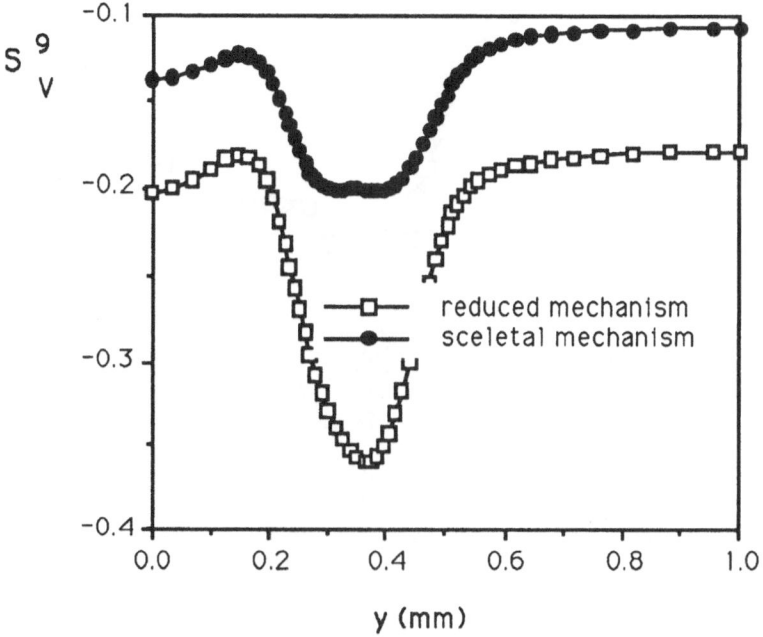

Figure 30: As Fig. 21, but for velocity.

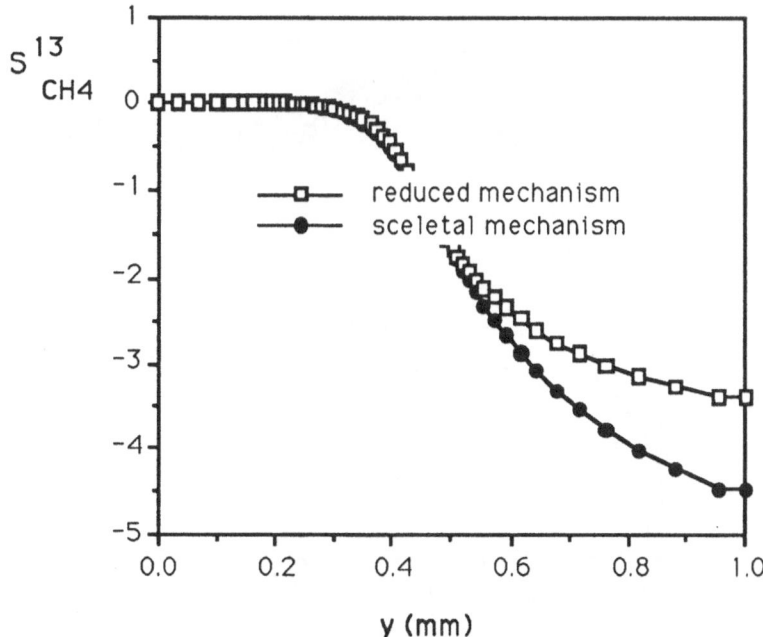

Figure 31: Computed profile of relative first-order sensitivity of the CH_4 mass fraction with respect to elementary step 13. Conditions and symbols as before.

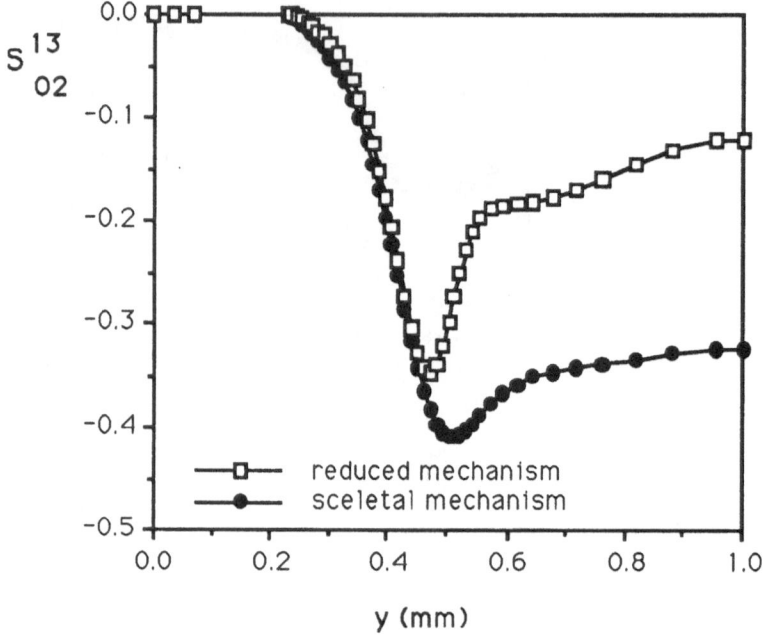

Figure 32: As Fig. 31, but for O_2.

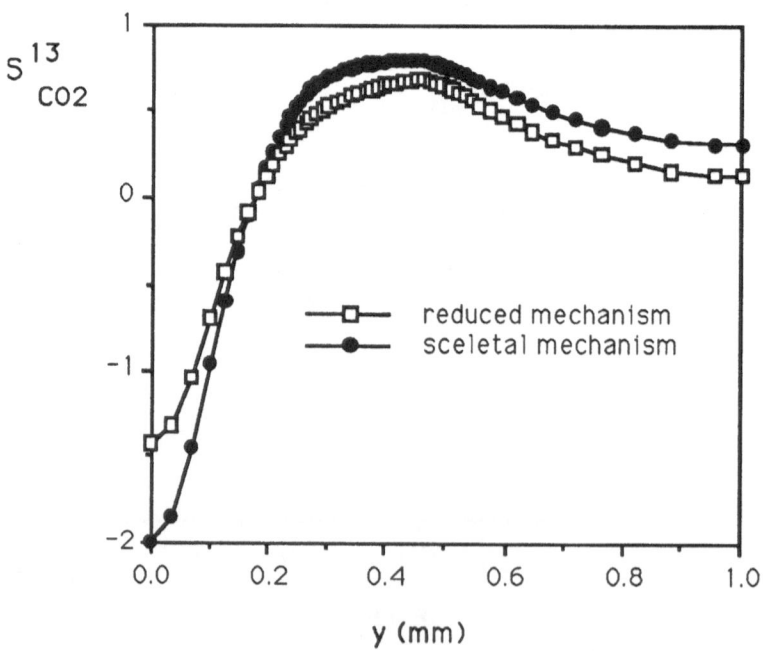

Figure 33: As Fig. 31, but for CO_2.

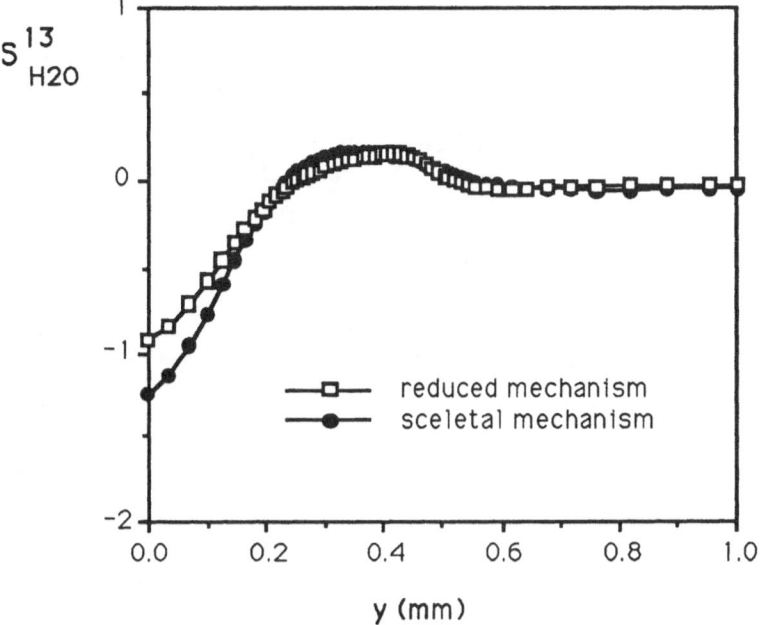

Figure 34: As Fig. 31, but for H_2O.

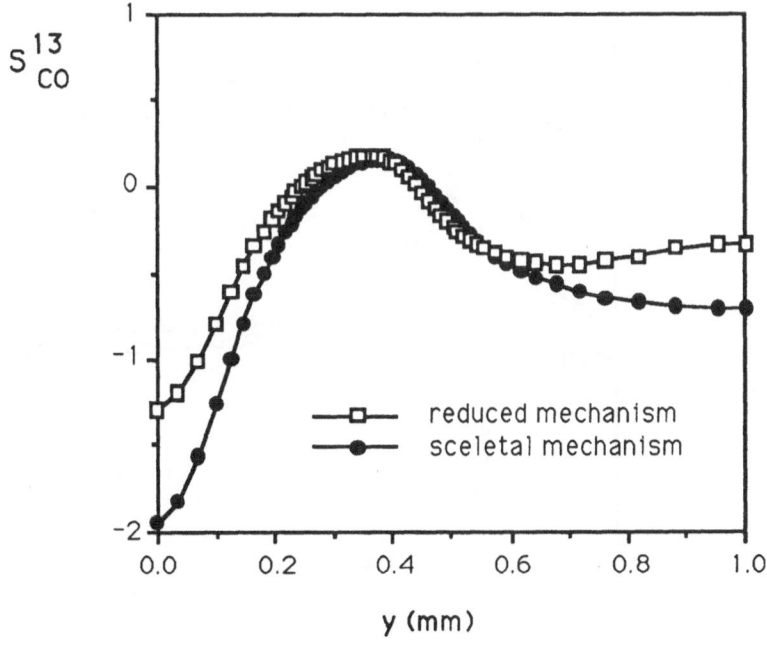

Figure 35: As Fig. 31, but for CO.

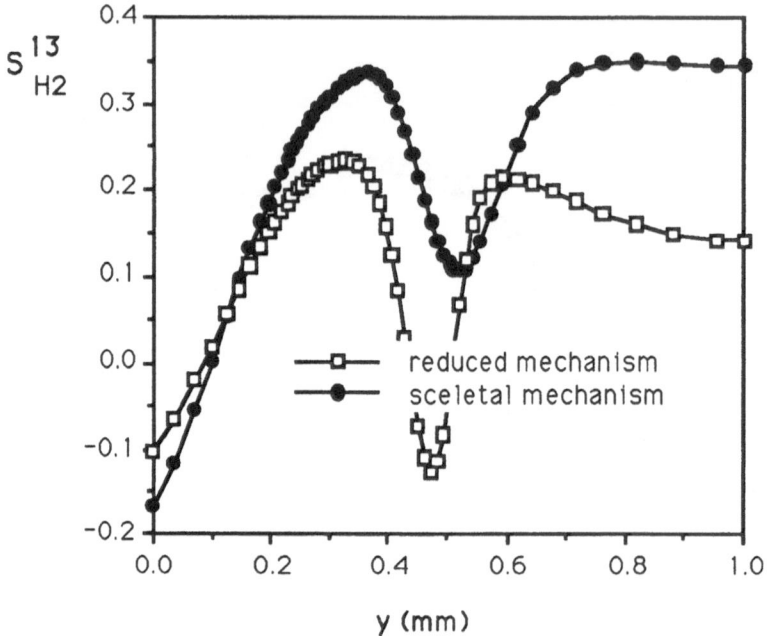

Figure 36: As Fig. 31, but for H_2.

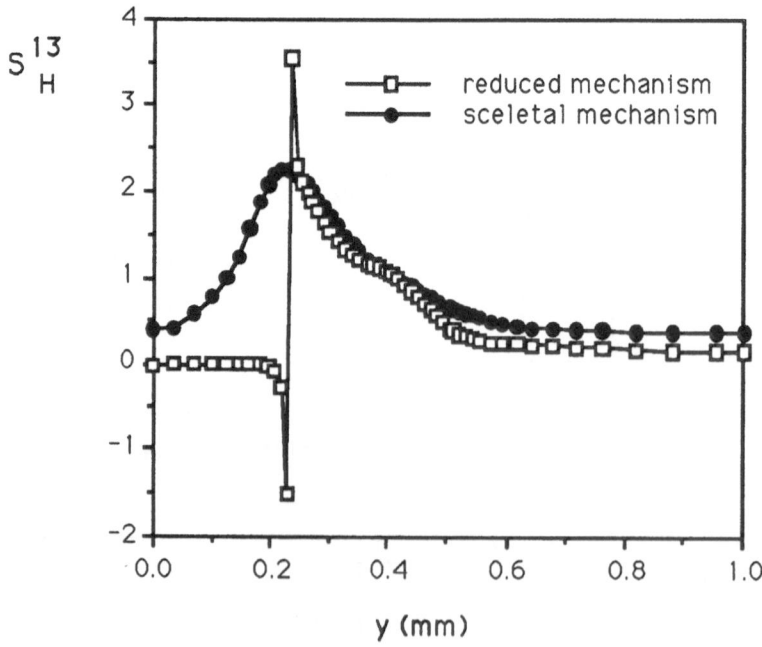

Figure 37: As Fig. 31, but for H.

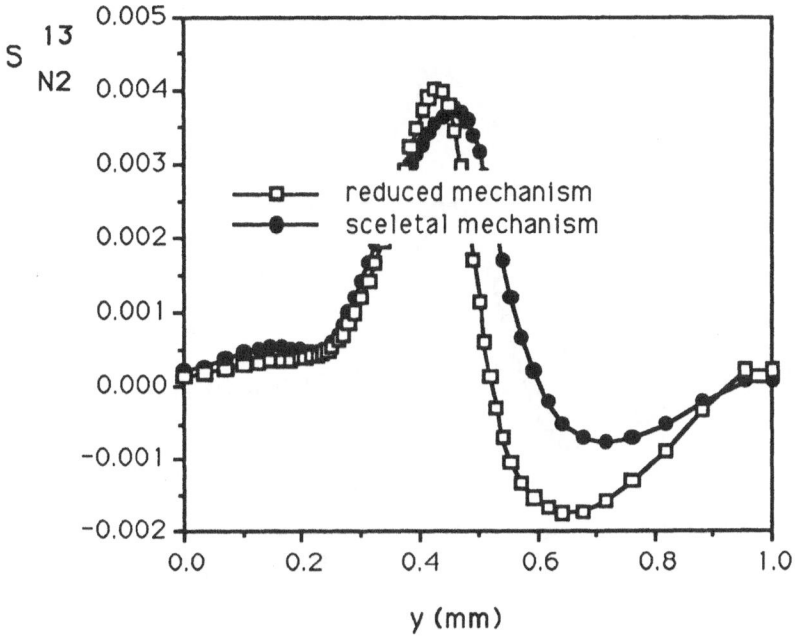

Figure 38: As Fig. 31, but for N_2.

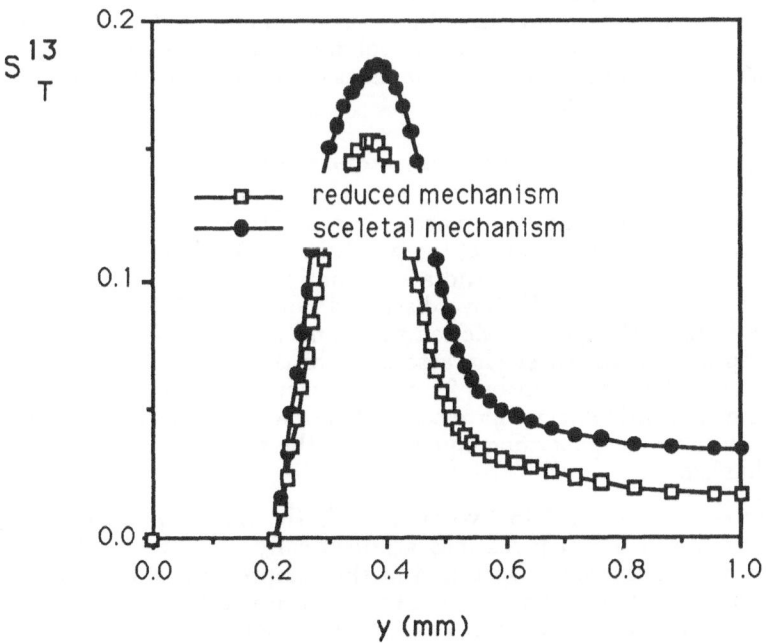

Figure 39: As Fig. 31, but for temperature.

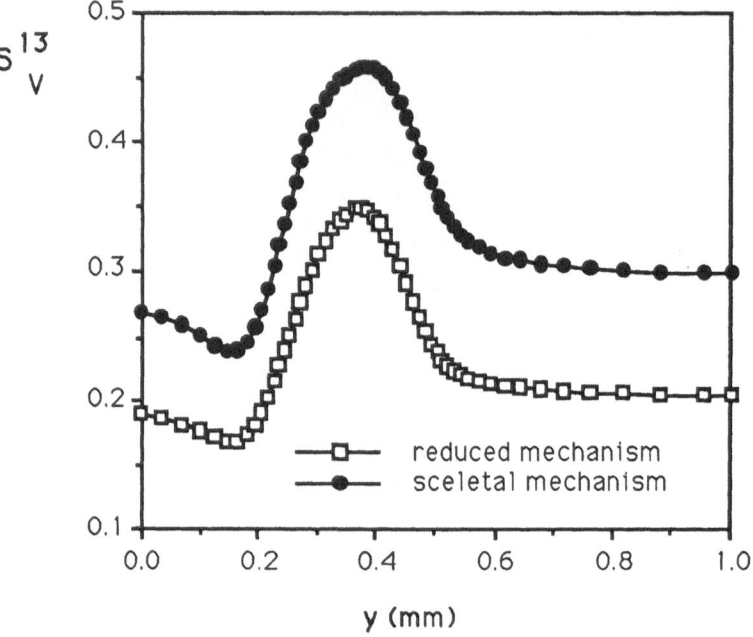

Figure 40: As Fig. 31, but for velocity.

It should be noted here that the arguments given in the previous paragraph are not exclusively the result of the work described in this chapter, but they are also based on the author's several years of experience in solving laminar combustion problems with Newton's method and both detailed mechanisms of elementary reactions and systematically reduced kinetic mechanisms.

9. Summary, Conclusions and Recommendations

In this chapter we have calculated atmospheric pressure, stoichiometric premixed laminar methane-air flames with an initial temperature of 300 Kelvin using both a detailed mechanism of elementary reactions and a systematically reduced mechanism consisting of only four global reactions. We have discussed numerical difficulties associated with the functional form of the global reaction rates and have applied first-order sensitivity analysis to identify and interpret characteristics of the reduced scheme which make it difficult to obtain numerical solutions for flame structures using this particular reduced reaction scheme. In particular, we have concluded that Newton's method may not be the ideal method to solve the governing equations on the basis of a systematically reduced reaction mechanism.

Further research is needed in two areas. Firstly, means must be found to bring systematically reduced mechanisms into a numerically convenient form: for instance, division by zero, as discussed in section 4 of this chapter, must be avoided. Secondly, numerical solution methods must be found which are not based on a "Newtonian-like search" for a solution vector in some multidimensional space, but guide the solution iteration-step by iteration-step or, alternatively, time-step by time-step towards the correct (time-independent) solution of a problem. With respect to the latter point, possibly semi implicit-explicit numerical schemes may offer an alternative to Newton's method.

Acknowledgement

The author would like to thank Dr. Henning Bockhorn for his assistance in the development of the sensitivity analysis for detailed kinetic mechanisms and its implementation into the computer program.

References

[1] Peters, N., "Numerical and Asymptotic Analysis of Systematically Reduced Reaction Schemes for Hydrocarbon Flames," in Numerical Simulation of Combustion Phenomena, R. Glowinski et al., Eds., Lecture Notes in Physics, Springer-Verlag, (1985), p. 90.

[2] Peters, N. and Williams, F. A., "The Asymptotic Structure of Stoichiometric Methane Air Flames," *Comb. and Flame*, **68**, (1987), p. 185.

[3] Paczko, G., Lefdal, P. M. and Peters, N., "Reduced Reaction Schemes for Methane, Methanol and Propane Flames," in 21st Symp. (Int.) on Comb., The Combustion Institute, Pittsburgh, (1986), p. 739.

[4] Bilger, R. W. and Kee, R. J., "Simplified Kinetics for Diffusion Flames of Methane in Air," *Joint Conference of the Western States and Japanese Sections of the Combustion Institute*, Honolulu, Hawaii, Nov. 1987, p. 277.

[5] Treviño, C. and Williams, F. A., "Asymptotic Analysis of the Structure and Extinction of Methane-Air Diffusion Flames," in Dynamics of Reactive Systems Part I: Flames, A. L. Kuhl et al., Eds., Progress in Astronautics and Aeronautics, Vol. 113, AIAA, (1988), p. 129.

[6] Rogg, B. and Williams, F. A., "Structures of Wet CO Flames with Full and Reduced Kinetic Mechanisms," in 22nd Symp. (Int.) on Comb., The Combustion Institute, Pittsburgh (1988), p. 1441.

[7] Jones, W.P. and Lindstedt, R. P. "The Calculation of the Structure of Laminar Counterflow Diffusion Flames Using a Global Reaction Mechanism," *Combust. Sci. Tech.*, **61**, (1988), p. 31.

[8] Seshadri, K. and Peters, N., "Asymptotic Structure and Extinction of Methane-Air Diffusion Flames," *Comb. and Flame*, **73**, (1988), p. 23.

[9] Rogg, B. and Peters, N., "The Asymptotic Structure of Weakly Strained Stoichiometric Methane-Air Flames," *Comb. and Flame*, **79**, (1990), p. 402.

[10] Chelliah, H. K. and Williams, F. A., "Aspects of the Structure and Extinction of Diffusion Flames in Methane-Oxygen-Nitrogen Systems," *Comb. and Flame*, **80**, (1990), p. 17.

[11] Peters, N. and Kee, R. J., "The Computation of Stretched Laminar Methane-Air Diffusion Flames Using a Reduced Four-Step Mechanism," *Comb. and Flame*, **68**, (1987), p. 17.

[12] Rogg, B., "Response and Flamelet Structure of Stretched Premixed Methane-Air Flames, *Comb. and Flame*, **73**, (1988), p. 45.

[13] Rabitz, H, Kramer, M. and Dacol, D., "Sensitivity Analysis in Chemical Kinetics," *Ann. Rev. Phys. Chem.*, **34**, (1983), p. 419.

[14] Yetter, R. A., Dryer, F. L. and Rabitz, H., "Some Interpretive Aspects of Elementary Sensitivity Gradients in Combustion Kinetics Modeling," *Comb. and Flame,* **59,** (1985), p. 107.

[15] Smooke, M. D., Rabitz, H., Reuven, Y. and Dryer, F. L., "Application of Sensitivity Analysis to Premixed Hydrogen-Air Flames," *Combust. Sci. Tech.,* **59,** (1988), p. 295.

[16] Bockhorn, H., "Sensitivity Analysis Based Reduction of Complex Reaction Mechanisms in Turbulent Non-Premixed Combustion," paper presented at the 23nd Symp. (Int.) on Comb., Orleans (France), (1990), in press.

[17] Rogg, B., "Modelling and Numerical Simulation of Premixed Turbulent Combustion in a Boundary Layer," in Proc. 7th Symp. on Turbulent Shear Flows, Stanford University (Stanford, CA), 21-23 August 1989, p. 26-1/1.

[18] W.C. Gardiner, Jr. (Ed.), "Combustion Chemistry," Springer-Verlag, New York (1984).

[19] Rogg, B., "Numerical Modelling and Computation of Reactive Stagnation-Point Flows," in Computers and Experiments in Fluid Flow, G.M. Carlomagno and C.A. Brebbia (Eds.), Springer-Verlag, Berlin-Heidelberg, (1989), p. 75.

[20] Rogg, B., "Numerical Analysis of Strained Premixed CH_4-Air Flames with Detailed Chemistry," in Numerical and Applied Mathematics, W.F. Ames (Ed.), Baltzer, (1989), p. 159.

[21] Rogg, B., "Adaptive Computational Methods for Time-Dependent Problems in Combustion Engineering," in NUMETA 90, Numerical Methods in Engineering, Vol.2, G.N. Pande and J. Middleton (Eds.), Elsevier, London-New York, (1990), p. 1129.

[22] Deuflhard, P., "A Modified Newton Method for the Solution of Ill-Conditioned Systems of Nonlinear Equations with Application to Multiple Shooting," *Numer. Math.,* **22,** (1974), p. 289.

[23] Smooke, M. D., "An Error Estimate for the Modified Newton Method with Application to the Solution of Nonlinear Two-Point Boundary Value Problems," *J. Opt. Theory and Appl.,* **39,**(1983), p. 489.

[24] Giovangigli, V. and Smooke, M.D., "Extinction of Strained Premixed Laminar Flames with Complex Chemistry," *Combust. Sci. Tech.,* **53,** (1987), p. 23.

CHAPTER 9

APPLICATION OF REDUCED CHEMICAL MECHANISMS
FOR PREDICTION OF
TURBULENT NONPREMIXED METHANE JET FLAMES

J.-Y. Chen and R.W. Dibble
Combustion Research Facility, Sandia National Laboratories
Livermore, CA

1. Introduction

Moment modeling is in widespread use for turbulent flows of aerodynamic interest where momentum transfer and pressure drop are the major features of interest. These moment models are less useful when they are extended to turbulent mixing flows. The additional complication of Arrhenius type chemical kinetic rates has yet to be satisfactorily incorporated into moment methods for turbulent reacting flows. Although attempts, such as the eddy-breakup model or an assumed pdf approach, have been proposed and studied extensively, the extension of moment methods to turbulent reacting and mixing flows remains unresolved. As an alternative, the probability density function (pdf) approach, which emerged a decade ago (see, e.g., Pope [1], O'Brien [2]), offers a promising method for treating chemical reactions in turbulent flows. Specifically, the chemical reaction terms appear exact in the governing equation for the single point pdf evolution; consequently, there is no modeling of the chemical source terms. However, modeling of the scalar dissipation terms is a crucial issue in the pdf method, since these terms represent the mixing process occurring at the molecular level. Closure of the mixing process requires statistical information from fluids at different locations, i.e., two-point correlations, which is not available from the single-point pdf method.

Models for the scalar dissipation terms are often referred to as *mixing* models. Due to the complex nature of the mixing process occurring in turbulent flows, development of mixing models has been based largely on *ad hoc* assumptions, such as those in Curl's [3] coalescence/dispersion model and its derivatives (Janicka *et al.* [4]; Pope [5]). Statistical information either from experiments or from direct numerical simulations of turbulent flows is needed to provide guidance for further development of mixing models. In spite of the primitive state of mixing models, there has been significant advancement in the numerical techniques for the pdf method. In particular, solution of the pdf equation by Monte Carlo simulation (Pope [6]) has made the method computationally feasible for multi-species problems. A recent review of progress in the pdf method is given by Pope [7]. Even with the development of Monte Carlo simulations, computer limitations are encountered when a nontrivial chemical mechanism is included.

Pope [6] has shown that the convergence rate of the statistics deduced from the Monte Carlo simulation is proportional to the square root of the number of representations used. Consequently, a large number of statistical representations (commonly referred to as particles) are needed to achieve accurate solutions; we typically use 1000 particles per grid cell. In principle, the effects of combustion on the joint scalar pdf can be calculated exactly by moving *each* Monte Carlo representative particle in the scalar space according to its chemical kinetic rate equations. The computation would be a formidable task for combustion processes with multiple species whose chemical rate equations are usually stiff. We estimate 1000 cpu hours on a Cray XMP-48 computer would be required for such simulations of jet flames with five scalars; such cpu requirements are not practical.

One strategy for reducing the enormous computational demand is *a priori* calculation of the integrals of the reaction rates for all possible chemical states and then storing the results in a 'look-up' table. The look-up table is finite when the domain of all possible chemical states is discretized. During the Monte Carlo simulations, the changes of the scalar properties are obtained from interpolations based on the previously generated look-up table, avoiding the need for repeated, and often redundant, time integration of the chemical kinetic rate equations (Pope and Correa [8], Chen and Kollmann [9]). This strategy is computationally very efficient. Typically, four hours of cpu time on a Cray XMP-48 are used for simulation of a turbulent jet flame up to thirty diameters downstream. Even with this improvement, computation time radically increases when the size of the look-up table exceeds the direct-access memory of the computer. The look-up table can be reduced in size by using a reduced chemical mechanism; one that retains salient, desirable features of the complete chemical mechanism. Accordingly, the search for such reduced mechanisms is an area of active research.

The recent development of reduced mechanisms for complex combustion processes (Peters and Kee [10], Peters and Williams [11], Bilger and Kee [12], Rogg and Williams [13]) is based on systematic reduction of complete chemical mechanisms rather than curve fits to limited experimental observations. Consequently, these newly developed reduced mechanisms contain many of the features of the entire chemical kinetic scheme. More importantly, for purposes here, these features of the full mechanism are obtained with a reduced number of scalars. For example, with the assumption of equal diffusivity and constant pressure, a four-step reduced mechanism having only five scalars (four reactive scalars and one conserved scalar) was reduced from a eighteen scalar, thirty-four step, mechanism for methane combustion. From a computational point of view, the number of steps in a mechanism is not nearly as important as the number of species included in the mechanism; the number of equations that are to be solved is related to the number of species rather than the number of chemical kinetic steps. Consequently, significant reduction in the computing effort is possible with mechanisms that contain a small number of species.

2. Four-Step Reduced Mechanisms

Development of both full and reduced mechanisms for combustion has been and is an area of active investigation. From a full mechanism, one can systematically develop reduced mechanisms by making steady state or partial equilibrium assumptions (e.g., Peters and Kee [10], Chen [14]). In this section, we will summarize two four-step reduced mechanisms proposed by Peters and Seshadri and by Bilger at the Workshop on Reduced Kinetic Mechanism and Asymptotic Approximations for Methane-Air Flames held on March 13-15, 1989, at UCSD, La Jolla, California. These reduced mechanisms are distilled from a larger mechanism (called *skeletal* by other authors) that has eighteen scalars and thirty-four reaction steps. The larger mechanism for methane-air combustion is summarized in Table 1.

Four-step Reduced Mechanism By Seshadri and Peters:

The four-step reduced mechanism of Seshadri and Peters can be derived using a computer program, developed by Chen [14], that eliminates user specified elementary reactions. Eliminating elementary steps 2f, 2b, 3f, 3b, 7, 13, 14, 17, 20, 23f, and 23b in Table 1, the resulting four-step mechanism is

$$CH_4 + 2H + H_2O = CO + 4H_2, \qquad (R1)$$

$$CO + H_2O = CO_2 + H_2, \qquad (R2)$$
$$2H + M = H_2 + M, \qquad (R3)$$
$$O_2 + 3H_2 = 2H + 2H_2O. \qquad (R4)$$

The global reaction rates are expressed as

$$w_{R1} = w_{10} + w_{11} + w_{12}$$

$$w_{R2} = w_9$$

$$w_{R3} = w_5 - w_{10} + w_{16} - w_{18} + w_{19} - w_{22} + w_{24} + w_{25}$$

$$w_{R4} = w_1 + w_6 + w_{18} + w_{22}.$$

Assuming that steps 3f and 3b are equilibrated leads to

$$[OH] = \frac{k_{3b}}{k_{3f}} \frac{[H_2O][H]}{[H_2]}.$$

Other intermediate species are obtained by assuming them to be in steady state. The results are

$$[O] = \frac{\sqrt{B^2 + 4AC} - B}{2A}$$

$$A = k_{13}b$$
$$B = bd + k_{13}(c - a)$$
$$C = ad + k_{18}c[O_2]$$

where

$$a = k_{1f}[H][O_2] + k_{2b}[OH][H] + k_{4f}[OH][OH]$$
$$b = k_{1b}[OH] + k_{2f}[H_2] + k_{4b}[H_2O]$$
$$c = (k_{10f} + k_{11f}[H] + k_{12f}[OH])[CH_4]$$
$$d = k_{10b}[H] + k_{11b}[H_2] + k_{12b}[H_2O] + k_{18}[O_2]$$

$$[CH_3] = \frac{(k_{10f} + k_{11f}[H] + k_{12f}[OH])[CH_4]}{k_{10b}[H] + k_{11b}[H_2] + k_{12b}[H_2O] + k_{13}[O] + k_{18}[O_2]}$$

$$[CH_3O] = \frac{k_{18}[CH_3][O_2]}{k_{19}[H] + k_{20}[M]}$$

$$[CH_2O] = \frac{k_{13}[O] + (k_{19}[H] + k_{20}[M])[CH_3O]}{k_{14f}[H] + k_{15f}[OH]}$$

$$[HCO] = \frac{k_{13}[CH_3][O] + (k_{19}[H] + k_{20}[M])[CH_3O]}{k_{16}[H] + k_{17}[M]}$$

$$[HO_2] = \frac{\sqrt{B^2 + 4AC} - B}{2A}$$

$$A = (2 - z)k_{21}$$
$$B = (1 - z)k_{23b}[H_2O] + (k_6 + k_7)[H] + k_8[OH]$$
$$C = k_{22b}[OH][OH][M]z + k_5[H][O_2][M^a]$$
$$z = \frac{k_{23f}[OH]}{k_{22f}[M] + k_{23f}[OH]}$$

$$[H_2O_2] = \frac{k_{21}[HO_2][HO_2] + k_{22b}[OH][OH][M] + k_{23b}[H_2O][HO_2]}{k_{22f}[M] + k_{23f}[OH]}$$

Four-step Reduced Mechanism By Bilger:

Steps 2f, 2b, 3f, 3b, 6, 12, 15, 17, 19, and 21 in Table 1 are used to eliminate intermediate species and the following reduced mechanism can be obtained:

$$CH_4 + 2H + H_2O = CO + 4H_2, \qquad (R'1)$$

$$CO + H_2O = CO_2 + H_2, \qquad (R'2)$$

$$2H_2 + O_2 + M = 2H_2O + M, \qquad (R'3)$$

$$O_2 + 3H_2 = 2H + 2H_2O. \qquad (R'4)$$

The global reaction rates are expressed as

$$w_{R'1} = w_{13} + w_{18}$$
$$\approx w_{13}$$
$$w_{R'2} = w_9$$
$$w_{R'3} = w_5 - w_{10} + w_{16} - w_{20} - w_{22} + w_{24} + w_{25}$$
$$\approx w_5 + w_{10b}$$
$$w_{R'4} = w_1 - w_7 - w_8 + w_{10} - w_{16} + w_{18} + w_{20} - w_{23} - w_{24} - w_{25}$$
$$\approx w_1 - w_{10b}.$$

Assuming that step 5 is equilibrated leads to

$$[OH] = \frac{k_{3b}}{k_{3f}} \frac{[H_2O][H]}{[H_2]}.$$

TABLE 1

CHEMICAL KINETIC MECHANISM

No.	Reaction	A_k	n_k	E_k
1f	$H + O_2 \rightarrow O + OH$	2.0×10^{14}	0.0	16800.0
1b	$O + OH \rightarrow O_2 + H$	1.575×10^{13}	0.0	690.0
2f	$O + H_2 \rightarrow OH + H$	1.8×10^{10}	1.0	8826.0
2b	$OH + H \rightarrow O + H_2$	8.0×10^{9}	1.0	6760.0
3f	$H_2 + OH \rightarrow H_2O + H$	1.17×10^{9}	1.3	3626.0
3b	$H_2O + H \rightarrow H_2 + OH$	5.09×10^{9}	1.3	18588.0
4a	$OH + OH \rightarrow H_2O + O$	6.0×10^{8}	1.3	0.0
4b	$H_2O + O \rightarrow OH + OH$	5.9×10^{9}	1.3	17029.0
5	$H + O_2 + M^a \rightarrow HO_2 + M^a$	2.3×10^{18}	-0.8	0.0
6	$H + HO_2 \rightarrow OH + OH$	1.5×10^{14}	0.0	1004.0
7	$H + HO_2 \rightarrow H_2 + O_2$	2.5×10^{13}	0.0	700.0
8	$OH + HO_2 \rightarrow H_2 + O_2$	2.0×10^{13}	0.0	1000.0
9f	$CO + OH \rightarrow CO_2 + H$	1.51×10^{7}	1.3	-758.0
9b	$CO_2 + H \rightarrow CO + OH$	1.57×10^{9}	1.3	22337.0
10f	$CH_4 (+M)^b \rightarrow CH_3 + H (+M)^b$	6.3×10^{14}	0.0	104000.0
10b	$CH_3 + H (+M)^b \rightarrow CH_4 (+M)^b$	5.20×10^{12}	0.0	-1310.0
11f	$CH_4 + H \rightarrow CH_3 + H_2$	2.2×10^{4}	3.0	8750.0
11b	$CH_3 + H_2 \rightarrow CH_4 + H$	9.57×10^{2}	3.0	8750.0
12f	$CH_4 + OH \rightarrow CH_3 + H_2O$	1.6×10^{6}	2.1	2460.0
12b	$CH_3 + H_2O \rightarrow CH_4 + OH$	3.02×10^{5}	2.1	17422.0
13	$CH_3 + O \rightarrow CH_2O + H$	6.8×10^{13}	0.0	0.0
14	$CH_2O + H \rightarrow HCO + H_2$	2.5×10^{13}	0.0	3991.0

TABLE 1 — Continued

CHEMICAL KINETIC MECHANISM

No.	Reaction	A_k	n_k	E_k
15	$CH_2O + OH \rightarrow HCO + H_2O$	3.0×10^{13}	0.0	1195.0
16	$HCO + H \rightarrow CO + H_2$	4.0×10^{13}	0.0	0.0
17	$HCO + M \rightarrow CO + H + M$	1.6×10^{14}	0.0	14700.0
18	$CH_3 + O_2 \rightarrow CH_3O + O$	7.0×10^{12}	0.0	25652.0
19	$CH_3O + H \rightarrow CH_2O + H_2$	2.0×10^{13}	0.0	0.0
20	$CH_3O + M \rightarrow CH_2O + H + M$	2.4×10^{13}	0.0	28812.0
21	$HO_2 + HO_2 \rightarrow H_2O_2 + O_2$	2.0×10^{12}	0.0	0.0
22f	$H_2O_2 + M \rightarrow OH + OH + M$	1.3×10^{17}	0.0	45500.0
22b	$OH + OH + M \rightarrow H_2O_2 + M$	9.86×10^{14}	0.0	-5070.0
23f	$H_2O_2 + OH \rightarrow H_2O + HO_2$	1.0×10^{13}	0.0	1800.0
23b	$H_2O + HO_2 \rightarrow H_2O_2 + OH$	2.86×10^{13}	0.0	32790.0
24	$OH + H + M^a \rightarrow H_2O + M^a$	2.2×10^{22}	-2.0	0.0
25	$H + H + M^a \rightarrow H_2 + M^a$	1.8×10^{18}	-1.0	0.0

Rate constants:

$$k_k = A_k T^{nk} \exp\left(-\frac{E_k}{RT}\right)$$

Units are moles, cubic centimeters, seconds, degrees Kelvin, and calories/mole.

Superscripts:

a *Third body efficiencies are:*
   CH_4: 6.5, H_2O; 6.5, CO_2: 1.5, H_2:1.0, CO: 0.75, O_2: 0.4, N_2: 0.4, all other species: 1.0.

b *High pressure value k_∞.* The pressure dependence is given by the Lindemann form $k = \dfrac{k_\infty}{\left(1 + \dfrac{\alpha RT}{p}\right)}$

   where $\alpha R = 0.517 \exp(-9000/T)$, with p in atm and T in degrees Kelvin.

Other intermediate species are obtained by assuming them to be in steady state. The results are

$$[O] = \frac{\sqrt{B^2 + 4AC} - B}{2A}$$

$$A = k_{13}$$

$$B = k_{11b}[H_2] + \frac{k_{13}(k_{11f}[H][CH_4] - k_{1f}[H][O_2] - k_{4f}[OH][OH])}{k_{1b}[OH] + k_{4b}[H_2O]}$$

$$C = \frac{k_{11b}[H_2](k_{1f}[H][O_2] + k_{4f}[OH][OH]) + k_{18}[O_2]k_{11f}[H][CH_4]}{k_{1b}[OH] + k_{4b}[H_2O]}$$

$$[CH_3] = \frac{k_{11f}[CH_4][H]}{k_{11b}[H_2] + k_{13}[O]}$$

$$[CH_3O] = \frac{k_{18}[CH_3][O_2]}{k_{19}[H] + k_{20}[M]}$$

$$[CH_2O] = \frac{k_{13f}[CH_3][O] + k_{18f}[CH_3][O_2]}{k_{14f}[H] + k_{15f}[OH]}$$

$$[HCO] = \frac{k_{13}[CH_3][O] + k_{18}[CH_3][O_2]}{k_{16}[H] + k_{17}[M]}$$

$$[HO_2] = \frac{\sqrt{B^2 + 4AC} - B}{2A}$$

$$A = 2k_{21}$$
$$B = k_{23b}[H_2O] + (k_6 + k_7)[H] + k_8[OH]$$
$$C = k_5[H][O_2][M]$$

$$[H_2O_2] = \frac{k_{21}[HO_2][HO_2] + k_{22b}[OH][OH][M] + k_{23b}[H_2O][HO_2]}{k_{22f}[M] + k_{23f}[OH]}.$$

Performance of the mechanisms:

Performance of these two reduced mechanisms was evaluated by comparing our calculated results for opposed *laminar* jet flames (Tsuji type) with those obtained with the complete 'skeletal' mechanism. Figure 1 shows a comparison of peak flame temperatures at various strain rates. In general, the Bilger mechanism gives higher (about 50 K) peak temperatures compared to Peters and Seshadri's mechanism, which matches well with the skeletal mechanism. The predicted extinction limits are $a \approx 460/s$ for all three mechanisms, where a is the magnitude of velocity gradient in the vicinity of a stagnation point. Comparison of temperature and species profiles are presented in Figures 2 to 4 for flames at $a = 300/s$. In Figure 2, the predicted temperature profiles agree reasonably well among the three mechanisms although some noticeable differences are observed, especially in rich mixtures. Likewise, the predicted major species in Figure 3 also show good agreement among the three mechanisms. Some of the important intermediate species are shown in Figure 4. For H_2, the mechanism by Peters and Seshadri predicts values about 50% higher than those from the skeletal mechanism in rich mixtures; while the mechanism by Bilger matches quite well with those from the skeletal mechanism. Both reduced mechanisms fail to predict CH_3 correctly. Noticeable differences are found in the predicted H atom profiles among the three mechanisms. Overall, the mechanism by Bilger gives more satisfactory results than that by Peters and Seshadri.

The large skeletal mechanism has eighteen scalars. A Monte Carlo simulation with eighteen scalars far exceeds our present computational capability. With the four-step reduced mechanisms, the total number of scalars is decreased to ten, including seven reactive chemical species, temperature, density, and pressure. Further reduction of the number of scalars by three, to seven, is possible if the concept of mixture fraction is utilized. With the assumption of *equal diffusivity* for species and enthalpy, the mixing process of a two-stream turbulent jet flame can be described by a conserved scalar, such as the mixture fraction f. The mixture fraction is the normalized mass fraction of an atomic element originating from one of the input streams, usually the fuel stream. We chose the mixture fraction to be unity in the fuel stream and zero in the outer coflowing stream. As a consequence of the equal diffusivity assumption, the mixture enthalpy and the atomic element populations are linear functions of the mixture fraction. Accordingly, given a mixture fraction, four relations can be derived; three for the atomic element concentrations and one for the mixture enthalpy. If we further assume constant-pressure combustion and the ideal gas law, two additional relations result. With these six relations, we need a total of five scalars (four reactive species and the mixture fraction) to determine thermodynamic properties of nonpremixed methane-air flames.

Interpolation Table:

In constructing an interpolation table for thermodynamic properties, we first define the allowable range of the five scalars. The limits of these ranges are obtained from the following constraints: (1) atomic species are conserved; (2) species must have non-negative concentrations; and (3) for mass balance, the radical species concentrations are neglected compared to those of major species. These extreme limits are summarized below and their detailed derivations can be found in Chen *et al.* [15].

(1) Limits for f:

$$f^{max} = 1, \qquad f^{min} = 0. \tag{2.1}$$

(2) Limits for n_{CH_4} at a given f:

$$n_{CH_4}^{max} = \frac{f}{W_{CH_4}}. \tag{2.2}$$

$$n_{CH_4}^{min} = max\{\frac{f}{W_{CH_4}} - \frac{2\psi(1-f)}{(W_{N_2} + \psi W_{O_2})} \ , \ 0 \ \}, \tag{2.3}$$

where the symbol $max\{a, b\}$ denotes selection of the larger of a and b.

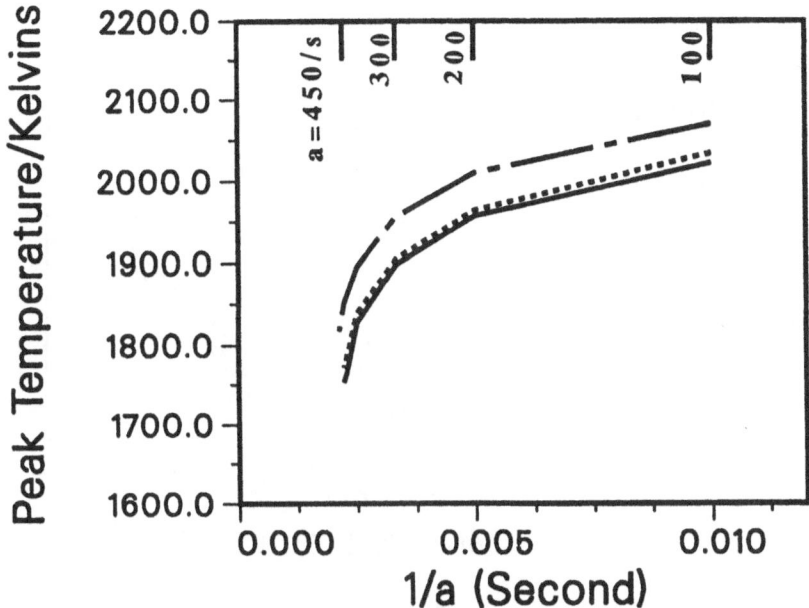

Fig. 1. Predicted peak temperatures for Tsuji type opposed laminar methane-air jet flames at various strain rates. Solid line: skeletal mechanism; dashdot line: Bilger's mechanism; dashed line: Peters and Seshadri's mechanism. Extinction limit of $a \approx 460/s$ was predicted by all three mechanisms.

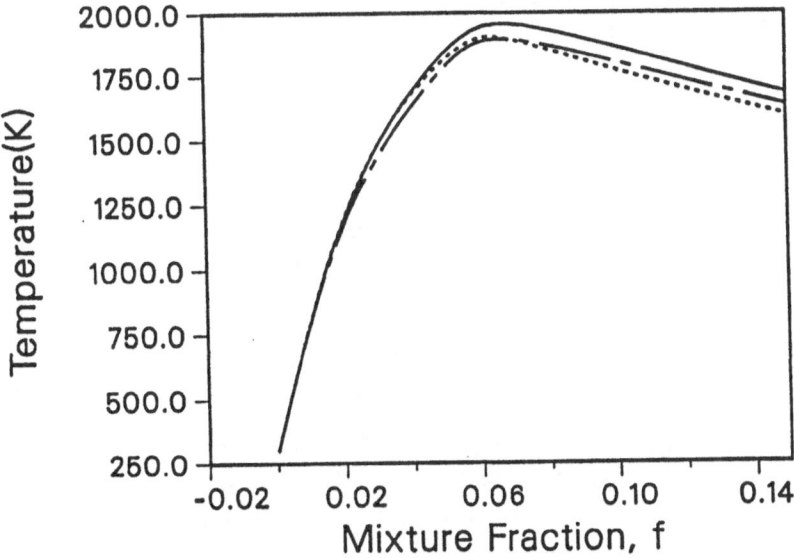

Fig. 2. Predicted temperatures versus mixture fraction for Tsuji type opposed laminar methane-air jet flame at a strain rate $a = 300/s$. Dashdot line: skeletal mechanism; solid line: Bilger's mechanism; dashed line: Peters and Seshadri's mechanism.

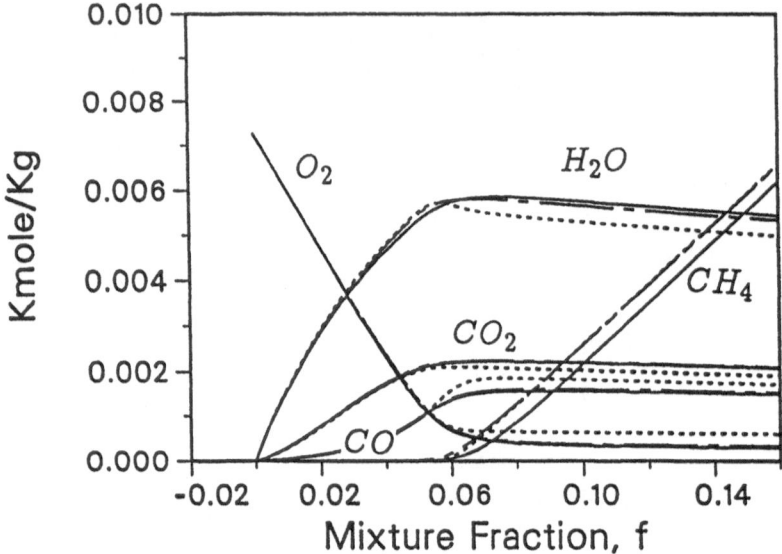

Fig. 3. Predicted major species versus mixture fraction for Tsuji type opposed laminar methane-air jet flame at a strain rate $a = 300/s$. Dashdot line: skeletal mechanism; solid line: Bilger's mechanism; dashed line: Peters and Seshadri's mechanism.

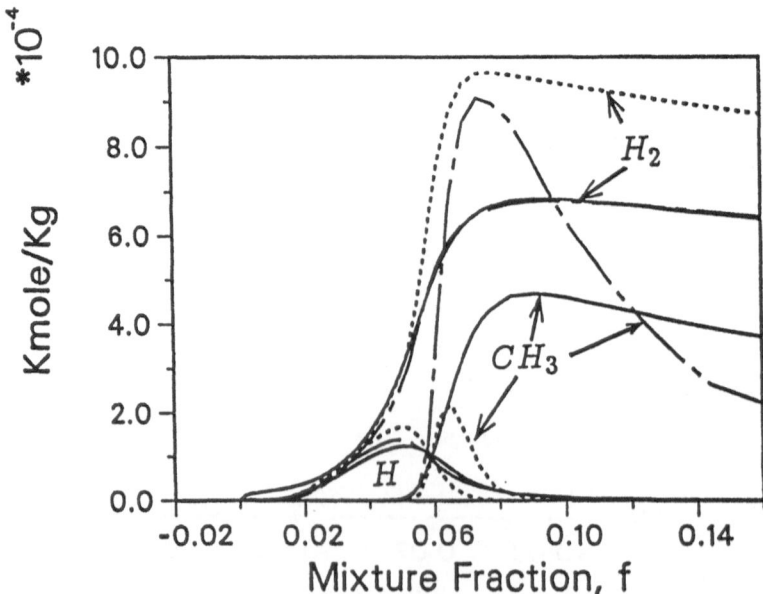

Fig. 4. Predicted intermediate species versus mixture fraction for Tsuji type opposed laminar methane-air jet flame at a strain rate $a = 300/s$. Dashdot line: skeletal mechanism; solid line: Bilger's mechanism; dashed line: Peters and Seshadri's mechanism.

(3) Limits for n_{CO} at a given (f, n_{CH_4}):

$$n_{CO}^{max} = \frac{f}{W_{CH_4}} - n_{CH_4} \equiv \Omega \tag{2.4}$$

$$n_{CO}^{min} = max\{2\Omega - \frac{2\psi(1-f)}{(W_{N_2} + \psi W_{O_2})} \quad, 0\}. \tag{2.5}$$

(4) Limits for n at a given (f, n_{CH_4}, n_{CO}):

$$n^{max} = \frac{(1+\psi)(1-f)}{(W_{N_2} + \psi W_{O_2})} + \frac{2f}{W_{CH_4}} - n_{CH_4} + \frac{1}{2}n_{CO} \tag{2.6}$$

$$n^{min} = \frac{(1-f)}{(W_{N_2} + \psi W_{O_2})} + \frac{3f}{W_{CH_4}} - 2n_{CH_4} + max\{\frac{\psi(1-f)}{(W_{N_2} + \psi W_{O_2})} - 2\Omega + \frac{1}{2}n_{CO} \quad, 0\}. \tag{2.7}$$

(5) Limits for n_H at a given (f, n_{CH_4}, n_{CO}, n):

$$n_H^{max} = \frac{4}{3}\{n - n_{CH_4} - \frac{\psi(1-f)}{(W_{N_2} + \psi W_{O_2})} - \Omega - \frac{1}{2}n_{CO}\}$$
$$- \frac{2}{3}max\{n_{CH_4} - n + \frac{(1-2\psi)(1-f)}{(W_{N_2} + \psi W_{O_2})} + 7\Omega - n_{CO} \quad, 0\}, \tag{2.8}$$

$$n_H^{min} = 0. \tag{2.9}$$

Compositions outside these *allowable* limits cannot occur. In a given flow, not all of the allowed composition may be accessible. The *accessible* domain contains all possible compositions that could occur in a particular flow (Pope [7]). Figures 5 to 8 illustrate both the *allowable* and the *accessible* domains of major species versus mixture fraction for a nonpremixed turbulent jet flame stabilized by a coflowing pilot flame of stoichiometric mixture. The accessible domain can be obtained by equations (2.1) to (2.9) except that equation (2.3) is now replaced by

$$n_{CH_4}^{min} = max\{\frac{f}{W_{CH_4}} - \frac{\psi(1-f)}{2(W_{N_2} + \psi W_{O_2})} \quad, 0\}. \tag{2.10}$$

Also shown in the figures are the compositions obtained from calculations of Tsuji-type opposed flow *laminar* flames at a high stretch rate, $a = 450/s$ (recall that $a \approx 460$ is extinction) and at a low stretch rate, $a = 5$. These laminar flame calculations provide a test of the applicability of the two reduced mechanisms only for a small fraction of the whole accessible domain. In order to further evaluate the performance of the reduced mechanisms, flows other than the opposed flow flame configuration are needed. One possibility is the well-stirred reactor flow in which the time scales of mixing and reaction are comparable, and thus a wider range of compositions might be tested.

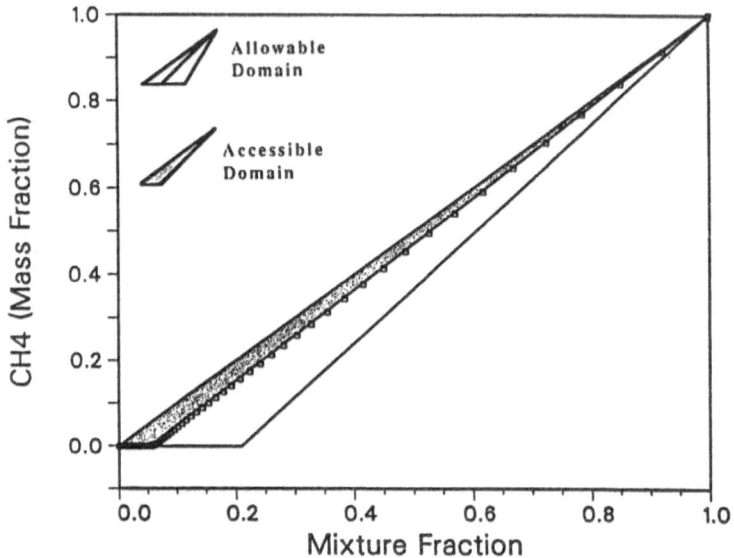

Fig. 5. Allowable and accessible domains for methane versus mixture fraction in methane-air jet flows with a pilot flame of burned stoichiometric mixture. Symbols are results from calculations for Tsuji type laminar opposed jet flames at $a = 450/s$ △ and at $a = 5/s$ ◯.

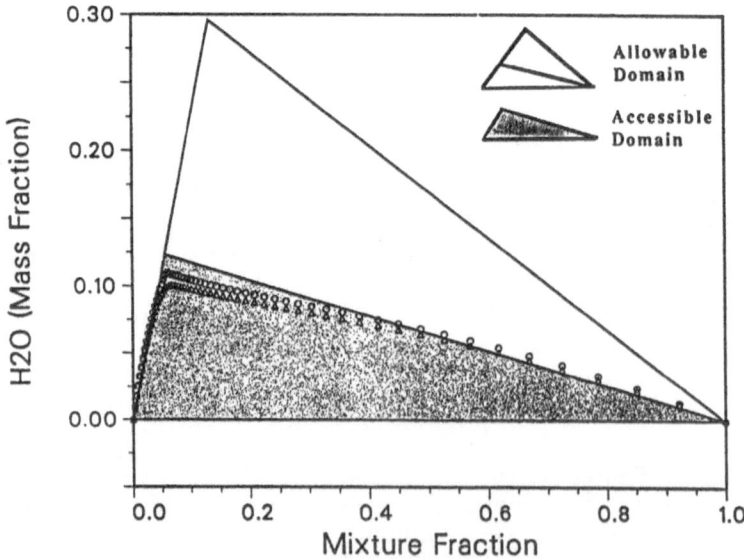

Fig. 6. Allowable and accessible domains for water versus mixture fraction in methane-air jet flows with a pilot flame of burned stoichiometric mixture. Symbols are results from calculations for Tsuji type laminar opposed jet flames at $a = 450/s$ △ and at $a = 5/s$ ◯.

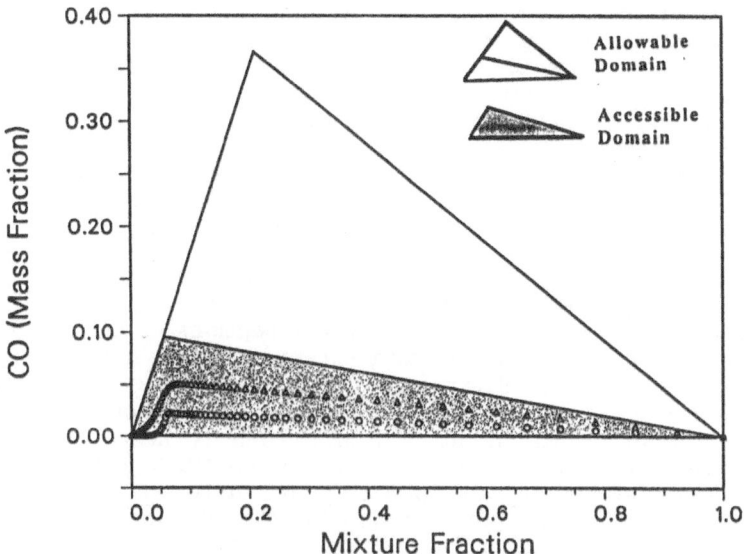

Fig. 7. Allowable and accessible domains for CO versus mixture fraction in methane-air jet flows with a pilot flame of burned stoichiometric mixture. Symbols are results from calculations for Tsuji type laminar opposed jet flames at $a = 450/s$ △ and at $a = 5/s$ ○.

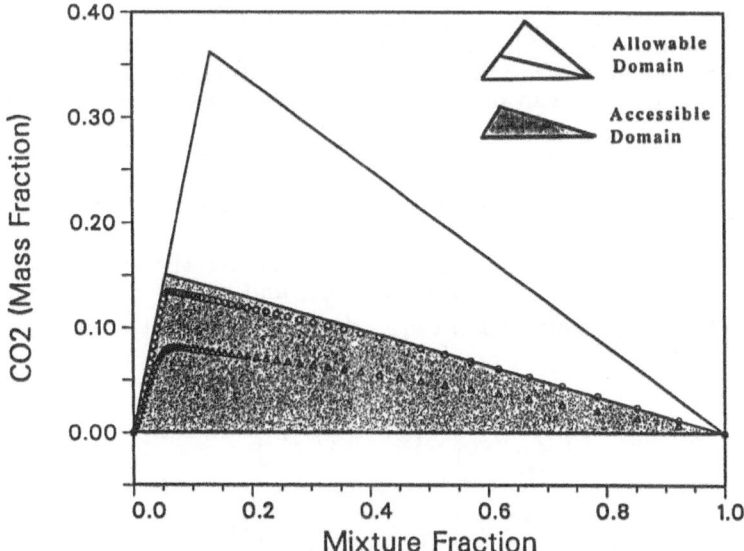

Fig. 8. Allowable and accessible domains for CO_2 versus mixture fraction in methane-air jet flows with a pilot flame of burned stoichiometric mixture. Symbols are results from calculations for Tsuji type laminar opposed jet flames at $a = 450/s$ △ and at $a = 5/s$ ○.

The effects of chemical reaction on the joint pdf are simulated by moving the representative particles through the allowable domain according to their chemical kinetic rate equations. In the allowable region, the reaction path and the speed along a path are all functions of the properties of the mixture. They do not depend on the velocity of the flow in which the reaction is taking place. Therefore, a 'look-up' table containing changes of reactive species for a given period of time can be constructed prior to the Monte Carlo simulations.

A look-up table for the entire *allowable* domain can be generated but it is wasteful as the compositions outside the *accessible* domain are not needed. Therefore, for better accuracy, look-up tables have been generated only for compositions in the *accessible* domain. The five dimensional domain for the reduced mechanisms is discretized in a five-scalar array $(f, n_{CH_4}, n_{CO}, n, n_H)$ with dimensions of (25,10,9,9,9). The grid point distributions in the array are uniform except in the f array which has dense grids clustered around the stoichiometric point. At each grid point in the domain, the look-up tables contain changes of four reactive scalars due to combustion for two time intervals, $10\mu s$ and $100\mu s$ (eight entries) and in addition, the density; hence a total of nine entries. The table consists of about 1.5 million words ($= (25*10*9*9*9)*9$). During Monte Carlo simulations, the scalar changes of each representative particle due to combustion are obtained from the look-up tables by an efficient multi-linear interpolation scheme. This interpolation is nontrivial and is described in detail by Chen *et al.* [15].

3. Turbulence Model and Scalar Pdf Model

The statistics of the mixing process and the chemical reactions occurring in the flow field are described in terms of the single point pdf $\widetilde{P}(\psi_1, .., \psi_n; \underline{x}, t)$ of the thermo-chemical variables (scalars) $\phi_1, .., \phi_n$. Here (ψ_i) is a composition variable and (ϕ_i) is the corresponding sample-space variable. The statistical properties of the fluctuating *velocity* field are simulated by means of a second-order closure model most recently described by Dibble *et al.* [16]. This second-order closure turbulence model provides a turbulent time scale and diffusivity for the pdf equation, and the pdf equation in turn allows calculation of the mean density that is required for the velocity closure.

The scalar variables $\phi_1, .., \phi_n$ are governed by transport equations of the general form

$$\rho D_t \phi_i = \partial_\alpha(\rho \Gamma_i \partial_\alpha \phi_i) + \rho S_i(\phi_1, .., \phi_n), \qquad i = 1, .. n \qquad (3.1)$$

where Γ_i denotes the molecular diffusivity, D_t denotes the substantial derivative, S_i denotes the chemical source term for scalar ϕ_i, and repeated Greek subscripts imply summation. The single point pdf $\widetilde{P}(\psi_1, .., \psi_n; \underline{x}, t)$ satisfies the following conservation equation

$$\bar{\rho}\partial_t \widetilde{P} + \bar{\rho}\widetilde{v}_\alpha \partial_\alpha \widetilde{P} + \bar{\rho} \sum_{i=1}^{N} \partial_{\psi_i}\left\{ S_i(\psi_1, .., \psi_N)\widetilde{P} \right\} =$$
$$- \partial_\alpha\left(\bar{\rho} < v''_\alpha \mid \phi_i = \psi_i > \widetilde{P} \right) - \bar{\rho} \sum_{i=1}^{N}\sum_{j=1}^{N} \partial^2_{\psi_i \psi_j}\left(< \varepsilon_{ij} \mid \phi_k = \psi_k > \widetilde{P} \right), \qquad (3.2)$$

where $v''_\alpha \equiv v_\alpha - \widetilde{v}_\alpha$ and $\varepsilon_{ij} \equiv \Gamma \, \nabla\phi_i \bullet \nabla\phi_j$, with $\Gamma_i = \Gamma_j = \Gamma$. The use of density-weighted variables (*e.g.*, $\widetilde{P}, \widetilde{v}_\alpha$) has formal advantages for flows with large density fluctuations (Bilger [17]). Since density ρ is a local function of the thermo-chemical scalars, the unweighted pdf P can be recovered from the density weighted pdf $\widetilde{P}$

$$P(\psi_1, .., \psi_n) = \frac{\bar{\rho}}{\rho(\psi_1, .., \psi_n)} \widetilde{P}(\psi_1, .., \psi_n) \qquad (3.3)$$

The chemical kinetic sources S_i act as convection velocities in the scalar space. They appear in closed form in equation (3.2), and thus no closure model is required, even for large chemical mechanisms.

The turbulent flux term $\partial_\alpha (< v''_\alpha \mid \phi_i = \psi_i > \widetilde{P})$ and the scalar dissipation rate $\sum_{i=1}^{N} \sum_{j=1}^{N} \partial^2_{\psi_i \psi_j} (< \varepsilon_{ij} \mid \phi_k = \psi_k > \widetilde{P})$ are unknowns and they are modeled in terms of 'known' variables. The closure expressions for the turbulent flux and the scalar dissipation terms were previously developed (Janicka et $al.$ [4], Dopazo [18]), and applied to turbulent nonpremixed jet flames (Chen and Kollmann [9]). The turbulent flux of the pdf is modeled by

$$-\bar{\rho} < v''_\alpha \mid \phi_i = \psi_i > \widetilde{P} \cong C_s \bar{\rho} \, \frac{\widetilde{k}}{\widetilde{\varepsilon}} \, \widetilde{v''_\alpha v''_\beta} \partial_\beta \widetilde{P}, \qquad (3.4)$$

in analogy with the gradient-flux model for statistical moments. The closure for the scalar dissipation of the pdf was based on pair-wise interaction of fluid parcels (see, e.g., [4,18]) and resulted in

$$-\sum_{i=1}^{N} \sum_{j=1}^{N} \partial^2_{\psi_i \psi_j} \left(< \varepsilon_{ij} \mid \phi_k = \psi_k > \widetilde{P} \right)$$
$$\cong \frac{C_D}{\tau} \left\{ \int_{D_N} \cdot \cdot \int d\psi' \int_{D_N} \cdot \cdot \int d\psi'' \; \widetilde{P}(\psi') \widetilde{P}(\psi'') \Theta(\psi', \psi'' \mid \psi) - \widetilde{P} \right\}, \qquad (3.5)$$

where

$$\Theta(\psi', \psi'' \mid \psi) = \prod_{i=1}^{N} \Theta_i(\psi'_i, \psi''_i \mid \psi_i)$$

and

$$\Theta(\psi'_i, \psi''_i \mid \psi_i) = \begin{cases} \frac{1}{|\psi''_i - \psi'_i|} & \text{for } \psi'_i \le \psi_i \le \psi''_i \text{ or } \psi''_i \le \psi_i \le \psi'_i; \\ 0 & \text{otherwise,} \end{cases} \qquad (3.6)$$

and $C_D = 6.0$. The scalar domain D_N is the set of all $allowable$ values of $\{\phi_1, .. \phi_n\}$ and τ denotes the time scale for this process (i.e., the reciprocal of mixing frequency). The present value for τ is given by

$$\tau = \frac{\widetilde{k}}{\widetilde{\varepsilon}} . \qquad (3.7)$$

The above closure model for the scalar dissipation, equation (3.5), is a simplified picture of the mixing process (Curl [3]). It is known that the present model (a modified coalescence/dispersion model as shown in equation (3.5)) satisfies the mathematical requirements for the pdf but is incorrect in the short time limit and leads to unbounded flatness and superskewness (standardized fourth and sixth moments, which are equal to 3 and 15 for a Gaussian pdf) values for decaying turbulence as time goes to infinity

(Pope [5], Kosaly [19], Kosaly and Givi [20]). The latter shortcoming may be remedied if another variable, such as the time interval between mixing events, is included in the pdf and the probability of sampling $\widetilde{P}$ is replaced by the probability of mixing in the integral term on the right hand side of equation (3.5). A recent study by Chen and Kollmann [21] shows that modification of current mixing models by age-biased sampling has little or no influence on the predicted scalar statistics up to fourth moment for turbulent reacting and nonreacting jet flows. A more serious problem is the fact that the mixing rate is determined from local properties of the turbulent flow field, and no information on distribution of energy or scalar intensities over large scales is used. This can lead to incorrect profiles for the scalar pdf in turbulent mixing layers, even though the mean of the scalar is in agreement. This pdf prediction failure is likely a consequence of the role of large scale structure not being reflected in the mixing model. An excellent example of this mode of model failure is clearly illustrated by Figure 24 in Koochesfahani and Dimotakis [22]. They showed that the model predicts the wrong peak positions in the pdf profiles which are believed to be strongly influenced by the large scale structure.

In addition to the above-mentioned shortcomings of the coalescence/dispersion mixing models, application of such models to reacting flows requires further consideration of the effects of combustion on mixing. For instance, the mixing frequency and the transition probability defined in equation (3.6) are independent of the reaction rates; hence, possible influence of chemical reactions on mixing is not included.

Implementation of Monte Carlo Simulation:

As the joint scalar pdf $\widetilde{P}$ is a function of space and composition, the total number of independent variables required to describe $\widetilde{P}$ is three plus the total number of composition scalars. Traditional finite difference techniques are not efficient in solving equation (3.2) with a large dimension. A stochastic approach (i.e., the Monte Carlo simulation) introduced by Pope [6] alleviates this enormous computational demand and problem solution is feasible with several scalars.

Formal mathematical representations of each term in equation (3.2) by their corresponding statistical Monte Carlo procedures can be found in the paper by Pope (1981). Here, we present illustrations to describe the stochastic procedures in simulating effects due to each term in equation (3.2). First, the effects of convection by mean velocity and turbulent flux (first two terms on LHS and the first term on RHS) on the joint pdf $\widetilde{P}$ are simulated by moving representative particles in the *physical* space, i.e., along the grid points, as shown in Figure 9. For a round jet configuration, let us denote the downstream and cross-stream directions by x and r respectively. If the pdf equation were solved by a three-point explicit finite difference scheme, the pdf at the downstream location $x + dx$ can be expressed as sum of the pdfs at upstream location x as

$$
\begin{aligned}
\widetilde{P}(\psi_1, ..\psi_n; x + dx, r_J) = {} & A(J+1)\widetilde{P}(\psi_1, ..\psi_n; x, r_{J+1}) \\
& + \quad A(J)\widetilde{P}(\psi_1, ..\psi_n; x, r_J) \qquad\qquad (3.8) \\
& + A(J-1)\widetilde{P}(\psi_1, ..\psi_n; x, r_{J-1})
\end{aligned}
$$

where $A(J+1)$, $A(J)$, and $A(J-1)$ denote the amount of contributions from cells at $J+1$, J and $J-1$. As the joint pdf is represented by a number of statistical events (particles), the pdf at the downstream location can be represented by a sum of statistical events from the three upstream cells weighted by their respective contributions. The particles are selected randomly (by calling a random number generator) from each upstream cell to form the new pdf at the downstream location.

Second, the influences of chemical reactions and of molecular mixing on the joint pdf are equivalent to moving the particles in the *composition* domain. As emphasized earlier, the effects of chemical processes on the joint pdf can be simulated exactly according to the kinetic rate equations and only the molecular mixing process requires modeling. However, for combustion problems, these two processes are intimately related, and the interactions between them must be properly simulated. Modeling of mixing and of interactions between chemical reaction and mixing are currently important and challenging issues for the pdf method (Chen *et al.* [15]). To demonstrate the stochastic procedures, the chemical reaction and mixing are treated separately, and hence, any potential interactions between them, except through the mean density changes, are not included. Simulation of the modified Curl's mixing model [3] by Janicka *et al.* [4] is performed by selecting randomly a pair of particles within a grid cell and mixing them up to random degree. The number of pairs to be selected is calculated by the product of the *mixing frequency* defined in equation (3.7) and the time spent during the marching step, dx/U. The chemical reactions then move all the particles to new positions in the composition space according to their rate equations. Figure 10 illustrates such procedures for a selected pair of particles undergoing molecular mixing and chemical reaction in the CH_4-O_2 composition space. Modifications and improvements of these Monte Carlo procedures to account for the interactions between chemical reactions and molecular mixing are currently being undertaken.

4. Numerical Modeling of Turbulent Methane Jet Flames:

The recent laser Raman measurements in nonpremixed turbulent methane jet flames by Dibble *et al.* [23] and by Masri *et al.* [24,25] provide the data base for model evaluation. Figure 11 shows a schematic of the experimental test facility. The burner has a central jet of methane, 7.2 mm in diameter, surrounded by a stoichiometric premixed and fully burned annulus of C_2H_2, H_2, and air with C/H ratio adjusted to that of methane. The outer diameter of the annulus is 18 mm and the lips are thin. During all the measurements, the coflowing air stream velocity and the pilot burnt gas velocity are maintained at 15 and 24 m/s, respectively. The adiabatic temperature of the pilot gas is 2600 K.

Four flames with increasing central jet velocities have been studied with time resolved laser Raman scattering. These new investigations provide the first detailed description of the nonequilibrium chemistry nature of these flames. We focus on two flames, Flame L and Flame B, with central jet velocities of 41 m/s and 48 m/s respectively. The nonequilibrium nature of the chemical kinetics is characterized by the ratio of time scales between turbulent mixing and chemical reaction (a Damköhler number). Increasing central jet velocity decreases the turbulent mixing time scale and leads to smaller Damköhler numbers. Local flame extinction can occur when the chemical kinetics are unable to keep up with the turbulent mixing processes. The experimental data show that both in Flame L and in Flame B, local flame extinction is observed. However, in Flame B the flame is almost totally extinguished at $x/D = 20$, where the peak of the *mean* temperature is only 900K. The flame is re-ignited further downstream, where the strain rate is decreased. It is interesting to note that, although the jet velocity in Flame B is only 7 m/s higher than that in Flame L, the flame exhibits quite different characteristics. Such a sharp change in the flame structure indicates a crossover point of competition between chemical reactions and turbulent mixing. Prediction of this crossover behavior is a severe test for turbulent combustion models.

Numerical modeling of Flame L and Flame B was carried out with 50 grid cells distributed from the centerline to a satisfactorily large radius and with 1000 particles per grid cell, a total of 50,000 particles for the Monte Carlo calculation. The numerical solutions are obtained by a parabolic marching scheme with a block-diagonal solver and a staggered grid system (Chen *et al.* [26]). In contrast to equilibrium chemical models,

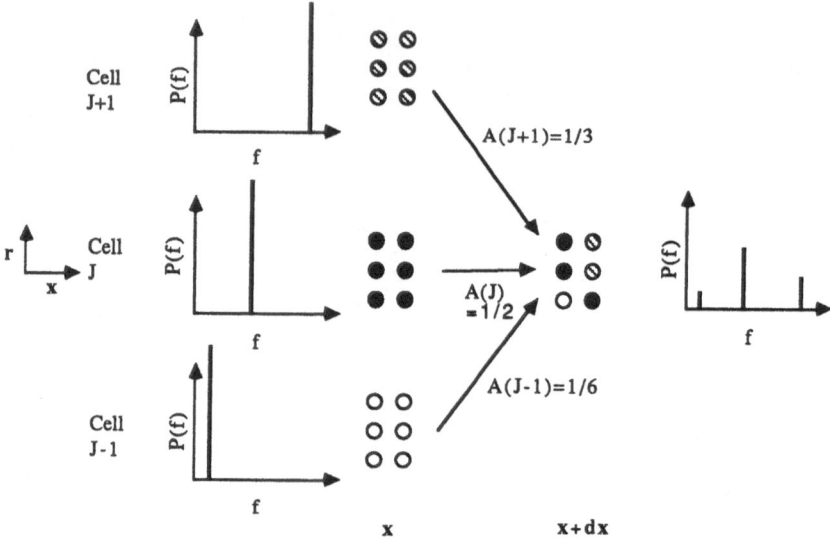

Fig. 9. Schematic of the stochastic procedure simulating the effects of convective terms on the joint scalar pdf.

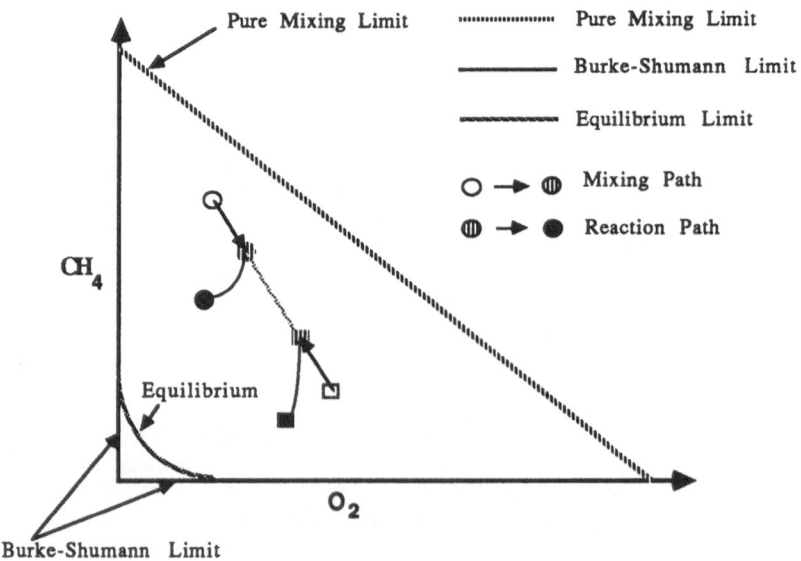

Fig. 10. Schematic of the stochastic procedure simulating the effects of chemical reaction and mixing on the joint scalar pdf.

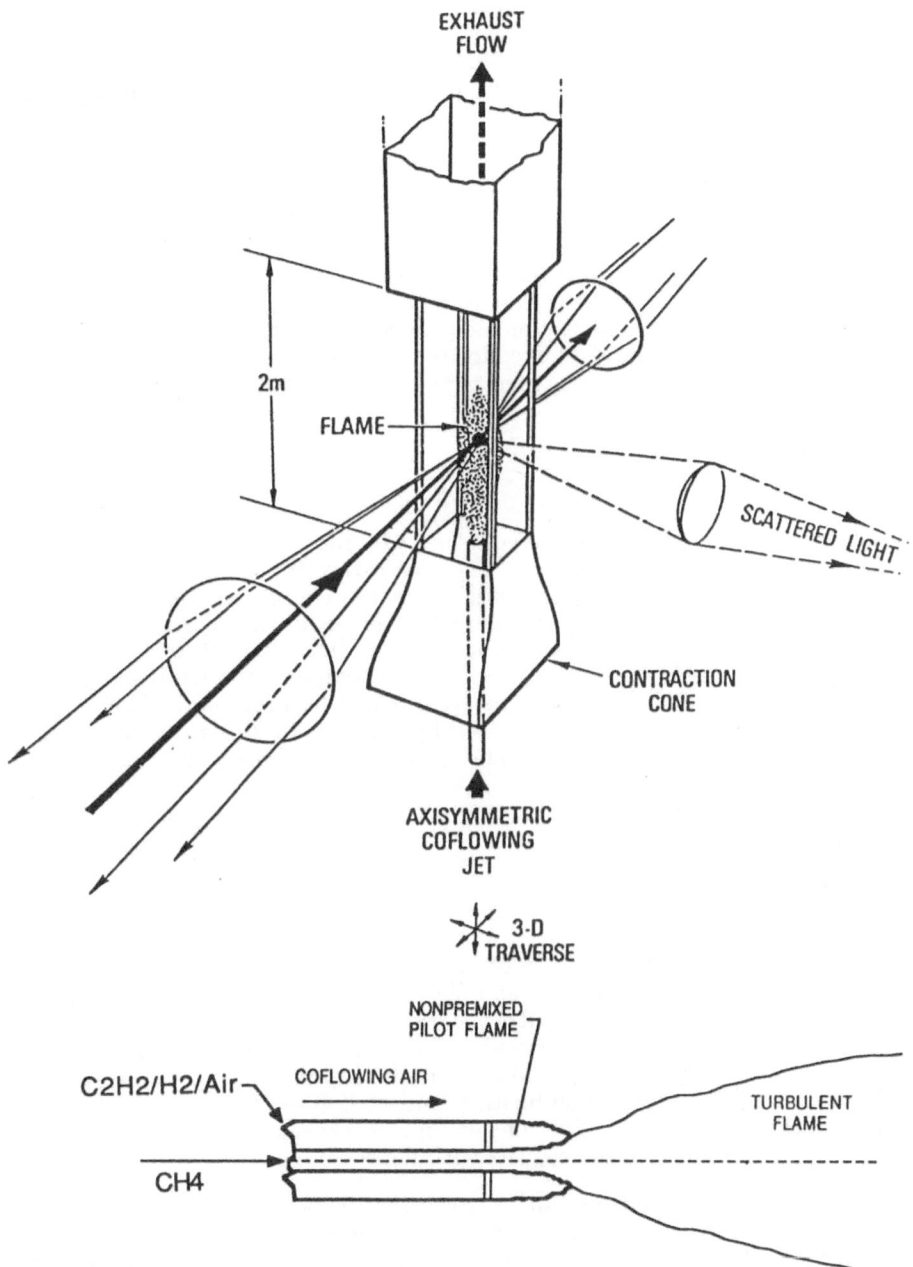

Fig. 11. Schematic of experimental test facility showing the Raman laser beam and details of the central jet.

the current method requires an ignition source for the flame to burn. In the Monte Carlo simulations, ignition is simulated by inserting particles with burned stoichiometric mixture into the jet at a location just downstream of the jet pipe exit (x/D = 0.03) and for the radial range given by the coflowing pilot flame of the experimental apparatus ($0.5 \leq r/D \leq 1.25$).

It should be noted that the ignition process in the numerical simulation uses the same fuel as the main jet flame (a stoichiometric mixture of methane and air), whereas the pilot flame in the experiment employs a mixture of H_2, C_2H_2 and air, which has the same atomic species as a stoichiometric methane flame, but has a much higher premixed flame speed needed for pilot flame stabilization. Consequently, the pilot flame has an adiabatic flame temperature that is 400 degrees higher than that for stoichiometric methane-air mixture. The use of such a pilot, in fact, introduces a third stream; the other two being nozzle fluid and coflowing air. As the model can only simulate two-stream mixing problems, we assume that the pilot flame temperature is that for stoichiometric methane-air mixture.

5. Results and Discussion

The solution of the pdf transport equation provides statistical information on the thermo-chemical scalars at a given point. However, we will present only a selected set of predicted statistical moments and joint pdf's that are compared with the experimental data. From these comparisons, we evaluate model performance and we then suggest areas for model improvements. Comparisons of predictions with experiments for the mean and variance of mixture fraction, density, and temperature will be considered first, and then the joint distribution among the scalars will be presented. The experimental results of Masri *et al.* [24,25] are represented by symbols in all figures, the calculations using the reduced mechanism of Peters and Seshadri by dashed lines, and the reduced mechanisms of Bilger by solid lines.

Mean and Variances

Mixture Fraction

The mixture fraction is a measure of extent of mixing. The radial profiles for the mean and variance of mixture fraction at x/D=20 for Flame B and Flame L presented in Figure 12 show good agreement with the experiment data. Noticeable, but small, differences in the predictions are observed between the two reduced mechanisms. The only difference between these two predictions is the input 'look-up' table, which determines the scalar changes due to combustion and therefore the density. We concluded that the chemical models are not different enough to cause significant differences in the mean density which in turn influences the turbulent mixing.

As is evident in Figure 12 from both the simulations and the experimental data, the overall mixing characteristics are not significantly altered when the central jet velocity is increased from 41 m/s to 48 m/s (from Flame L to Flame B). A closer comparison shows that in Flame B the fuel mixes with air slightly faster than in Flame L as indicated by the lower mean mixture fraction $\tilde{f}$ at various axial locations. This trend is consistent with the observation in nonreacting flows that turbulent transport in the radial direction is increased with the strain rate.

Density

The profiles for the mean density $\bar{\rho}$ and its variance, presented in Figure 13, are of great importance because fluid dynamics and chemical reactions are linked primarily

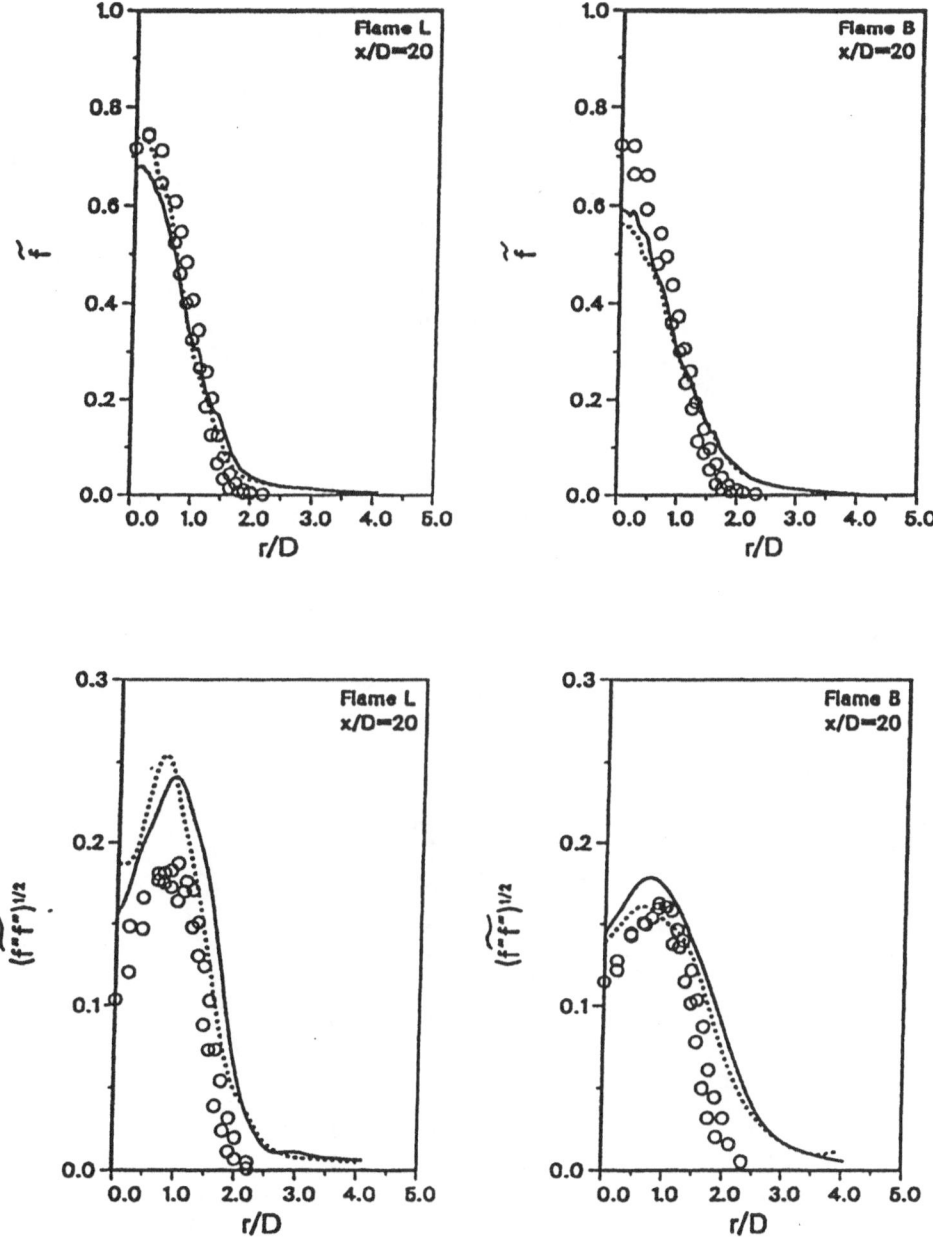

Fig. 12. Radial profiles of mean mixture fraction and its variance at x/D=20 for Flame B and Flame L. Solid line: Bilger's mechanism; dashed line: Peters and Seshadri's mechanism; symbols ◯ data from Masri *et al.* (1988a) and (1988b).

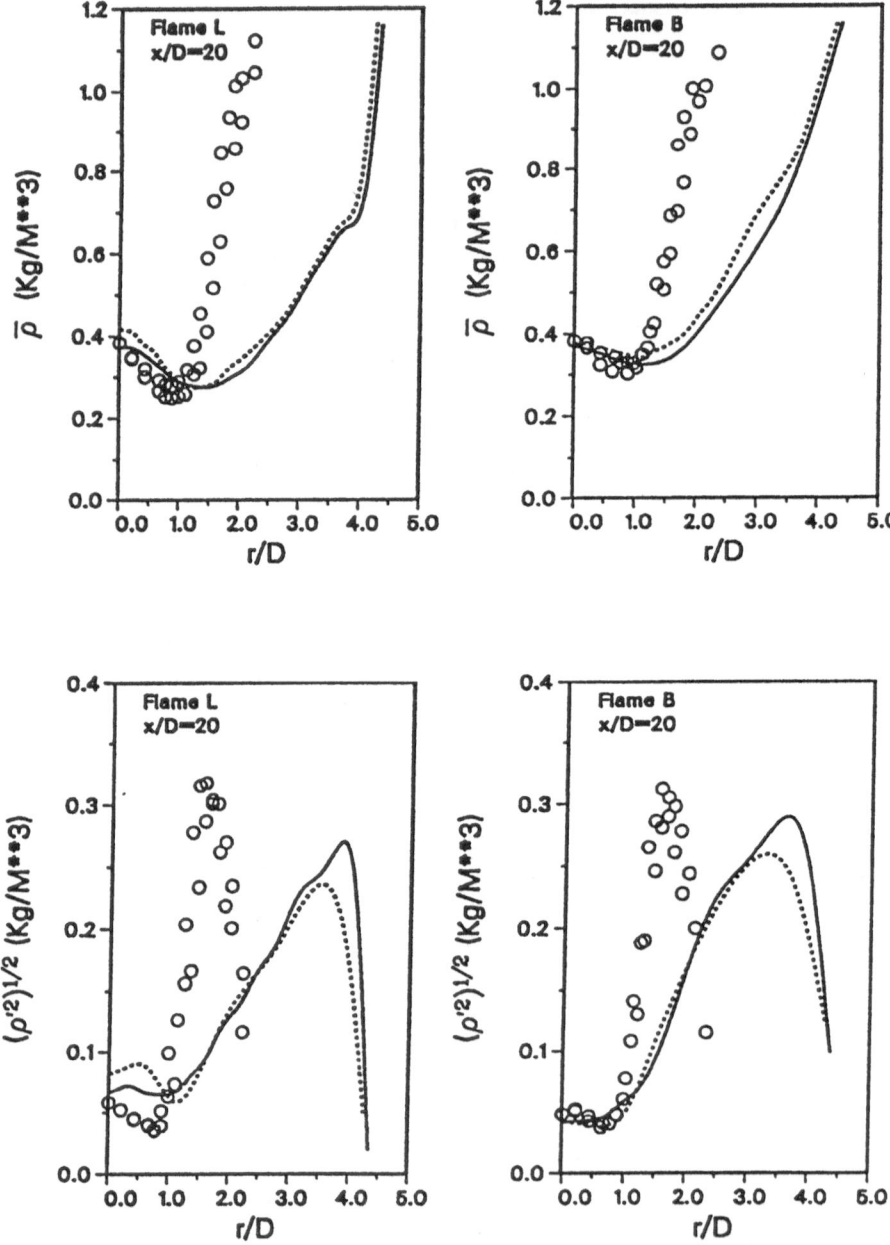

Fig. 13. Radial profiles of mean density and its variance at x/D=20 for Flame B and Flame L. Solid line: Bilger's mechanism; dashed line: Peters and Seshadri's mechanism; symbols ◯ data from Masri *et al.* (1988a) and (1988b).

through the density. Compared to the experimental data, the predicted mean density profiles are much broader. The observed discrepancies are large and the reasons for them are as follows.

In the fast chemistry limit, the density becomes solely a function of the mixture fraction. For methane-air combustion, the stoichiometric value of the mixture fraction is $f_{sto.}$=0.055. Density changes rapidly from 1.2 kg/m^3 to 0.2 kg/m^3 as f increases from 0 to 0.055 creating a steep gradient in the $\rho - f$ plane of about -18 kg/m^3. Therefore, in the lean side of the flame, a change of 1% in the mixture fraction causes a change of 0.18kg/m^3 (15% of air density) in the density. However, in the rich part of the flame, the corresponding density change is only about 0.01kg/m^3 (1% of air density). Hence, even for flows with equilibrium chemistry, an accurate prediction of the mixture fraction is prerequisite for accurate density predictions in the lean part of the flame.

As shown above, the predicted mean mixture fraction profiles compare reasonably well with experiments but noticeable differences (more than few percent) are observed. The observed large discrepancies between the calculated and the measured mean density profiles can be caused both by the inaccurate predictions of the mixture fraction and by the nonequilibrium chemistry. To exclude the effects of inaccurate mixture fraction predictions, the fluctuation of mixture fraction and the mean density are plotted versus the mean mixture fraction as shown in Figure 14. As seen from the $f'' - \tilde{f}$ plot, the turbulent fluctuation levels have been reasonably well predicted, especially for lean mixtures. If the nonequilibrium chemistry is correctly predicted by the current model, the predicted mean density profiles should agree with the experimental data in the $\rho - \tilde{f}$ plane. As evident from the plot, large discrepancies exist between the predicted mean density and experimental data, implying that the nonequilibrium chemical kinetics are not well captured by the current model. As will be shown in a later section, frequent local flame extinction is observed in both Flame L and Flame B; whereas, the model predicts a rare occurrence of local flame extinction. When the flame is locally extinguished, the instantaneous density increases significantly. As a result, the mean densities are high compared to model predictions for flames with high probability of local extinction.

Temperature

An important thermo-chemical property is temperature. In the experiments of Masri *et al.* [24,25], temperatures were deduced from the sum of all measured species concentrations with the assumption of the ideal gas law. The radial profiles of mean temperatures at x/D=20 are presented in Figure 15. Comparisons of these plots show that the peak temperatures are well predicted by the model for both Flame L and Flame B. However, due to the less frequent local extinction predicted by the model, the temperature profiles are much broader than those from experiments. Figure 16 presents a comparison of predicted and measured mean temperature profiles at various axial downstream locations for Flame B. The experimental data indicate that the peak mean temperature decreases from x/D=10 to x/D=20 and then increases from x/D=30 to x/D=50. A physical picture of the dynamics of the flame can be described as follows. At x/D=20, the flame experiences intense mixing and hence local extinction occurs frequently, resulting in low mean temperatures. The extinguished reactants are re-ignited by hot burned products at locations further downstream, where turbulent mixing is less intense. As a result of this reignition process, the mean peak temperature increases. The numerical model on the other hand predicts a continuously decreasing trend in temperature indicating the extinction-reignition process is not captured, suggesting an area for model improvement.

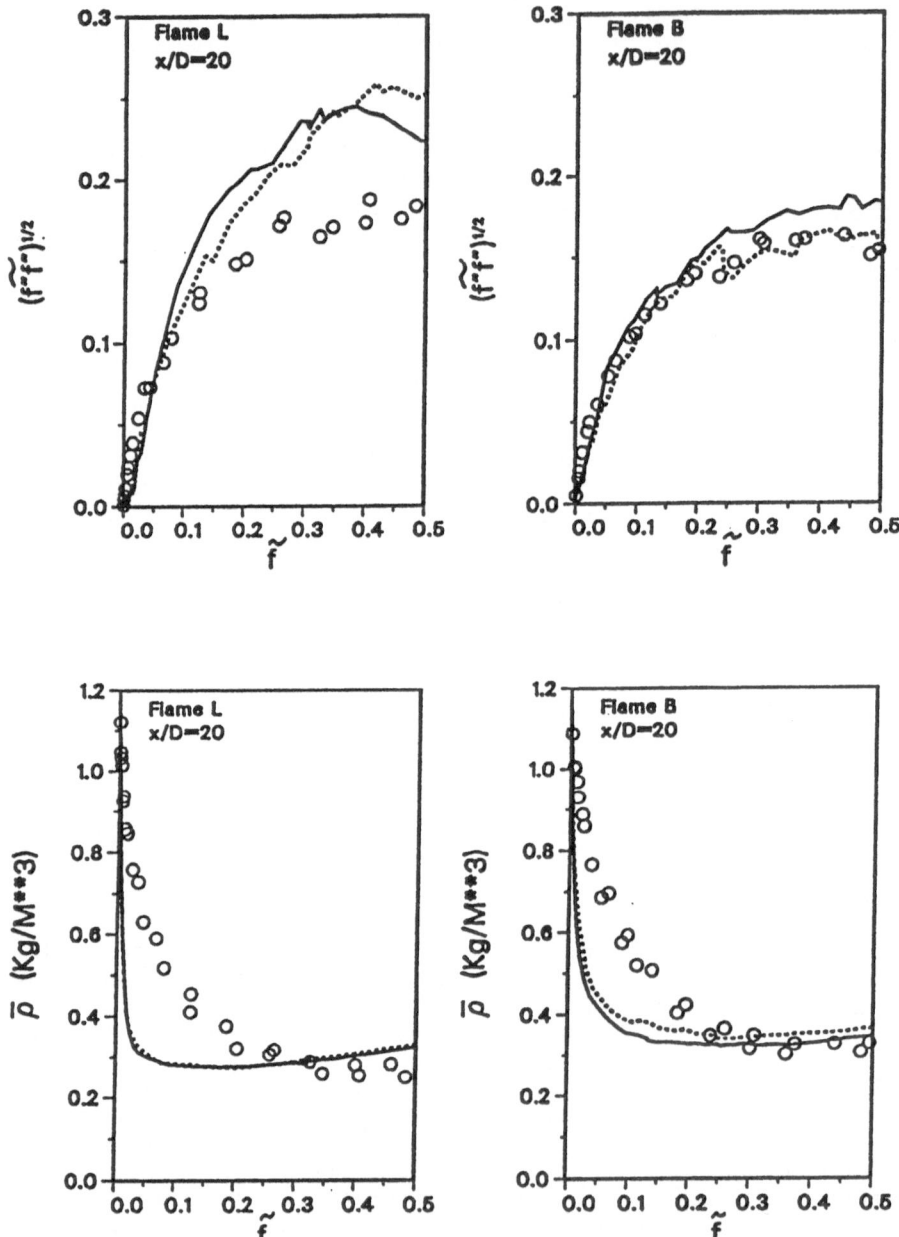

Fig. 14. Profiles of mixture fraction fluctuation and mean density versus mean mixture fraction at x/D=20 for Flame B and Flame L. Solid line: Bilger's mechanism; dashed line: Peters and Seshadri's mechanism; symbols ◯ data from Masri *et al.* (1988a) and (1988b).

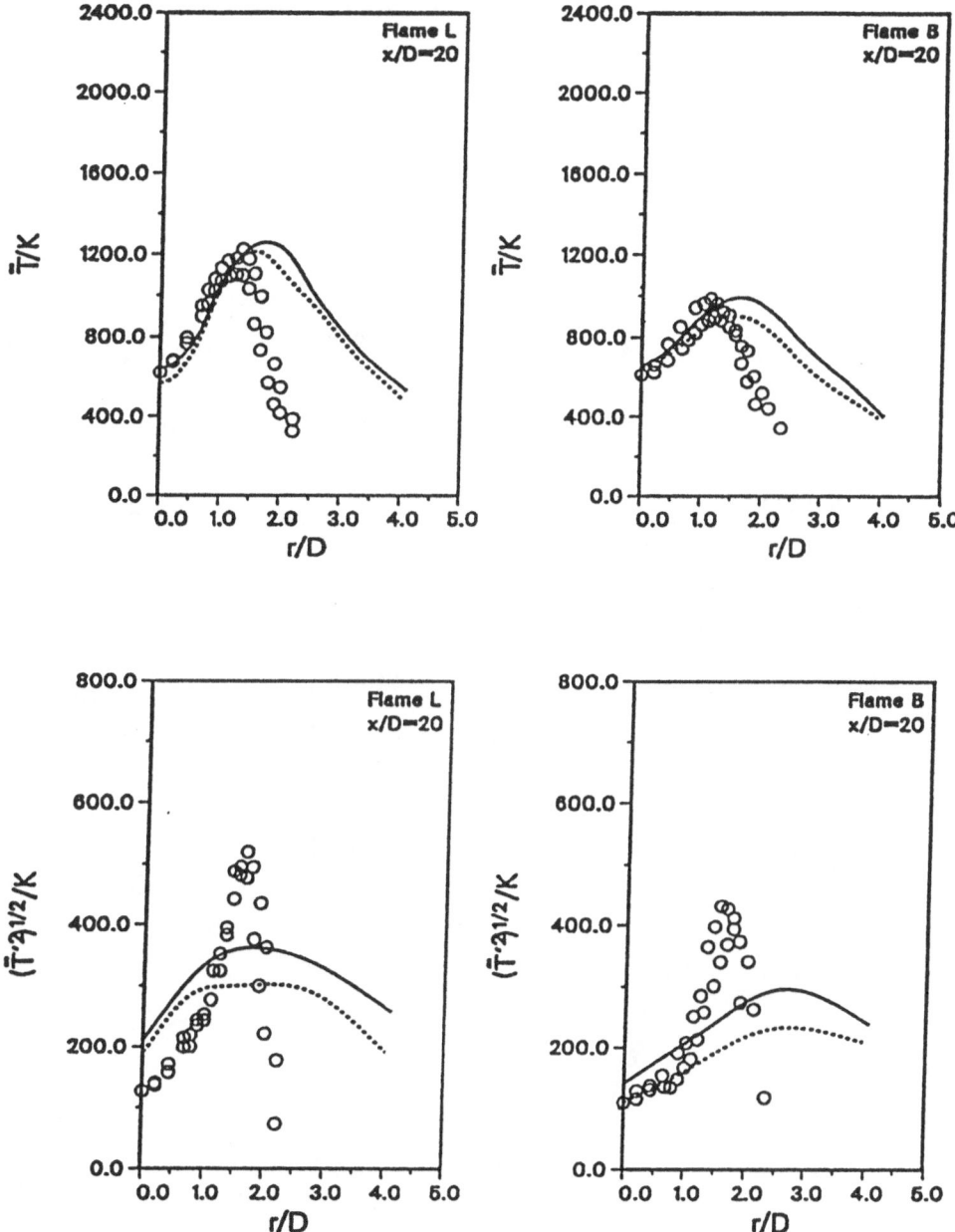

Fig. 15. Radial profiles of mean temperature and its variance at x/D=20 for Flame B and Flame L. Solid line: Bilger's mechanism; dashed line: Peters and Seshadri's mechanism; symbols ○ data from Masri *et al.* (1988a) and (1988b).

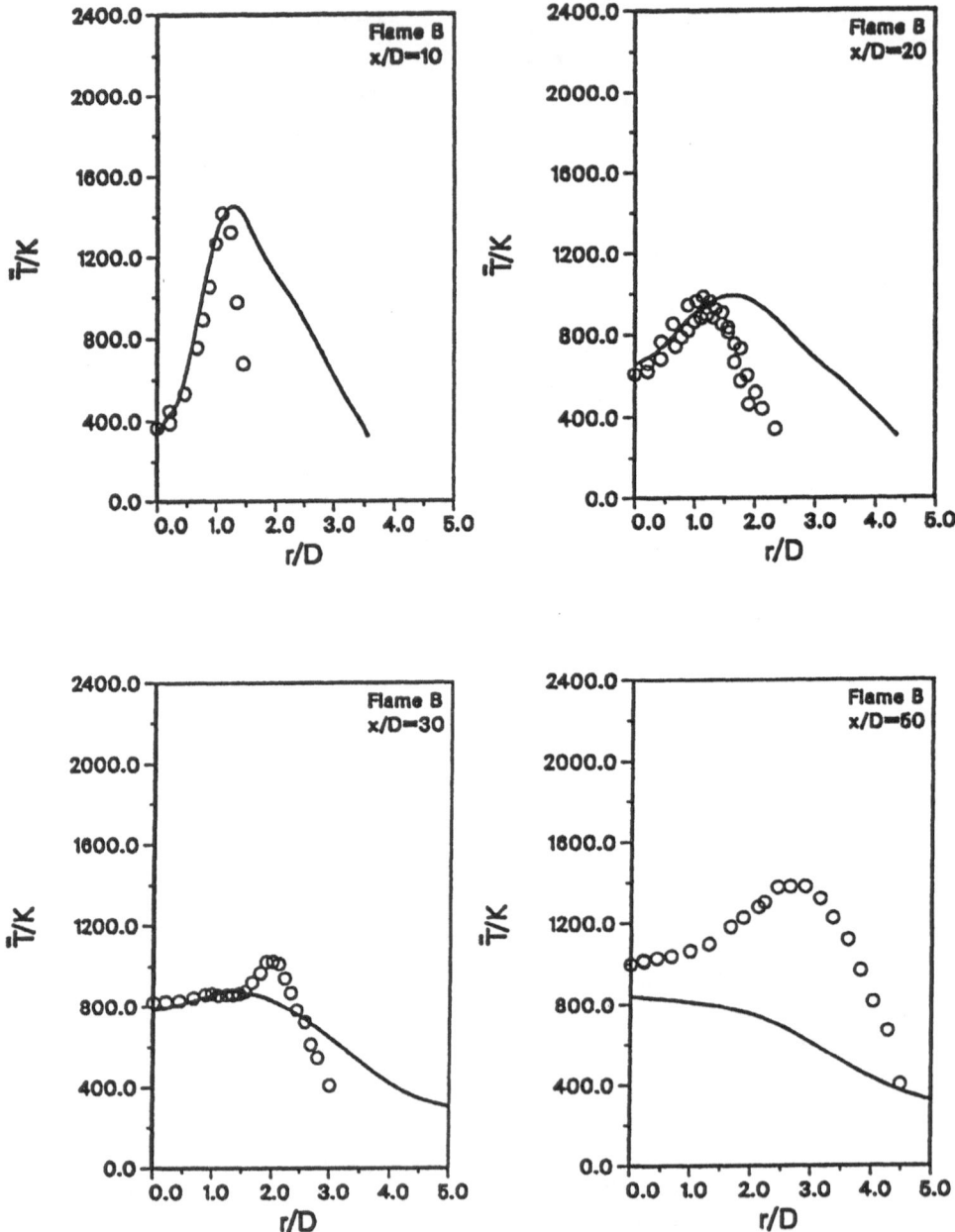

Fig. 16. Radial profiles of mean temperature at x/D=10, 20, 30, 50 for Flame B. Solid line: Bilger's mechanism; symbols ◯ data from Masri *et al.* (1988a) and (1988b).

Joint Pdf Comparisons

One of the major advantages of the pdf method is the ability to generate an ensemble of data points which can be directly compared to the ensemble of experimental data points obtained by pulsed laser diagnostics. Comparison of predictions and experimental measurements of joint statistics among the scalars, such as temperature and the mixture fraction, provides not only a valuable assessment of the models but also gives new insight leading to future improvements in the model and, as we shall see, in the experiment as well. The pdf's are presented in the form of scatter plots together with flamelet results taken from the calculation of laminar opposed jet flames as shown in Figures 17 to 19. The solid line corresponds to the strain rate a=450 s^{-1}, the dashdot line to a=300 s^{-1}, and the broken line to a=100 s^{-1}. Inclusion of these results allows us to explore the concepts of treating this turbulent flame as an ensemble of laminar flamelets (Liew et $al.$ [27], Peters [28]). The calculated pdfs are represented by 1,000 points per grid cell and the experiments provide approximately 1,250 laser shots at a given spatial position. Comparisons of predictions and the data will be presented for both Flame L and Flame B with different chemical models. A high density of scatter points corresponds to a large probability of finding the system in this region; hence, the pdf has large values in the region. Conversely, regions with low density of scatter points are associated with low values of the pdf.

Joint pdf between T and f

Figure 17 presents the joint statistics between temperature and the mixture fraction for both flames together with the corresponding experimental data. Compared to the experiment, the model predicts pdfs with small variance and much less frequent low temperature points. As the low temperature statistics are an indication of local flame extinction, the strong bimodal nature in the measured pdfs suggest that the flame is either burned or extinguished. Comparison shows that the model does not predict such a clear bimodal nature but instead a distributed pattern is predicted. As the jet velocity is increased, lower temperatures are observed in the experimental data. Consistent with the experimental data, the model also predicts lower temperatures as the strain rate is increased.

The experimental (T, f) scatter data allow us to evaluate the flamelet approach for modeling the present methane turbulent flames. According to the laminar flamelet concepts for turbulent nonpremixed combustion (Peters [28]), combustion takes place in narrow zones, called strained laminar flames, and local flame extinction occurs when the local mixing rate exceeds a critical value. The frequency of flame extinction depends on the mixture fraction dissipation rate and its statistical distribution. An important consequence of these concepts is that turbulent flames can be described entirely by local variables irrespective of the past history of the flame. Some developments of such models for nonpremixed turbulent flames with the pdf method have been reported by Haworth et $al.$ [29]. If such a model is employed here, then the burning flames would give a scatter pattern lying in a region bounded by the solid line (a=450 s^{-1}) and the highest temperatures. One clear example of local extinction would be represented by the 300K line for all values of the mixture fraction. As is evident in Figure 17, the experimental data indicate a high probability of occurrence for the instantaneous temperatures to fall between these two regimes.

Joint pdf between O$_2$ and CH$_4$ mass fractions

The joint pdf of O$_2$ and CH$_4$ mass fraction shown in Figure 18 allows us to examine the coexistence of fuel and oxidizer. For nonpremixed flames, infinitely fast and irreversible single step reaction does not allow coexistence of fuel and oxidizer. Consequently, the only allowable states are either on the Y_{O_2} axis or on the Y_{CH_4} axis. The

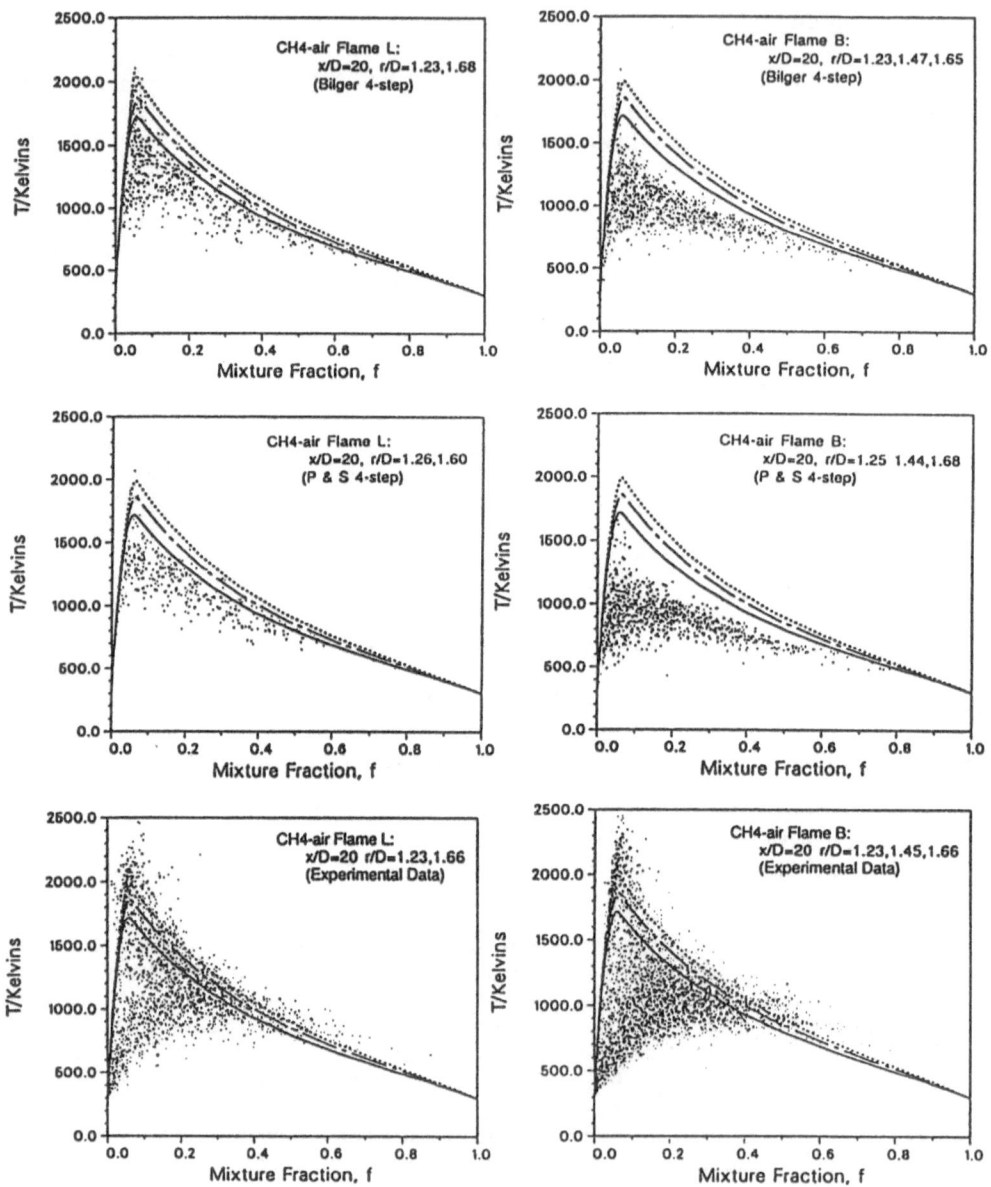

Fig. 17. Joint pdf scatter plots for temperature versus mixture fraction. Lines denote results from calculations for opposed laminar jet diffusion flames; solid lines with a=450 s^{-1}, dashdot lines with a=300 s^{-1}, and dashed lines with a=100 s^{-1}.

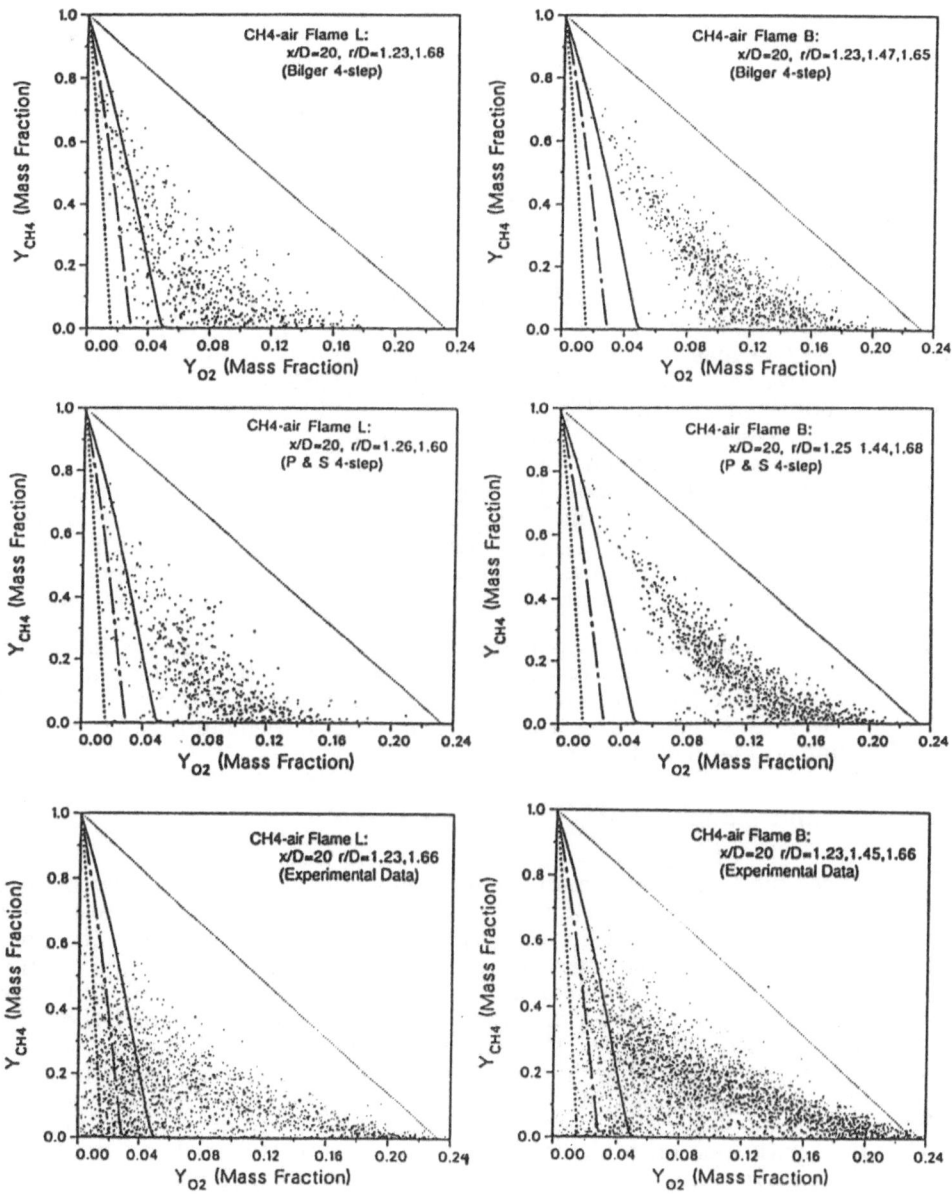

Fig. 18. Joint pdf scatter plots for Y_{CH_4} versus Y_{O_2}. Lines denote results from calculations for opposed laminar jet diffusion flames; solid lines with a=450 s^{-1}, dashdot lines with a=300 s^{-1}, and dashed lines with a=100 s^{-1}. Dotted lines denote pure mixing limit.

opposite limit of pure mixing without reaction is indicated by the dotted line in the graphs. In addition, the flamelet states corresponding to three strain rates are included for evaluation of the flamelet approach. Compared to the predictions, experiments indicate again a larger scatter. Consistent with the experimental data, the scatter pattern shifts toward the pure mixing line as the jet velocity increases. It is interesting to note that for both Flame L and Flame B, both models predict rare occurrence of the coexistence of CH_4 and O_2 in the region bounded by the a=100 s^{-1} dashed line and the Y_{CH_4} axis, while the experiments show a more frequent occurrence.

Joint pdf between CO mass fraction and f

Distribution of CO versus f in the joint pdf domain can provide valuable information for assessing the performance of the chemical models. Figure 19 presents the joint pdf between the CO mass fraction and f showing the consistent trend of decreasing variance with increasing strain rate in the predicted scatter plots. Comparison of the two chemical models reveals that Bilger's mechanism is more sensitive to the jet velocity than Peters and Seshadri's mechanism. It is worth noting that the predicted scatter of CO mass fractions are confined within the flamelet limits. In contrast to the predictions, the experimental data exhibit a significant probability of CO concentrations much larger than those allowed in the low strain rate flamelets. This result points to CO measurements as a crucial factor in separating one model from another. It is suspected that experimental errors are responsible for the high CO points, and experiments are in progress with the object of improving CO measurements (Starner *et al.* [30])

We conclude this section of comparison between predictions and experimental data by presenting predicted joint pdf's of CO with CH_4 and with T from a three-dimensional view as shown in Figure 20. The complex structure of these joint pdf's certainly does not support the traditional approach of assuming a certain shape for the joint pdf's, such as the joint normal distribution.

6. Summary and Conclusions

Two reduced mechanisms for methane-air combustion, one proposed by Bilger and the other proposed by Peters and Seshadri, have been investigated by incorporating each one into a pdf model that is solved by Monte Carlo simulation. The performance of either reduced mechanism is found to be reasonably good for predicting Tsuji-type *laminar* opposed flow flames. For combustion in turbulent flows, the accessible domain is shown to contain a wider range of scalar compositions than explored those found in the laminar opposed jet flames. Further evaluation of the performance of reduced mechanisms for those compositions outside the opposed jet flame regime is needed.

Monte Carlo simulations of turbulent nonpremixed methane-air jet flames have been performed with look-up tables generated by using the two reduced mechanisms. Flow conditions corresponding to high and low strain rates (Flame B and Flame L) were selected to study the ability of the model to predict the significant nonequilibrium chemistry effects in these flames. Extensive comparison of predictions and experimental data reveals that the model is capable of predicting the global trend of further departure from chemical equilibrium when the jet velocity is increased. However, the calculated CO concentrations are too low compared to the data and no distinct bimodal pdfs are predicted. Furthermore, the predictions give a decreasing mean peak temperatures in the axial direction while the experimental data indicate a decrease-increase trend. These deficiencies are likely due to the limitation of current mixing models.

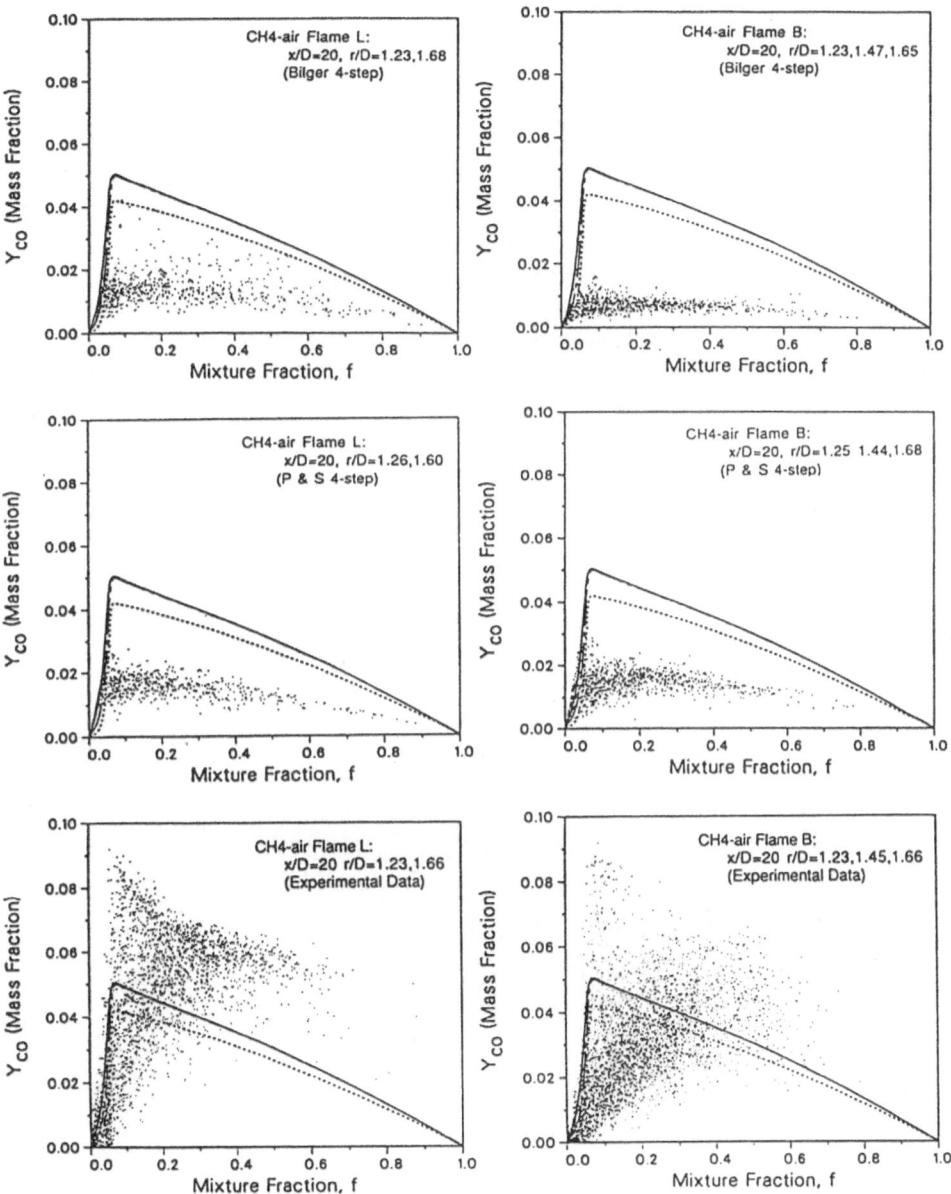

Fig. 19. Joint pdf scatter plots for Y_{CO} versus mixture fraction. Lines denote results from calculations for opposed laminar jet diffusion flames; solid lines with a=450 s^{-1}, dashdot lines with a=300 s^{-1}, and dashed lines with a=100 s^{-1}.

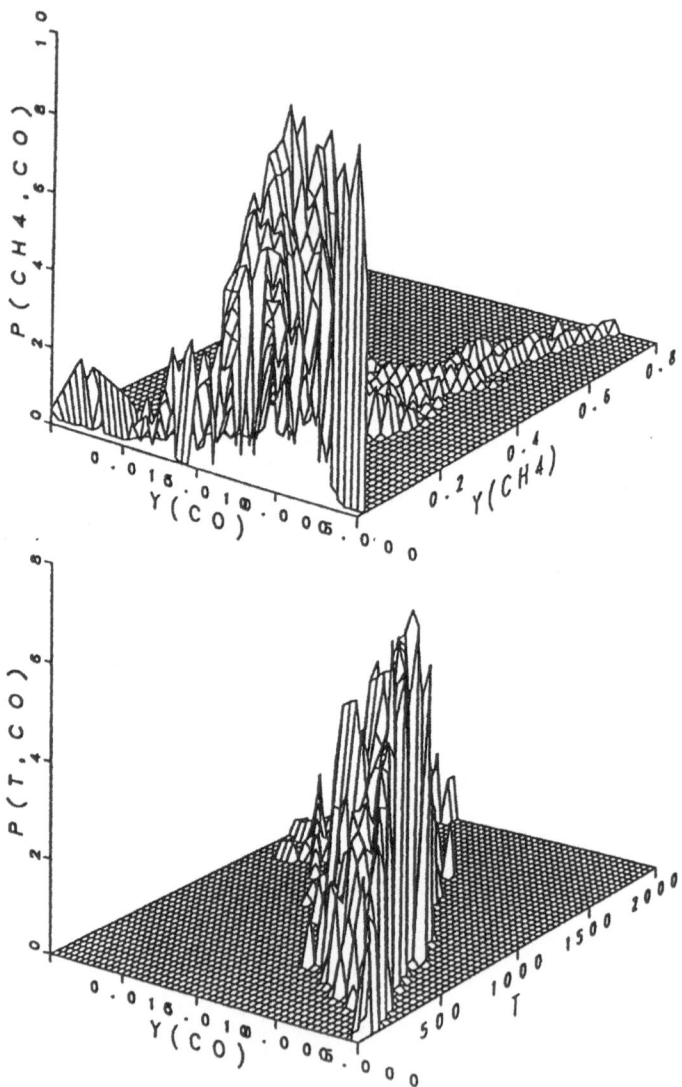

Fig. 20. Predicted joint pdfs for Y_{CO} versus Y_{CH4} and for Y_{CO} versus T in a three-dimensional view showing complex patterns of the joint pdfs.

225

References

[1] Pope, S.B., "The probability approach to the modeling of turbulent reacting flows," *Combust. Flame*, **27**, (1976), p. 299.

[2] O'Brien, E.E.,"The Probability density function (pdf) approach to reacting turbulent flows,"In Libby, P.A. and Williams, F.A. (Ed.). *Turbulent Reacting Flows* , Springer-Verlag, Berlin, Chap. 5, (1980), p. 185.

[3] Curl, R.L.,"Dispersed phase mixing: I. Theory and effects of simple reactors," *A.I.Ch.E.J.*, **9**, (1963), p. 175.

[4] Janicka, J., Kolbe, W., and Kollmann, W.,"Closure of the transport equation for the probability density function scalar field,"*J. Non-equilib. Thermodyn.* **4**, (1979), p. 47.

[5] Pope, S.B.,"An improved turbulent mixing model,"*Combust. Sci. and Technol.*, **28**, (1982), p. 131.

[6] Pope, S.B.,"A Monte Carlo method for the pdf equations of turbulent reactive flow,"*Combust. Sci. and Technol.*, **25**, (1981), p. 159.

[7] Pope, S.B.,"Pdf methods for turbulent reactive flows,"*Prog. Energy Combust. Sci.*, **11**, (1985), p. 119.

[8] Pope, S.B. and Correa, S.M.,"Joint pdf calculations of a nonequilibrium turbulent diffusion flame,"*Twenty-First Symp. (Int.) Combust./The Combustion Institute, Pa.*, (1986), p. 1341.

[9] Chen, J.-Y. and Kollmann, W.,"Pdf modeling of chemical nonequilibrium effects in turbulent nonpremixed hydrocarbon flames,"*Twenty-Second, Symp. (Int.) Combust./The Combustion Institute, Pa.*, (1988), p. 645.

[10] Peters, N. and Kee. R.J.,"The computation of stretched laminar methane-air diffusion flames using a reduced four-step mechanism,"*Combust. Flame*, **68**, (1987), p. 17.

[11] Peters, N. and Williams, F.A.,"The asymptotic structure of stoichiometric methane-air flames,"*Combust. Flame*, **68**, (1987), p. 185.

[12] Bilger, R.W. and Kee, R.J.,"Simplified kinetics for diffusion flames of methane in air,"*Western States Fall Meeting/The Combustion Institute*, (1987), Paper no. 87-85.

[13] Rogg, B. and Williams, F.A.,"Structures of wet CO flames with full and reduced kinetic mechanisms. *Twenty-Second Symp. (Int.) on Combust./The Combustion Institute, Pa.*, (1988), p. 1441.

[14] Chen, J.-Y.,"A general procedure for constructing reduced reaction mechanisms with given independent relations,"*Combust. Sci. and Technol.*, **57**, (1988), p. 89.

[15] Chen, J.-Y., Kollmann, W., Dibble, R.W.,"Pdf modeling of turbulent nonpremixed methane jet flames,"*Combust. Sci. and Technol.*, **64**, (1989), p. 315.

[16] Dibble, R.W., Kollmann, W., Farshchi, M., Schefer, R. W.,"Second-order closure for turbulent nonpremixed flames: scalar dissipation and heat release effects," Twenty-First Symp. (Int.) Combust./The Combustion Institute, Pa., (1986), p. 1329.

[17] Bilger, R.W.,"A note on Favre averaging in variable density flows," *Combust. Sci. and Technol.*, **11**, (1975), p. 215.

[18] Dopazo, C., "Relaxation of initial probability density functions in the turbulent convection of scalar fields," *Phys. Fluids* **22**, (1979), p. 20.

[19] Kosaly, G.,"Theoretical remarks on a phenomenological model of turbulent mixing," *Combust. Sci. and Technol.*, **49**, (1986), p. 227.

[20] Kosaly, G. and Givi, P.,"Modeling of turbulent molecular mixing," *Combust. Flame*, **70**, (1987), p. 101.

[21] Chen, J.-Y. and Kollmann, W.,"Mixing models for turbulent flows with exothermic reactions," *Seventh Symposium on Turbulent Shear Flows*, Stanford University, August 21-23, 1989.

[22] Koochesfahani, M.M. and Dimotakis, P.E.,"Mixing and chemical reactions in a turbulent liquid mixing layer," *J. Fluid Mech.*, **170**, (1986), p. 83.

[23] Dibble, R.W., Masri, A.R., and Bilger, R.W.,"The spontaneous Raman scattering technique applied to nonpremixed flames of methane," *Combust. Flame*, **67**, (1987), p. 189.

[24] Masri, A.R., Bilger, R.W., and Dibble, R.W.,"Turbulent nonpremixed flames of methane near extinction: mean structure from Raman measurements," *Combust. Flame*, **71**, (1988), p. 245.

[25] Masri, A.R., Bilger, R.W., and Dibble, R.W.,"Turbulent nonpremixed flames of methane near extinction: probability density functions," *Combust. Flame*, **73**, (1988), p. 261.

[26] Chen, J.-Y., Kollmann, W., and Dibble, R.W.,"Numerical computation of turbulent free-shear flows using a block-tridiagonal solver for a staggered grid system," *Proceedings of the Eighteenth Annual Pittsburgh Conference*, Instrument Society of America, **18**, (1987), p. 1833.

[27] Liew, S.K., Bray, K.N.C., and Moss, J.B.,"A stretched laminar flamelet model of turbulent nonpremixed combustion," *Combust. Flame*, **56**, (1984), p. 199.

[28] Peters, N.,"Laminar flamelet concepts in turbulent combustion," *Twenty-First Symp. (Int.) Combust./The Combustion Institute*, Pa., (1986), p. 1231.

[29] Haworth, D.C., Drake, M.C., Pope, S.B., and Blint, R.J.,"The importance of time-dependent flame structures in stretched laminar flamelet models for turbulent jet diffusion flames," *Twenty-Second Symp. (Int.) Combust./The Combustion Institute*, Pa., (1988), p. 589.

[30] Starner, S.H., Barlow, R.S., and Dibble, R.W., (1990) in preparation.

CHAPTER 10

CONVENTIONAL ASYMPTOTICS AND COMPUTATIONAL SINGULAR PERTURBATION FOR SIMPLIFIED KINETICS MODELLING

S. H. Lam and D. A. Goussis
Department of Mechanical and Aerospace Engineering
Princeton University, Princeton, NJ 08544

1. Introduction

Generally speaking, the chemical kinetics of any reaction system can potentially be very complex. Even though the investigator may only be interested in a few species, the reaction model almost always involves a much larger number of species. Some of the species involved are often referred to as *radicals* with a special meaning: they are usually reactants of *low* concentrations which are believed to be important *intermediaries* in the whole reaction scheme. A large number of elementary reactions can occur among the species; some of these reactions are fast and some are slow. The aim of simplified kinetics modelling is to derive the simplest reaction system which nevertheless retains the essential features of the full reaction system. The conventional technique [1] is to systematically apply the so-called *steady-state* approximation to the appropriate radicals, the *partial-equilibrium* approximation to the fast reactions, and to ignore the very slow (and therefore unimportant) reactions. The investigator is responsible for identifying what are the appropriate radicals, which are the fast elementary reactions, and which are the very slow ones by making intelligent order of magnitude estimates using information gathered from detailed examination of available data. A skilled and knowledgeable chemical kineticist is usually needed, and the results obtained are expected to be valid only in some limited domain of initial and operating conditions in some limited interval of time. The successful derivation of a simplified chemical kinetics model by conventional methodology thus depends considerably on the experience, intuition and judgement of the investigator.

In the present paper, we present a summary, plus some new recent developments, of the theory of Computational Singular Perturbation [2], [3], [4], [5] (CSP), a general method for non-linear boundary layer type stiff equations particularly appropriate for chemical kinetics problems. A hypothetical reaction system is studied and the appropriate *simplified kinetics models* are derived using both the conventional method and the CSP method, showing clearly their comparative strengths and weaknesses. The goal of the paper is to show that, given the relevant database of elementary reaction rates, the derivation of time-resolved simplified kinetics models for large reaction systems can be routinely accomplished using numerical data generated by CSP.

2. Preliminary Discussions

We are interested in a certain reaction system with initial conditions in a certain temperature and pressure range. A comprehensive and up-to-date database containing all possibly relevant elementary reactions and their rates is assumed available. The *full mechanism* of this reaction system can readily be constructed from this database and consists of R elementary reactions and N species. The species concentrations are represented by the *column* vector y:

$$\mathbf{y} = [y^1, y^2, \ldots, y^N]^T, \tag{2.1}$$

where $[., ., \ldots]$ is a *row* vector and its transpose $[., ., \ldots]^T$ is a column vector. For the sake of simplicity, we consider here only isothermal and homogeneous reaction systems. With this qualification, the governing chemical kinetics equations for $\mathbf{y}$ is written formally as follows:

$$\frac{d\mathbf{y}}{dt} = \mathbf{g}(\mathbf{y}), \tag{2.2}$$

where the column vector $\mathbf{g}$ is the *global reaction rate* and consists of contributions from each of the R elementary reactions:

$$\mathbf{g} = [g^1, g^2, \ldots, g^N]^T = \sum_{r=1}^{R} \mathbf{S}_r F^r, \tag{2.3}$$

where $\mathbf{S}_r$ and F^r are the *stoichiometric* (column) vector and the *reaction rate* of the rth elementary reactions, respectively. For complex reaction systems [6], Eq.(2.2) is usually stiff, and the values of N and R may be quite large and in general not equal.

The basic idea of CSP is to project the R terms in Eq.(2.3) into N linearly independent *modes*, and group the N modes into a *fast* group and a *slow* group. For most chemical kinetics problems, the fast modes are usually of the boundary layer type, and their amplitudes decay rapidly with time. The appropriate simplified kinetics model is then straightforwardly obtained when the contribution of the fast group becomes "sufficiently small" and is neglected. The process of projecting and grouping of the terms is accomplished by the use of a special set of *basis vectors*.

3. CSP in Summary

Since $\mathbf{g}$ is a N-dimensional vector, it can always be expressed in terms of any set of N linearly independent basis vectors, $(\mathbf{a}_1, \mathbf{a}_2, \ldots, \mathbf{a}_N)$. For the moment, let us assume that a trial set of N linearly independent basis (column) vectors $(\mathbf{a}_1(t), \mathbf{a}_2(t), \ldots, \mathbf{a}_N(t))$ has *somehow* been chosen. The corresponding set of (row) vectors $(\mathbf{b}^1(t), \mathbf{b}^2(t), \ldots, \mathbf{b}^N(t))$ satisfying the orthonormal relations:

$$\mathbf{b}^i \bullet \mathbf{a}_k = \delta_k^i, \qquad\qquad i, k = 1, 2, \ldots, N, \tag{3.1}$$

where δ_k^i is the N by N identity matrix, can readily be computed. Eq.(2.3) can be alternatively expressed in terms of these basis vectors:

$$\mathbf{g} = \sum_{i=1}^{N} \mathbf{a}_i f^i, \tag{3.2}$$

where f^i is the "amplitude" of $\mathbf{g}$ in the "direction" of $\mathbf{a}_i$. We interpret each of the additive terms in Eq.(3.2) as representing a *reaction mode* or simply a *mode*. Therefore, $\mathbf{a}_i$ and f^i are the *effective stoichiometric vector* and the *effective reaction rate* of the ith mode, respectively. Taking the dot product of $\mathbf{b}^k$ with $\mathbf{g}$, we obtain, using Eqs.(2.3), (3.1) and (3.2):

$$
\begin{aligned}
f^k &\equiv \mathbf{b}^k \bullet \mathbf{g} & &\text{(3.3a)} \\
&= \mathbf{b}^k \bullet \sum_{r=1}^{R} \mathbf{S}_r F^r & & \\
&= \sum_{r=1}^{R} B_r^k F^r, & k = 1, 2, \ldots, N, &\quad\text{(3.3b)}
\end{aligned}
$$

where B_r^k is:

$$B_r^k \equiv \mathbf{b}^k \bullet \mathbf{S}_r, \qquad k = 1, 2, \ldots, N, \qquad r = 1, 2, \ldots, R. \tag{3.4}$$

It is seen from Eq.(3.3b) that f^k is some linear combination of the F^r's. Summarizing: once a full set of basis vectors (either the $\mathbf{a}_i$'s or the $\mathbf{b}^i$'s) is somehow chosen, the "other" set of basis vectors and the f^i's can readily be computed from Eqs.(3.1) and (3.3), respectively. Eq.(3.2) is exact and is an alternative representation of Eq.(2.3).

It is important to note here that each basis vector $\mathbf{a}_i$ may contain an arbitrary time-dependent scale factor. Without loss of generality, we shall assume that all the $\mathbf{a}_i$ vectors have been appropriately scaled such that their orders of magnitude are relatively constant with time. Consequently, the order of magnitude of the contribution of the ith mode in Eq.(3.2) is primarily dependent on the order of magnitude of f^i. How does the choice of the basis vectors influence the time evolution of the f^i's? To find out, we differentiate Eq.(3.3a) with respect to time to obtain, with the help of Eqs.(3.2) and (2.2):

$$\frac{df^i}{dt} = \sum_{k=1}^{N} \Lambda_k^i f^k, \qquad i = 1, 2, \ldots, N, \tag{3.5}$$

where

$$\Lambda_k^i \equiv \Phi_k^i + \frac{d\mathbf{b}^i}{dt} \bullet \mathbf{a}_k = \Phi_k^i - \mathbf{b}^i \bullet \frac{d\mathbf{a}_k}{dt}, \qquad i, k = 1, 2, \ldots, N, \tag{3.6}$$

$$\Phi_k^i \equiv \mathbf{b}^i \bullet \mathbf{J} \bullet \mathbf{a}_k, \qquad i, k = 1, 2, \ldots, N, \tag{3.7}$$

and $\mathbf{J}$ is the Jacobian of $\mathbf{g}$ with respect to $\mathbf{y}$:

$$\mathbf{J} = J_k^i \equiv \frac{\partial g^i}{\partial y^k}, \qquad i, k = 1, 2, \ldots . N. \tag{3.8}$$

Hence the time evolution of the f^i's is seen to be controlled solely by Λ_k^i which, according to Eq.(3.6), is completely determined by $\mathbf{a}_i(t)$, $\mathbf{b}^i(t)$ (and their time derivatives) in addition to $\mathbf{J}$. It is useful to note that J_k^i and Φ_k^i are similar matrices, but Λ_k^i and Φ_k^i are not (unless the $\mathbf{a}_i$'s and $\mathbf{b}^i$'s are constant vectors). Note that if Λ_k^i is diagonal, the modes would be completely uncoupled from each other.

If the given problem were linear, $\mathbf{J}$ would be a constant matrix, and the obvious choice for basis vectors would be the (constant) right and left eigen-vectors of $\mathbf{J}$ defined by:

$$\beta^i \bullet \mathbf{J} = \lambda(i)\beta^i, \qquad i = 1, 2, \ldots, N, \tag{3.9a}$$

$$\mathbf{J} \bullet \alpha_i = \alpha_i \lambda(i), \qquad i = 1, 2, \ldots, N. \tag{3.9b}$$

The resulting Λ_k^i would be the diagonal matrix (or the Jordan form) of the eigen-values, $\lambda(i)$'s. Consequently, the amplitude of each of the uncoupled modes, f^i, would evolve with its own characteristic time scale. If $\lambda(i)$ is essentially real and negative, the ith mode is said to be of the boundary layer type, and f^i would decay exponentially toward zero and become eventually exhausted for some $t \gg |1/\lambda(i)|$. An obvious algorithm for the linear case is: whenever the amplitude of the currently fastest mode falls below some user-specified threshold, the term representing that mode could be dropped from Eq.(3.2), yielding a less stiff equation to be integrated for larger time. In other words, terms representing exhausted boundary layers can be "neglected" to yield a simplified and less stiff equation for the next boundary layers.

In essence, CSP simply extends the above described algorithm to non-linear problems. Unlike the linear case, it is now not obvious how to choose the basis vectors. However, even though $\mathbf{J}$ is no longer a constant matrix, its left and right eigen-vectors, β^i and α_i, are nevertheless always defined, and can be computed at any time from Eqs.(3.9). Let us denote the reciprocal of the absolute value of $\lambda(i)$ by $\tau(i)$, call the $\tau(i)$'s the current *time scales* of the reaction modes, and order them in ascending magnitudes: $\tau(1) \leq \tau(2) \leq \ldots \leq \tau(N)$. These eigen-vectors and eigen-values of $\mathbf{J}$ can either be used directly (provided that $\mathbf{J}$ is non-defective) or serve as a guide for choosing a set of *trial* basis vectors. Note that the main impact of non-linearity is that even if eigen-vectors of $\mathbf{J}$ were used directly as the trial set, the resulting Λ_k^i still would *not* be diagonal. The non-zero off-diagonal elements cause *mixing* of the modes, and as a consequence the fast modes may not decay and become small as in the linear case. In the present paper, we shall focus our efforts to derive a "refinement" algorithm which can generate from any reasonable trial set of basis vectors an improved set which has less mode mixing than before. As shall be demonstrated later in §5, these CSP generated basis vectors and other data can be used in chemical kinetics problems to deduce physically meaningful information such as the global stoichiometry and reaction rates of simplified kinetics models.

We can manipulate Eqs.(3.1), (3.6) and (3.7) to yield:

$$\frac{d\mathbf{b}^i}{dt} + \mathbf{b}^i \bullet \mathbf{J} = \sum_{k=1}^{N} \Lambda_k^i \mathbf{b}^k, \quad i = 1, 2, \ldots, N, \tag{3.10a}$$

$$-\frac{d\mathbf{a}_i}{dt} + \mathbf{J} \bullet \mathbf{a}_i = \sum_{k=1}^{N} \mathbf{a}_k \Lambda_i^k, \quad i = 1, 2, \ldots, N, \tag{3.10b}$$

which may be considered the governing differential equations for the $\mathbf{a}_i$'s and $\mathbf{b}^i$'s for any desired Λ_k^i. Since no restriction has been placed on the initial conditions, it is clear that the set of basis vectors corresponding to a given Λ_k^i is not unique. One simple-minded approach would be to choose Λ_k^i to be the diagonal (or the Jordan form) eigen-value matrix of $\mathbf{J}$, and compute for the corresponding uncoupled $\mathbf{a}_i$'s and $\mathbf{b}^i$'s using Eqs.(3.10) with some appropriately chosen initial conditions. It suffices to state here that this approach is fundamentally flawed but a detailed discussion of its shortcomings is beyond the scope of this paper.

Let τ^* be a user-specified time resolution of interest. Thus reaction modes with $\tau(i) < \tau^*$ are considered fast, otherwise they are considered slow. We assume that, for the time interval of interest and guided by the eigen-vectors of $\mathbf{J}$ in some vague sense, the first M basis vectors of the chosen set of trial basis vectors span a M-dimensional fast subspace, and that the rest of the $N - M$ basis vectors span the remaining slow subspace. The left and right eigen-vectors of $\mathbf{J}$ at $t = t_0$ or any other reasonably intelligent choice can be used as the (constant) trial basis vectors for $t \geq t_0$. As mentioned previously, the Λ_k^i computed from any trial set will in general have non-zero off-diagonal elements. We shall presently derive an algorithm which can reduce the amount of mode mixing of any reasonably chosen trial basis vectors.

We shall use indices m and n to refer to the fast subspace ($m, n = 1, \ldots, M$), and I and J to refer to the slow subspace ($I, J = M + 1, \ldots, N$). Indices i and k shall continue to refer to the whole N-dimensional space ($i, k = 1, 2, \ldots, N$).

The right-hand side of Eq.(3.5) can be formally divided into a fast and a slow group.

Writing out the fast and the slow equations separately, we have:

$$\frac{df^m}{dt} = \sum_{n=1}^{M} \Lambda_n^m f^n + \sum_{J=M+1}^{N} \Lambda_J^m f^J, \qquad m = 1, 2, \ldots, M, \tag{3.11a}$$

$$\frac{df^I}{dt} = \sum_{n=1}^{M} \Lambda_n^I f^n + \sum_{J=M+1}^{N} \Lambda_J^I f^J, \qquad I = M + 1, \ldots, N. \tag{3.11b}$$

The question is: how do the fast f^m's behave as time increases? Eq.(3.11a) can be rewritten as follows:

$$\frac{df^m}{dt} = \sum_{n=1}^{M} \omega_n^m(M)(f^n - f_\infty^n(M)), \qquad m = 1, 2, \ldots, M, \tag{3.12}$$

where $\omega_n^m(M)$ is the M by M principal submatrix of Λ_k^i:

$$\omega_n^m(M) \equiv \Lambda_n^m, \qquad m, n = 1, 2, \ldots, M, \tag{3.13}$$

and

$$f_\infty^m \equiv -\sum_{J=M+1}^{N} p_J^m(M) f^J, \qquad m = 1, 2, \ldots, M, \tag{3.14}$$

$$p_J^m(M) \equiv \sum_{n=1}^{M} \tau_n^m(M) \Lambda_J^n, \qquad m = 1, 2, \ldots, M,$$
$$J = M + 1, \ldots, N, \tag{3.15}$$

and $\tau_n^m(M)$ is the inverse of $\omega_n^m(M)$:

$$\sum_{k=1}^{M} \omega_k^m(M) \tau_n^k(M) = \sum_{k=1}^{M} \tau_k^m(M) \omega_n^k(M) = \delta_n^m, \qquad m, n = 1, 2, \ldots, M.$$

The impact of having non-zero off-diagonal elements of the matrix Λ_k^m can be seen from Eq.(3.12): when $\lambda(m)$ is essentially real and negative, f^m no longer decays exponentially to zero as the corresponding boundary layer is exited. Eq.(3.12) shows that f^m decays exponentially with the fast time scale $\tau(m)$ only initially. In a transient period of order $\tau(M)$, f^m tends rapidly toward $f_\infty^m(M)$; thereafter, f^m simply tries to follow the corresponding $f_\infty^m(M)$ which evolves with the slower time scale $\tau(M+1)$ as time increases. At this point, there is no theoretical assurance that $f_\infty^m(M)$ is in any sense small or negligible. Thus, so long as $\Lambda_J^m(M)$ has non-zero elements, there is mixing of the slow modes with the fast modes, and this mixing prevents the amplitude of the fast modes from continuing their rapid decay to become "small" when $t \gg \tau(M)$.

Eq.(3.12) suggests the use of a new f_o^m to replace f^m:

$$f_o^m \equiv f^m - f_\infty^m(M), \qquad m = 1, 2, \ldots, M. \tag{3.16}$$

We shall show presently that f_o^m is expected to be small in some asymptotic sense when $t \gg \tau(M)$. Eqs.(3.11) can be rewritten as follows:

$$\frac{df_o^m}{dt} = \sum_{n=1}^{M} \omega_n^m(M)(f_o^n - f_{o,\infty}^n(M)), \qquad m = 1, 2, \ldots, M, \tag{3.17a}$$

$$\frac{df^I}{dt} = \sum_{J=M+1}^{N} \Omega_J^I(M) f^J + \sum_{m=1}^{M} \Lambda_m^I f_o^m, \qquad I = M + 1, \ldots, N, \tag{3.17b}$$

where

$$f_{o,\infty}^n(M) \equiv \sum_{m=1}^{M} \tau_m^n(M)\frac{df_\infty^m(M)}{dt}, \qquad n = 1, 2, \ldots, M, \tag{3.18}$$

$$\Omega_J^I(M) \equiv \Lambda_J^I - \sum_{m,n=1}^{M} \Lambda_m^I \tau_n^m(M)\Lambda_J^n, \qquad I, J = M+1, \ldots, N. \tag{3.19}$$

Note that Eqs.(3.12) and (3.17a) differ only in their forcing term, $f_\infty^n(M)$ versus $f_{o,\infty}^n$; hence $f_o^m(M)$ is expected to tend to $f_{o,\infty}^m(M)$ for $t \gg \tau(M)$. Eq.(3.18) clearly shows that $f_{o,\infty}^m(M)$ is smaller than $f_\infty^m(M)$ for $t \gg \tau(M)$ by the factor $\varepsilon(M)$:

$$\varepsilon(M) \equiv \frac{\tau(M)}{\tau(M+1)}. \tag{3.20}$$

Thus f_o^m is a "purer" fast mode than f^m. In the asymptotic limit of vanishingly small $\varepsilon(M)$, f_o^m is now theoretically small for $t \gg \tau(M)$.

Inspection of Eq.(3.11b) or (3.17b) indicates that in general there is also mixing of the fast modes with the slow modes so long as $\Lambda_m^I(M)$ has non-zero elements. For obvious reasons, this mixing should be as weak as possible.

The above analysis suggests that the set of basis vectors $\mathbf{a}_i^o(M)$ and $\mathbf{b}_o^i(M)$ defined below is an "improvement" over the original trial set, $\mathbf{a}_i$ and $\mathbf{b}^i$:

$$\mathbf{b}_o^m(M) \equiv \mathbf{b}^m + \sum_{J=M+1}^{N} p_J^m(M)\mathbf{b}^J, \qquad m = 1, 2, \ldots, M, \tag{3.21a}$$

$$\mathbf{a}_J^o(M) \equiv \mathbf{a}_J - \sum_{n=1}^{M} \mathbf{a}_n p_J^n(M), \qquad J = M+1, \ldots, N, \tag{3.21b}$$

$$\mathbf{b}_o^I(M) \equiv \mathbf{b}^I - \sum_{n=1}^{M} q_n^I(M)\mathbf{b}_o^n(M), \qquad I = M+1, \ldots, N, \tag{3.21c}$$

$$\mathbf{a}_m^o(M) \equiv \mathbf{a}_m + \sum_{J=M+1}^{N} \mathbf{a}_J^o(M)q_m^J(M), \qquad m = 1, 2, \ldots, M, \tag{3.21d}$$

where $p_J^m(M)$ was previously given by (3.15), and $q_m^I(M)$ is given by:

$$q_m^I(M) \equiv \sum_{n=1}^{M} \Lambda_n^I \tau_m^n(M), \qquad m = 1, 2, \ldots, M, \quad I = M+1, \ldots, N. \tag{3.22}$$

Eq.(3.21a) is derived by a conventional singular perturbation analysis on Eq.(3.17a), and Eq.(3.21c) is derived by a similar analysis on Eq.(3.17b). Eqs.(3.21b,d) are derived by insisting that the new set of basis vectors satisfy the orthonormal relation (see Eq.(3.1)):

$$\mathbf{b}_o^i(M) \bullet \mathbf{a}_k^o(M) = \delta_k^i, \qquad i, k = 1, 2, \ldots, N.$$

Eqs.(3.21) provide an algorithm to generate a new set of basis vectors from an old set. We shall call the new set, $\mathbf{b}_o^i(M)$ and $\mathbf{a}_k^o(M)$, the *refined basis vectors*. Note that the refinement algorithm can be recursively applied. The use of refined $\mathbf{b}_o^m(M)$ and $\mathbf{a}_J^o(M)$

improves (*i.e.* made smaller by $O(\varepsilon(M))$) the order of magnitude of the residual amplitudes of the fast modes. The use of refined $\mathbf{b}_o^I(M)$ and $\mathbf{a}_m^o(M)$ obtained with $q_m^I(M)$ given by Eq.(3.22) purifies the slow modes.

In terms of these new refined basis vectors, a one-parameter family of alternative representation of $\mathbf{g}$ (with M as the parameter) is obtained:

$$\mathbf{g} = \sum_{k=1}^{N} \mathbf{a}_k^o(M) f_o^k \qquad (3.23a)$$

$$= \sum_{m=1}^{M} \mathbf{a}_m^o f_o^m + \sum_{J=M+1}^{N} \mathbf{a}_J^o(M) f_o^J \qquad (3.23b)$$

where

$$f_o^i = \mathbf{b}_o^i(M) \bullet \mathbf{g}, \qquad i = 1, 2, \ldots, N. \qquad (3.23c)$$

Eq.(3.23b) is exact and is mathematically identical to Eqs.(2.3) or (3.2). It is expected to be "better" because the use of refined basis vectors enables the fast f_o^m's to asymptote to smaller values for $t \gg \tau(M)$.

When the residual amplitudes of the f_o^m's become sufficiently small in comparison to some user-specified threshold, their contributions to Eq.(3.23b) may be dropped to yield the leading-order CSP-derived simplified model:

$$\frac{d\mathbf{y}}{dt} = \mathbf{g} \approx \sum_{J=M+1}^{N} \mathbf{a}_J^o(M) f_o^J, \qquad t \gg \tau(M). \qquad (3.24)$$

The initial condition for the simplified model must satisfy:

$$f_o^m = \mathbf{b}_o^m(M) \bullet \mathbf{g} \approx 0, \qquad m = 1, 2, \ldots, M. \qquad (3.25)$$

When computing for the numerical solution using the exact formulation, Eqs.(3.23), the mode amplitudes f_o^i's are most easily evaluated using Eq.(3.23c). However, as the amplitudes of the fast modes decay and become small, the values of the small f_o^m's so evaluated are likely to be very inaccurate because of possible round-off errors. Fortunately, Eq.(3.17a) provides an alternative route to evaluate them when they are near exhaustion. Applying conventional singular perturbation techniques to Eq.(3.17a), the following analytical approximation is straightforwardly obtained:

$$f_o^m \approx f_{o,asym}^m \equiv f_{o,\infty}^m(M) + \sum_{n=1}^{M} \tau_n^m(M) \frac{df_{o,\infty}^n(M)}{dt} + O(\varepsilon(M)^3), \qquad (3.26)$$

which is valid for $t \gg \tau(M)$. Theoretically, $f_{o,asym}^m$, the particular solution of Eq.(3.17a), is of order $O(\varepsilon(M))$. The two-term representation given above is accurate to order $O(\varepsilon(M)^3)$. Thus, a more accurate CSP-derived simplified model than Eq.(3.24) is obtained by using Eq.(3.26) in Eq.(3.23b):

$$\frac{d\mathbf{y}}{dt} = \mathbf{g} \approx \sum_{m=1}^{M} \mathbf{a}_m^o f_{o,asym}^m + \sum_{J=M+1}^{N} \mathbf{a}_J^o(M) f_o^J, \qquad t \gg \tau(M). \qquad (3.27)$$

The accuracy of the approximate $\mathbf{y}$ so generated is $O(\varepsilon(M)^3)$. It can be shown that Eqs.(3.25) are approximate *integrals of motion* of Eq.(3.27). Computationally, the time derivatives needed in the two-term representation of $f^m_{o,asym}$ can be approximately evaluated (*e.g.* by the use of finite differences), and the value of $f^m_{o,asym}$ so computed can be used to assess the accuracy of the CSP model. We have also experimented with the inclusion of the homogeneous solution of Eq.(3.17a) in $f^m_{o,asym}$ and obtained significant improvement in accuracy.

The idea of refining the fast basis vectors was first pointed out in reference [2]. Additional expositions on CSP can be found in references [3], [4] and [5]. An example is worked out in this paper using both the conventional method and CSP in §4 and §5. The derivation of the specific refinement algorithm presented above (Eqs.(3.21)) and other details are beyond the scope of the present paper and will be reported later in a separate paper.

4.   The Conventional Method at Work

The following hypothetical chemical reaction system is used as a vehicle for our discussions:

$$\#1: \qquad 2y^1 \rightleftharpoons y^2, \qquad F^1 = F^1(\mathbf{y}), \qquad (4.1a)$$

$$\#2: \quad y^1 + y^5 \rightleftharpoons y^3 + y^4, \qquad F^2 = F^2(\mathbf{y}), \qquad (4.1b)$$

$$\#3: \quad y^1 + y^3 \rightleftharpoons y^2 + y^4, \qquad F^3 = F^3(\mathbf{y}), \qquad (4.1c)$$

where the reaction rates satisfy the Law of Mass Action:

$$F^1 \;=\; k_{1f}y^1y^1 - k_{1b}y^2, \qquad (4.2a)$$

$$F^2 \;=\; k_{2f}y^1y^5 - k_{2b}y^3y^4, \qquad (4.2b)$$

$$F^3 \;=\; k_{3f}y^1y^3 - k_{3b}y^2y^4, \qquad (4.2c)$$

and k_{rf} and k_{rb} are the forward and backward reaction rates of the rth reaction, respectively. A hypothetical system is used here to more starkly highlight the role of experience and intuition.

The governing equations of the above reaction system are:

$$\frac{dy^1}{dt} \;=\; -2F^1 - F^2 - F^3 , \qquad (4.3a)$$

$$\frac{dy^2}{dt} \;=\; F^1 \qquad\;\; +F^3 , \qquad (4.3b)$$

$$\frac{dy^3}{dt} \;=\; \qquad F^2 \;\; -F^3 , \qquad (4.3c)$$

$$\frac{dy^4}{dt} \;=\; \qquad F^2 \;\; +F^3 , \qquad (4.3d)$$

$$\frac{dy^5}{dt} \;=\; \qquad -F^2 , \qquad (4.3e)$$

A conventional analysis would proceed as follows [7], [8]:

(i) The collection of elementary reactions (Eqs.(4.1)) is examined and found to involve only two atomic species. Let C_1 and C_2 represent the total amount of each atomic species:

$$C_1 \equiv y^1 + 2y^2 + y^3, \qquad C_2 \equiv y^3 + y^4 + 2y^5 . \qquad (4.4)$$

Using Eqs.(4.3) and (4.4), we can easily show that:

$$\frac{dC_1}{dt} = 0, \qquad \frac{dC_2}{dt} = 0 . \qquad (4.5)$$

Eqs.(4.5) indicate that the reaction system respects the physical law of conservation of atomic species.

(ii) The rate constants and the initial conditions are examined and species y^1 and y^4 are declared radicals (based on experience and/or intuition). The assumptions that $\frac{dy^1}{dt}$ and $\frac{dy^4}{dt}$ are small in some sense are then made. Applying the steady-state approximation to y^1, we have:

$$g^1(\mathbf{y}) = -2F^1 - F^2 - F^3 \approx 0 . \qquad (4.6)$$

(iii) Applying the steady-state approximation to y^4, we have:

$$g^4(\mathbf{y}) = F^2 + F^3 \approx 0 . \qquad (4.7)$$

(iv) Substituting Eqs.(4.6) and (4.7) into Eqs.(4.3), we obtain:

$$\frac{dy^1}{dt} \approx 0, \qquad \text{(see Eq.(4.12a) below)} \qquad (4.8a)$$

$$\frac{dy^2}{dt} \approx -F^2, \qquad (4.8b)$$

$$\frac{dy^3}{dt} \approx 2F^2, \qquad (4.8c)$$

$$\frac{dy^4}{dt} \approx 0, \qquad \text{(see Eq.(4.12b) below)} \qquad (4.8d)$$

$$\frac{dy^5}{dt} \approx -F^2, \qquad (4.8e)$$

Eqs.(4.8) represent the following approximate one-step chemical reaction system for large t:

$$y^2 + y^5 \rightleftharpoons 2y^3, \qquad \text{reaction rate} \approx F^2, \qquad (4.9)$$

where F^2 depends on y^1 and y^4 in addition to y^3 and y^5. It is interesting to note that the stoichiometric coefficients appearing in Eq.(4.9) are all rational numbers or integers; this is characteristic of simplified kinetics models derived by the use of steady-state approximations.

(v) It is well known that Eqs.(4.8a) and (4.8d) do *not* mean y^1 and y^4 are constants; they merely indicate that the net time rate of change of y^1 and y^4 are small in comparison to their forward and reverse contributions. To proceed further, Eqs.(4.6) and (4.7) are solved algebraically for y^1 and y^4 in terms of the other species, y^2, y^3 and y^5. In general, this step is not routine, and frequently additional assumptions and restrictions are needed [8] simply to overcome the algebraic difficulties here. For the present problem, straightforward algebraic manipulations yield:

$$y^1 \approx \sqrt{\frac{k_{1b}y^2}{k_{1f}}}, \qquad (4.10a)$$

$$y^4 \approx \left(\frac{k_{2f}y^5 + k_{3f}y^3}{k_{2b}y^3 + k_{3b}y^2}\right) \sqrt{\frac{k_{1b}y^2}{k_{1f}}} . \qquad (4.10b)$$

Using Eqs.(4.10) in Eq.(4.2b), we obtain F^2 as a function of y^2, y^3 and y^5 only:

$$F^2 \approx \frac{\sqrt{\frac{k_{1b}y^2}{k_{1f}}}}{k_{2b}y^3 + k_{3b}y^2}(k_{2f}k_{3b}y^2y^5 - k_{3f}k_{2b}y^3y^3) . \qquad (4.11)$$

Eq.(4.11) can now be used with Eqs.(4.8b,c,e) to compute for y^2, y^3 and y^5; the values of y^1 and y^4 are then computed from Eqs.(4.10). It is possible to algebraically manipulate Eqs.(4.4) and (4.10) to express all elements of the y vector in terms of any single species such as y^2 (plus C_1 and C_2) so that F^2 can be expressed as a function of y^2 only. Eq.(4.8b) then becomes a single equation for a single unknown.

The above results can be alternatively presented as follows. Taking the logarithmic time derivatives of Eqs.(4.10) and then solving for $\frac{dy^1}{dt}$ and $\frac{dy^4}{dt}$ with the help of Eqs.(4.8b,c,e), we have:

$$\frac{dy^1}{dt} \approx \frac{y^1}{2y^2}\frac{dy^2}{dt} \approx -\left\{\frac{y^1}{2y^2}\right\}F^2 \equiv -c_1 F^2, \qquad (4.12a)$$

$$\frac{dy^4}{dt} \approx -\left\{y^4\left(\frac{k_{2f} - 2k_{3f}}{k_{2f}y^5 + k_{3f}y^3} + \frac{2k_{2b} - k_{3b}}{k_{2b}y^3 + k_{3b}y^2} + \frac{1}{2y^2}\right)\right\}F^2$$

$$\equiv -c_2 F^2. \qquad (4.12b)$$

Eqs.(4.12) can be used to replace Eqs.(4.8a) and (4.8d). Hence, the one-step chemical reaction system represented by Eqs.(4.9) and (4.10) can be alternatively represented by:

$$c_1 y^1 + y^2 + c_2 y^4 + y^5 \;\rightleftharpoons\; 2y^3, \qquad (4.13a)$$

$$\text{reaction rate} \approx F^2, \qquad (4.13b)$$

showing more clearly that y^1 and y^4 are involved in the stoichiometry of this one-step reaction. It is now easy to show that the above approximate results are "slightly" inconsistent with Eqs.(4.5). Differentiating Eq.(4.4) with respect to time and using Eqs.(4.12) and (4.8b,c,e), we obtain:

$$\frac{dC_1}{dt} \approx -c_1 F^2 \neq 0, \qquad \frac{dC_2}{dt} \approx -c_2 F^2 \neq 0, \qquad (4.14)$$

i.e. this approximate model does not exactly conserve the total amount of atomic species. This "error" can be explained as follows: the steady-state approximation correctly gave the leading approximation to the stoichiometric coefficients of the one-step reaction model as (0, -1, 2, 0, -1), but provided only the small corrections (c_1 and c_2) for the two zeros. To be fully consistent, the other three stoichiometric coefficients also need small corrections (See Eq.(4.20) later). This minor inconsistency does not appear to be widely recognized. Formally, the simplified model derived using the steady-state approximation (Eqs.(4.9), (4.10)) is valid only when c_1 and c_2 as defined above are asymptotically small.

We shall re-analyze the same problem by retaining the steady-state approximation on y^1 but replacing the steady-state approximation on y^4 by the partial-equilibrium approximation on F^1. We shall show that using the more careful procedure [1] normally reserved for partial-equilibrium approximations, the inconsistency mentioned above can be avoided.

The steps after (ii) are:

(vi) The steady-state approximation on y^1 again yields Eq.(4.6). Applying the partial equilibrium approximation to F^1, we have:

$$F^1(\mathbf{y}) = k_{1f}y^1y^1 - k_{1b}y^2 \approx 0. \tag{4.15}$$

Thus Eqs.(4.7) and (4.10) remain valid. Consequently, the starting points of this and the previous analysis are identical.

(vii) We now rewrite Eqs.(4.3) in the following form:

$$\frac{dy^1}{dt} = g^1 , \tag{4.16a}$$

$$\frac{dy^2}{dt} = -g^1 -F^1 -F^2 , \tag{4.16b}$$

$$\frac{dy^3}{dt} = g^1 +2F^1+2F^2, \tag{4.16c}$$

$$\frac{dy^4}{dt} = -g^1 -2F^1, \tag{4.16d}$$

$$\frac{dy^5}{dt} = -F^2 . \tag{4.16e}$$

Instead of step (iv) which would have neglected g^1 and F^1, we algebraically eliminate them:

$$\frac{1}{2}\frac{dy^1}{dt} + \frac{dy^2}{dt} - \frac{1}{2}\frac{dy^4}{dt} = -F^2, \tag{4.17a}$$

$$\frac{dy^3}{dt} + \frac{dy^4}{dt} = 2F^2, \tag{4.17b}$$

$$\frac{dy^5}{dt} = -F^2. \tag{4.17c}$$

No approximation has so far been applied in arriving at Eqs.(4.17). The above three exact differential equations are to be supplemented by the two approximate algebraic relations, Eqs.(4.10) or Eqs.(4.6) and (4.7) or (4.15).

(viii) Eqs.(4.10) can be used to directly eliminate y^1 and y^4 from Eqs.(4.17). Alternatively, we can differentiate Eqs.(4.6) and (4.7) with respect to time and eliminate the time derivatives of y^1 and y^4 from Eqs.(4.17). Using the latter approach, we obtain, after considerable algebra:

$$\frac{dy^1}{dt} \approx -c_1(1 - c_3)F^2, \tag{4.18a}$$

$$\frac{dy^2}{dt} \approx -(1 - c_3)F^2, \tag{4.18b}$$

$$\frac{dy^3}{dt} \approx (2 + c_1)(1 - c_3)F^2, \tag{4.18c}$$

$$\frac{dy^4}{dt} \approx (2c_3 - c_1(1 - c_3))F^2, \tag{4.18d}$$

$$\frac{dy^5}{dt} \approx -F^2, \tag{4.18e}$$

where

$$c_3 \equiv \frac{e_2 - 2e_4 + e_5 + c_1(e_1 + e_3 - e_4)}{(2 + c_1)(e_3 - e_4) + c_1 e_1 + e_2}, \quad e_i \equiv \frac{\partial g^4}{\partial y^i}. \tag{4.19}$$

Eqs.(4.18) represent the following approximate one-step chemical reaction system for large t:

$$c_1(1 - c_3)y^1 + (1 - c_3)y^2 + y^5$$
$$\rightleftharpoons (2 + c_1)(1 - c_3)y^3 + (2c_3 - c_1(1 - c_3))y^4, \tag{4.20a}$$
$$\text{reaction rate} \approx F^2 = k_{2f}y^1 y^5 - k_{2b}y^3 y^4. \tag{4.20b}$$

Eq.(4.20a) is seen to be different from both Eq.(4.9) and Eq.(4.13a). It can be shown that when c_1 and c_3 are small in comparison to unity, c_1, c_2 and c_3 are approximately related by:

$$c_3 \equiv \frac{c_1 - c_2}{2}, \quad c_1 \ll 1, \quad c_2 \ll 1. \tag{4.21}$$

Eq.(4.20a) becomes:

$$c_1 y^1 + (1 - \frac{c_1}{2} + \frac{c_2}{2})y^2 + c_2 y^4 + y^5 \rightleftharpoons (2 + c_2)y^3. \tag{4.22}$$

When both c_1 and c_2 are small, Eqs.(4.9), (4.13a) and (4.22) are competitive approximate reaction models for the same problem, but Eq.(4.22) is clearly the superior model because it alone is consistent with the conservation of atomic species, Eqs.(4.5). When neither c_1 nor c_2 is small, Eq.(4.20a) is the only correct approximate one-step reaction model fully consistent with Eqs.(4.5). In contrast to Eq.(4.9), the stoichiometric coefficients in Eq.(4.20a) and (4.22) are mostly irrational numbers.

In the existing literature [1], a distinction is usually made between the steady-state approximation and the partial-equilibrium approximation. According to the above presentation, partial-equilibrium approximation appears to be the more general procedure: it includes the steady-state approximation as a special case when the approximation of total neglect of the time derivatives of the identified radicals in Eqs.(4.17) can be justified.

5.   The CSP Method at Work

We shall work out the same examples with CSP.

1. The eigen-vectors of $\mathbf{J}$ at $t = 0$ can be used as trial basis vectors for $t \geq 0 : \mathbf{a}_i(t) = \alpha_i(t = 0)$ and $\mathbf{b}^i(t) = \beta^i(t = 0)$. With this choice of constant trial basis vectors, we have $\Lambda_k^i = \Phi_k^i$ which is diagonal only at $t = 0$. In general, the refined basis vectors are not constant and will evolve with time.

2. The CSP computation commences with $M = 0$ at $t = 0$, and any standard (non-stiff) ODE solver can be used. M is incremented whenever the currently fastest active mode in the slow group falls below some user-specified threshold: the mode is promoted into the fast group and declared exhausted at the same time. The integration of Eq.(3.24) or Eq.(3.27) now allows a larger integration time step to be taken.

3. If the refined $\mathbf{b}_o^1$ is found to be approximately proportional to $[1, 0, 0, 0, 0]$ when $t \gg \tau(1)$, then y^1 is a radical which reaches steady-state first (because $f_o^1 = \mathbf{b}_o^1 \bullet \mathbf{g} \approx g^1(\mathbf{y})$; see Eq.(4.6)). If $f_o^2 \approx (\mathbf{b}_o^2 \bullet \mathbf{S}_1)F^1(\mathbf{y})$ when $t \gg \tau(2)$, then elementary reaction #1 is very fast and reaches partial-equilibrium next (see Eq.(4.15)). If, instead, $f_o^2 \approx (\mathbf{b}_o^1 \bullet \mathbf{S}_1)F^1(\mathbf{y}) + (\mathbf{b}_o^2 \bullet \mathbf{S}_3)F^3(\mathbf{y})$ when $t \gg \tau(2)$, then elementary reactions #1 and #3 quickly reach partial-equilibrium with each other next. Whatever $\mathbf{b}_o^1$ and $\mathbf{b}_o^2$ turn out to be, some physically meaningful interpretations for $f_o^1 \approx 0$ and $f_o^2 \approx 0$ may be obtained.

4. When $M = 2$, the effective stoichiometric coefficients of the one-step (see below) simplified kinetics model represented by Eq.(3.24) are given by the elements of $\mathbf{a}_3^o$, and the corresponding effective reaction rate is $f_o^3 = \mathbf{b}_o^3 \bullet \mathbf{g}$.

5. It can easily be established computationally that the rank of the matrix $\mathbf{J}$ is 3, indicating that there are two zero eigen-values. Thus reaction modes #4 and #5 have identically zero reaction rates, and thus represent some physically interesting conservation laws. Because these modes are never active, the maximum value for M for this problem is 2, at which point the reaction system has an one-step model. It can easily be shown that C_1 and C_2 always satisfy Eqs.(4.5).

What if we were interested in the time interval $\tau(2) \gg t \gg \tau(1)$? The CSP data generated in the time interval with $M = 1$ readily provides the corresponding two-step reaction model. If in the same time interval, it is found numerically that the contribution of f_o^3 (for each component of $\mathbf{g}$) is below some user-specified accuracy threshold in comparison to that of f_o^2, then f_o^3 is dormant (i.e. reaction mode #2 is not important) and can be neglected to yield a one-step model.

If, at any time, the value of $f_{o,asym}^m$ of one of the M exhausted fast modes rises above the user-specified threshold, that mode can simply be declared active again. If the value of one of the slow eigen-values of $\mathbf{J}$ is positive, then that mode in question is potentially explosive. Interesting information such as ignition delays and chain-branching mechanisms can readily be derived from the CSP data of explosive modes.

In the language of CSP, the conventional method presented in §4 relied on the experience and intuition of the investigator in the sample problem area in certain ranges of initial and operating conditions to choose the following set of trial basis vectors:

$$\mathbf{a}_1 = [1, -1, 1, -1, 0]^T, \qquad \mathbf{b}^1 = [1, 0, 0, 0, 0], \qquad (5.1a)$$

$$\mathbf{a}_2 = [0, -1, 2, -2, 0]^T, \qquad \mathbf{b}^2 = [0, 1, 1, 0, 1], \qquad (5.1b)$$

$$\mathbf{a}_3 = [0, 1, -2, 0, 1]^T, \qquad \mathbf{b}^3 = [0, 0, 0, 0, 1]. \qquad (5.1c)$$

The one-step model given by Eq.(4.9) can be obtained by CSP using the trial set without refinement, while that given by Eqs.(4.20) can be obtained by using the refined set.

6. Discussion

The essential feature of the CSP method is that it is completely algorithmic and programmable. Unlike the conventional method which depends critically on the investigator's experience and intuition in identifying and applying the appropriate approximations and on the success of the subsequent problem-specific algebraic manipulations (e.g. to solve for the concentrations of the radicals from highly non-linear algebraic equations), CSP recasts all chemical kinetics problems into a universal standard form. The specifics of the problem are completely contained in the single N by N matrix $\mathbf{J}$ from which all further results are derived. The reaction system is decomposed into N reaction modes divided into a fast and a slow group using basis vectors refined from an appropriately chosen trial set. With the help of the refined basis vectors, the reaction modes can be classified at any time as being either *exhausted, active* or *dormant*. Exhausted reaction modes are fast reaction modes which were once dominant but are now sufficiently spent to be ignored, active reaction modes are slow reaction modes which are mainly responsible for the currently observed activities, and dormant reaction modes are the remaining slow reaction modes which are not contributing significantly. Conventional asymptotic methodology is used to analyze the long time behavior of the f_o^m's in the fast group, taking advantage of the fact that the relevant equation is in a universal and particularly simple form (Eq.(3.17a)) to obtain the leading approximations.

The left and right eigen-vectors of $\mathbf{J}$ at $t = 0$ are always available to be the initial trial basis vectors for $t \geq 0$. From the programming point of view, *it is straightforward to use the freshly computed set of refined basis vectors at the end of every integration time-step to be the new trial set of constant basis vectors for the next time step.* Recent numerical results have shown that updating of $\mathbf{a}_i$ and $\mathbf{b}^i$ with Eqs.(3.21a,b,c,d) at every time step improves the accuracy of Eqs.(3.24) or (3.27) by nearly another order of $\varepsilon(M)$.

The refined mode amplitude f_o^i is given by:

$$f_o^i = \sum_{r=1}^{R} B_{o,r}^i(M) F^r(\mathbf{y}), \qquad i = 1, 2, \ldots, N, \tag{6.1}$$

where

$$B_{o,r}^i(M) \equiv \mathbf{b}_o^i(M) \bullet \mathbf{S}_r, \qquad i = 1, 2, \ldots, N, \qquad r = 1, 2, \ldots, R. \tag{6.2}$$

Information on the degree of participation of the rth elementary reaction (and its rate constants) toward the ith mode amplitude can be obtained from Eq.(6.1). The relevance of elementary reactions not included in the original calculation can also be similarly assessed. For the exhausted modes, we have

$$f_o^m = \sum_{r=1}^{R} B_{o,r}^m(M) F^r(\mathbf{y}) \approx 0, \qquad m = 1, 2, \ldots, M. \tag{6.3}$$

These are the CSP-derived approximate algebraic relations between the state variables; *i.e.* they are the equations of state of the radicals. For the active modes, the effective stoichiometry of the Ith mode is $\mathbf{a}_I^o$, and its effective reaction rate is f_o^I. These results together is the CSP-derived simplified kinetics model.

The small contribution to $\mathbf{g}$ of the exhausted fast modes can either be ignored (see Eq.(3.24)) or be included with good accuracy (see Eq.(3.26)). Non-stiff integration algorithms can then be used, and the integration time step Δt can be increased each time a fast mode is declared exhausted. Exhausted modes can be declared active again when the user-specified accuracy threshold is breached. The elementary reactions which do not participate significantly in the exhausted and active modes in the time interval of interest can be identified by examining Eq.(6.1); the "reduced mechanism" of the reaction system can then be easily determined [5] by their removal.

The role of CSP in chemical kinetics modelling is clear. For sufficiently simple problems for which the identities of the appropriate radicals and the fast reactions are well known, and the resulting algebraic equations are amendable to the needed manipulations, the conventional analysis is the method of choice because the results are analytical. The minor defect of the steady-state approximation pointed out earlier can easily be remedied.

It should be noted that for sufficiently simple problems the CSP algorithm can be carried out analytically if so desired. For sufficiently large and complex problems for which little is known and few guidelines exist, numerical CSP data can be used to deduce most of the information normally expected from a conventional asymptotic analysis. After using CSP to identify the available simplifications in the time interval of interest, one may follow up with conventional asymptotic analyses to obtain selected additional analytical insights. It is interesting to note that in our sample calculations [5], most of the exhausted reaction modes can indeed be cleanly associated with either the steady-state approximation for radicals or the partial-equilibrium approximation for fast elementary reactions, or both. However, there are also some ambiguous cases when neither seems applicable.

From the point of view of asymptotics, the CSP method removes the need for non-dimensionalization of variables, order of magnitude estimates, identification of small parameters, consistency checks for assumed forms of expansions, and the various labor-intensive and problem-specific manipulations in the derivation of simplified models for boundary-layer type non-linear O.D.E. problems. The myriad asymptotic procedures have been formalized into a straightforward and programmable algorithm. For linear problems, the algorithm reduces to standard eigen-analysis. For non-linear problems of the boundary layer type (*i.e.* when all the fast $\lambda(i)$'s are essentially real and negative), the algorithm "derives" the simplified inner and outer (*i.e.* fast and slow) equations, explicitly accounting for the leading order effects of the "rotation" of the local fast subspace (spanned by the M fast basis vectors) due to the non-linearities of $\mathbf{g}$.

The CSP user supplies the database of elementary reactions and their rates, specifies the thresholds of accuracy desired for each unknown in the simplified kinetics model, the time scale(s) of interest, τ^* (the desired time resolution of the numerical printouts), and the initial conditions. The main raw data generated by CSP are the refined basis vectors $\mathbf{a}_i^o(M)$'s, $\mathbf{b}_o^i(M)$'s, the time scales $\tau(i)$'s, the number of exhausted reaction modes M, and the number of active and dormant reaction modes (including the identification of conservation laws, if any), all as a function of time. It is a relatively simple matter to deduce from the above raw data most of the information normally expected from a conventional analysis of a reaction system. If the problem under study is insufficiently stiff or if the user-specified threshold of accuracy is too stringent, the contributions of spent fast reaction modes would simply refuse to fall below the threshold.

The CSP algorithm described in this paper has been programmed (CSP8) and tested, and the results have been excellent even when the separation of the fast and slow time scales is only moderate. It is straightforward to include the energy equation by assigning one of the elements of the y vector to be temperature. However, if spatial diffusive terms are included on the right-hand side of Eq.(2.2) making it into a system of partial differential equations (PDE), many new theoretical issues arise. While discretized PDE systems can be treated as finite dimensional ODE systems by the present CSP algorithm, the formal generalization of CSP concepts to infinite dimensional PDE systems is a significant step and is being explored at the present time.

Acknowledgement

This work is supported by NASA Langley's Aerothermodynamics Branch, Space Systems Division.

References

[1] Williams, F.A., *Combustion Theory, The Fundamental Theory of Chemically Reacting Systems*, 2nd edition, Benjamin/Cummings Pub. Co., Menlo Park, Ca., 1985.

[2] Lam, S. H., "Singular Perturbation for Stiff Equations Using Numerical Methods", in *Recent Advances in the Aerospace Sciences*, Corrado Casci, Ed., Plenum Press, New York and London, pp.3-20, 1985.

[3] Lam, S. H. and Goussis, D. A., "Basic Theory and Demonstrations of Computational Singular Perturbation for Stiff Equations", *12th IMACS World Congress on Scientific Computation*, Paris, France, July 18-22, 1988; IMACS Transactions on Scientific Computing '88, Numerical and Applied Mathematics,C. Brezinski, Ed., J. C. Baltzer A. G. Scientific Publishing Co., pp. 487-492, November, 1988.

[4] Lam, S. H. and Goussis, D. A., "Understanding Complex Chemical Kinetics With Computational Singular Perturbation", *Proceedings of the 22nd International Symposium on Combustion*, pp.931-941, University of Washington, Seattle, Washington, August 14-19, 1988.

[5] Lam, S. H., Goussis, D. A., and Konopka D., "Time-Resolved Simplified Chemical Kinetics Modelling Using Computational Singular Perturbation", AIAA 27th Aerospace Sciences Meeting, (AIAA 89-0575), Reno, Nevada, January 9-12, 1989. See also: Goussis, D. A., S. H. Lam and Gnoffo, P.A., "Reduced and Simplified Chemical Kinetics for Air Dissociation Using Computational Singular Perturbation", AIAA 28th Aerospace Sciences Meeting, (AIAA 90-0644), Reno, Nevada, January 9-12, 1990.

[6] Frenklach, M., "Modeling of Large Reaction Systems", in Complex Chemical Reaction Systems, Mathematical Modelling and Simulation, Eds. J. Warnatz and W. Jäger in Chemical Physics **47**, Springer Verlag, 1987.

[7] Peters, N. and Williams F. A., "The Asymptotic Structure of Stoichiometric Methane-Air Flames", Combust. Flame, **68**, pp. 185-207, 1987.

[8] Seshadri, K. and Peters, N., "The Inner Structure of Methane-Air Flames", presented at the Workshop on Reduced Kinetic Mechanisms and Asymptotic Approximations for Methane-Air Flames, UCSD, La Jolla, Ca., March 13-14, 1989. See also: Peters, N., "Numerical and Asymptotic Analysis of Systematically Reduced Reaction Schemes For Hydrocarbon Flames", presented at the Symposium on Numerical Simulation of Combustion Phenomena, Sophia - Antipolis, France, May 21-24, 1985.

Index

Lecture Notes in Physics

W. C. Gardiner Jr. (Ed.)

Combustion Chemistry

1984. XIII, 509 pp. 164 figs. Hardcover DM 198,–
ISBN 3-540-90963-X

W. Bartknecht

Dust Explosions

Course, Prevention, Protection

With a Contribution by G. Zwahlen

With a Preface by H. Brauer

Translated from German by R. E. Bruderer, G. N. Kirby,
R. Siwek

1989. X, 270 pp. 295 figs. (some in color), 26 tabs.
Hardcover DM 168,– ISBN 3-540-50100-2

Deutsche Ausgabe:
W. Bartknecht, **Staubexplosionen.** 1987. DM 168,–
ISBN 3-540-16243-7

D. L. Andrews

Lasers
in Chemistry

2nd ed. 1990. XII, 188 pp. 113 figs.
Softcover DM 68,–
ISBN 3-540-51777-4

Springer-Verlag
Berlin
Heidelberg
New York
London
Paris
Tokyo
Hong Kong
Barcelona
Budapest